# No. 311 (Czechoslovak) Squadron in RAF Bomber Command 1940 - 1942

# No. 311 (Czechoslovak) Squadron in RAF Bomber Command 1940 - 1942

'Never Regard Their Numbers'

Steve C. Smith

from

www.aviationbooks.org

First published 2024 by Aviation Books Ltd., CF47 8RY, Wales, UK.

Publisher's email address for correspondence: aviationbooksuk@gmail.com

Copyright 2024 © Steve C. Smith

The right of Steve Smith to be identified as Author of this work is asserted by him in accordance with the Copyright, Designs and Patents Act 1988.

All rights reserved. No part of this publication may be reproduced, stored in a retrieval system, transmitted in any form or by any means, electronic, mechanical, or photocopied, recorded or otherwise, without the written permission of the copyright owners.

Cover design: Topics - The Creative Partnership www.topicsdesign.co.uk

Photographs and other illustrations of these incidents are available on the internet, but it is difficult in most cases to establish their original source, especially as a large number of photographs appear on numerous different websites. The author has sought permission to use the images in every case, although this has not been possible in all, due to broken links on old websites, or publishers having gone out of business. Where the source is clear, but no answer has been forthcoming, the author has assessed the historical value of each picture when deciding whether or not to include it. By definition, every contemporary image used is more than one hundred years old, and the advice received is that their use is permissible under copyright law. This has been carried out in good faith.

Any queries or objections to the use of any images or other material should be referred to the publisher.

A CIP catalogue reference for this book is available from the British Library.

ISBN 978 1915335395

This book is dedicated to old friends who were lost during the writing of this book.

*Jim Broadrib*

*Bob Body*

Also, to my wonderful wife, Jill. Her courage and resolve is a constant inspiration to me. Despite her pain, she continues to support me day after day. Thank you darling.

# Contents

Acknowledgements .................................................................................................. 8
Preface .................................................................................................................... 8
July 1940 : Out of the Frying Pan, Into the Fire ..................................................... 8
August 1940 : Finding your Feet! ........................................................................... 11
September 1940 : Operations ................................................................................. 17
October 1940 : Striking Back ................................................................................. 27
November 1940 : Back to Basics ............................................................................ 44
December 1940 : Back on Operations .................................................................... 48
January 1941 : New Year, Fresh Start .................................................................... 57
February 1941 : Operations Intensify ..................................................................... 63
March 1941 : A Loss Free Month ........................................................................... 76
April 1941 : Keeping Up the Pressure .................................................................... 93
May 1941 : Switch to Germany .............................................................................. 108
June 1941 : German Production ............................................................................. 122
July 1941 : Keeping the Pressure On! ..................................................................... 139
August 1941 : English Summer - Rain! .................................................................. 161
September 1941 : Italian Skies ............................................................................... 177
October 1941 : The Weather Sets In ....................................................................... 193
November 1941 : Under the Spotlight .................................................................... 209
December 1941 : A Not So Merry Christmas ......................................................... 224
January 1942 : Those Bloody Toads Again ............................................................ 234
February 1942 : The Balloon Goes Up ................................................................... 247
March 1942 : Germany Bound ............................................................................... 254
April 1942 : The Beginning of the End .................................................................. 274
Ground Crews: The Role and Dangers They Faced ............................................... 292
311 Squadron Prisoners of War .............................................................................. 296
311 Squadron Roll of Honour ................................................................................. 297
Remembering .......................................................................................................... 301

# Acknowledgements

This book owes its existence to the unwavering support, encouragement, and patience of Pavel Vančata, a respected author and historian. His guidance has been invaluable in shaping this book.

I would also like to thank the following for their help: fellow enthusiasts who have given up their time to help me with this book, most are amateurs like me, passionate about Bomber Command, and not a university degree amongst us!

As usual, there was unstinting help from New Zealander, thanks Chris Newey. I want to thank and acknowledge, in particular, some old friends, Simon Hepworth, Bob Collis and Graham Howard and a special mention to Di Ablewhite, who kindly allowed me to bombard her with requests for help. Also, for their assistance and support, Hans Nauta, Joop Hendrix, Andy Long, Dr Theo Boiten, Dave Homewood, Kelvin Youngs, Karen Evans, and Bill Chorley. I also want to acknowledge the help of four excellent websites/forums that have often answered many difficult questions: RAF Commands and 12 O'Clock High, WW2Talk and the excellent Aircrew Remembered website.

Finally, let me apologise in advance for the inevitable errors. Until I started this book, I was unaware that the Squadron Operational Records Book for 311 (Czech) Squadron held at the National Archives recorded a fraction of the information on the squadron. Perhaps, due to the language problems, the usually British adjutant recorded the mere basics. Being ever resourceful, the Czechs kept their own meticulous records, which give a fascinating insight into their time with R.A.F Bomber Command

Throughout this book, I have strived to present the facts using primary source material, a practice I have been guided in by Pavel Vančata, a leading expert on the squadron. While the opinions expressed are my own, I trust that this squadron history offers a fair and honest view. I also want to pre-emptively apologise for any errors that may have slipped through.

# Preface

Number 311(Czech) Squadron was the first and only bomber squadron to serve in R.A.F Bomber Command. The bomber group chosen to home the squadron was No.3 Group, whose squadrons were based in rural East Anglia.

Unlike its R.A.F counterparts, most squadron personnel had already seen war first-hand. The occupation of their homeland and then the conflict in France during the German invasion of May 1940, they had personally witnessed the brutality of the Nazis. The squadron was blessed with exceptional airmen who had served in the Czechoslovak Air Force and fleetingly the Armée de l'Air française. Equipped with the sturdy Vickers Wellington bomber, the Czechs eagerly set about mastering their new aircraft and, equally important, the English language. Highly motivated and with an ardent desire to prove themselves, the Czechs were deemed ready to undertake their first operations. However, political pressure may have resulted in the squadron's appearance too soon, as losses quickly rose. Withdrawn from the front line, a period of intense training followed and guided by some skilled R.A.F aircrew. Within a few months, they were back operational and undertaking raids over France, Germany, and Italy. The dedication of the Aircrews was matched by the ground crews and backroom staff, who, day after day and regardless of the weather, applied their trades with equal resolve. The Czechs participated in all the major campaigns between 1940 and 1942. Throughout that time, some outstanding commanding officers and flight commanders applied their trade and forged a reputation that still exists today. A period of heavy losses and friction between senior R.A.F officers and the Czechs ultimately resulted in the squadron's transfer to R.A.F Coastal Command.

Steve C. Smith

# July 1940 : Out of the Frying Pan, Into the Fire

For many of the 200 young Czech airmen who arrived at Liverpool docks on July 9th 1940, the relative peace and friendly smiles were a welcome change from what they had experienced back in their homeland and France. Many of these brave young men started their fight in 1938 when Hitler turned his attention to the Sudetenland and undermined the then-new Czechoslovak state. Having already shown their resolute spirit, they were arriving in England to continue their fight, having sailed from Port Vendres aboard one of the last evacuation vessels to leave for England.

Wing Commander Karel Mareš Toman, a 42-year-old son of a railway worker, commanded the party. A veteran of the Great War, he had learnt to fly in 1925. He was no stranger to danger and was undoubtedly the right man to lead the motley collection of men. During his escape from Czechoslovakia, he changed his name to Karel Toman. Having arrived in Liverpool, the group were transported to Beeston Castle, perched on a rocky sandstone crag 350 feet above the Cheshire Plain. Built in 1220, it offered few creature comforts, but it was a welcome refuge to rest and take stock in the early summer sunshine. Government bureaucracy was not long in arriving, even in wartime Britain, when British Authorities paraded and inspected the Czechs. The stay at Beeston Castle was brief. The men were again on the move, and their destination was Innsworth Lane, Gloucestershire. Here, the catalyst of 311 (Czechoslovak) Squadron was formed, and the rather disorganised group of men were equipped and issued with R.A.F uniforms. On July 17th, they were once again on the move, this time to Cosford Depot, where the officers received commissions in the R.A.F and other ranks were sworn in. They were now officially members of the Royal Air Force. While at Cosford, those already trained pilots went back to basics with tuition on the Link Trainer. They may have had flying experience, but they first had to meet certain criteria of the R.A.F. Learning the English language was a major issue, essential if the squadron was to be an effective contributor to the bomber offensive. Also important was a working knowledge of the King's Regulations. The Czechs, showing typical aptitude and determination, spent many hours studying both in the classroom and in private when not training. The arrival of the experienced Wing Commander John Griffiths DFC was a welcome appointment. Canadian born Griffiths was known for his bravery and steady leadership qualities. One of his first tasks was to organise 311 (Czech) Squadrons move to R.A.F Honington, Suffolk which was planned for July 29th. The R.A.F Honington Station Records Book reported the following on this date.

> *A new squadron, No.311, composed of Czech-Slovakians was formed with Wing Commander Griffiths as Officer Commanding.*

R.A.F Honington was the home of 9 Squadron who were equipped with the sturdy and reliable Vickers Wellington. Construction of Honington Airfield, which was undertaken by John Laing & Son, began in 1935, and the airfield was opened on May 3rd, 1937. It was located just six miles from Thetford and had two grass runways of 1,400 yards, plus four Type C hangars. The technical and administrative sites and barracks were located two miles from the airfield close to the village of Honington. Both the station and the squadron would come under the control of No.3 Group, one of Bomber Command's premier groups. At Honington 311 Squadron began the slow process of finding its feet and forging its own identity, a challenging process given the language barrier. The squadron was uniquely positioned, having two commanding officers, Wing Commander Karel Mareš Toman and Wing Commander John Griffiths DFC. Toman would command the Czech airmen, now numbering 269, while Griffiths took care of the running and administrational duties. Another experienced pilot would ably assist them, Flight Lieutenant Percy Charles 'Pick' Pickard DFC, a tall, blonde, well-spoken Englishman. He would oversee the squadron's flying training. Bombing and Gunnery training fell to Flight Lieutenant Anthony Browne and Pilot Officer Arthur Roman, who had just finished a spell as Gunnery Leader with 38 Squadron. In charge of navigation were Squadron Leaders Norman Samuels DFC and Henry Graham. Another experienced pilot arrived on July 28th, Flying Officer Michael Earle, a very capable New Zealander fresh from a hectic tour with 38 Squadron based at R.A.F Marham. He would fill the position of Squadron Adjutant. Another 38 Squadron veteran was Flight Lieutenant Richard MacFadden DFC. Like Earle, he had just completed a tour and was initially ordered to R.A.F Lossiemouth for instructional duties. His recent experience however was considered better suited at R.A.F Honington and 311 Squadron. He would take over 'B' Flight's training. With these experienced officers now in place, an intensive training period followed. On July 30th, the Air Officer Commanding 3 Group AVM J.E.A Baldwin

CB, OBE, DSO visited the squadron and met both commanding officers, senior crews and other ranks, welcoming them to the Group and Bomber Command.

With R.A.F Honington being an operational station, flying training would be difficult given the day-to-day activities of 9 Squadron. It was decided that flying training would be undertaken at Honington's satellite at Breckland heath, six miles north-east of Thetford and south-east of the village of East Wretham. The site, was like others had been hurriedly chosen and developed. Hastily erected huts, tents, and a makeshift telephone line were all the airfield had to offer. Living conditions were at first primitive, but given the glorious weather, it was manageable. With two grass runways of 1,400 yards and 1,880 yards, it was an ideal location to train in the summer months. However, this would quickly change as summer turned to autumn.

# August 1940 : Finding your Feet!

*The Czechoslovak president Dr. Eduard Beneš talking to both flying and ground personnel of 311 Squadron during his first visit to Honington on 6th August 1940. L-R: gen. Antonín Hasal-Nižborský, W/Cdr J. F. Griffiths DFC, A/W/Cdr Karel Mareš-Toman, Air Gunner P/O Jan Fürbach, Wireless Operator P/O František Doležal, Fitter Sgt František Bartůněk listening to Dr. Beneš, Fitter Armourer Sgt Ferdinand Kopecký, Armourer Sgt Alois Konopický, Navigator P/O Vilém Konštacký. Note the mixture of uniforms, NCOs are already wearing RAF uniforms, while officers are still in their French uniforms. The photo was taken in front of Wellington Mk.I L4332 received from No. 214 Squadron. (Pavel Vančata)*

August 1st 1940, found R.A.F Honington on 'Yellow' Alert, the Battle of Britain was reaching its peak. All the while, 311 Squadron was working tirelessly to become operational. The Squadron Operations Record Book does not record any activity for August. Thankfully, R.A.F Honington Station records give a brief insight. On the 1st, the squadron was visited by Wing Commander Cyril Douglas Adams from 3 Group H.Q. He, in turn, was followed by AVM C.F Portal CB, DSO, MC and the AOC 3 Group AVM J.E.A Baldwin CB, OBE, DSO. The Czechs were becoming popular! They were so popular that the AOC returned the following day. On the 3rd, 9 Squadron were ordered to prepare nine Wellingtons for operations that night. The Czechs could only look on enviously for the day they were operational. Their mood was lightened with the arrival of the squadron's first Vickers Wellington, L4332, a Mk.I via 214 Squadron. This was followed by Wellington L4338 the following day, again an ex-Stradishall-based 214 machine.

It was soon apparent that additional men would be required to bring the squadron up to its operational strength and maintain that level. Aircrew and ground crews were in short supply, and any prospect of new arrivals escaping via the Middle East and North Africa was now almost impossible. To augment the shortfall, volunteers were recruited from the then relatively small Czechoslovak Army. The Czechoslovakian armament industry was one of Europe's leading manufacturers producing high quality weapons and Czechoslovakia had been one of the most industrialised countries in the world, exporting quality products comparable, if not better than other manufacturing countries. Given the country's rich history of manufacturing, there were a large number of skilled individuals of fighting age to choose from. Those successful would be divided into categories, aircrew and groundcrew. Those selected for aircrew were despatched to No.2 Signals School, Yatesbury for wireless training. While those considered more suitable for groundcrews, many of whom had engineering and mechanical backgrounds, were sent to No.6 Training School in Hednesford, Staffordshire. Those with a background or an aptitude for electronics were sent to No.14 School of Technical Training at Henlow and

*Vickers Wellington Mk.Ic R1378 KX-K ( Via John Costin)*

further training in the delicate art of instrument repairs at No.12 School of Technical Training at Melksham. On the 6th, the squadron had a number of esteemed visitors. These included the Czechoslovak President Dr Edvard Beneš, who inspected and spoke to a selection of aircrew. There followed an almost production line of senior R.A.F officers, Air Marshal Sir W.G.S. Mitchell, KCB, CBE, DSO, MC, AFC, Inspector General of the R.A.F, Group Captain F. Beaumont. Frank Beaumont of the Air Ministry, had an excellent relationship with the Czechoslovaks. From December 1935 to May 1938 he served as British Air Attaché in Prague and in the summer of 1940 he led negotiations on behalf of the British side for the integration of Czechoslovak airmen into the RAF. Also from the Air Ministry, was Wing Commander Sir Louis Greig KBE, CVO, mentor and adviser to King George VI. Lastly, a return visit by the Group AOC AVM J.E.A Baldwin CB, OBE, DSO. R.A.F Honington was on almost continuous Anti-Invasion Alert for the next week. 9 Squadron operated on the 6th, 9th and 12th, while the airfield received the occasional diverted or damaged aircraft, all the while, 311 Squadron busied itself. It was not until August 13th 1940 that the first training flights are believed to have started. Initially allocated Avro Ansons and aging Wellington Mk.Is there was some initial disappointment, but eager to start training the squadron set about making the best out of what aircraft were available.

On the 16th, Squadron Leader Geoffrey Amison MiD arrived having just completed a tour with 38 Squadron based at R.A.F Marham. Tall, good-looking with a deep scar on his cheek, the 27-year old Amison joined the R.A.F in 1932 and learned to fly at No.3 Training School, Grantham. In 1937, Amison found himself stationed with No.2 Armoured Car Company based near Ramla, Palestine. While in Palestine, he was awarded a MiD in recognition of distinguished services rendered during this period. More recently he had commanded 'A' Flight of 38 Squadron. Arriving on the same day was an E.N.S.A Party led by Lady Peel, the Canadian born Beatrice Lillie. Cast members included stage and film stars John Gielgud and Ivy St Helier. The performance was held in the N.A.A.F.I. and was a roaring success. On the 18th, R.A.F Honington was bombed by a lone raider who accurately dropped a number of high explosives and incendiaries at around 16:15hrs taking the whole station unawares. Superficial damage was caused to a number of buildings, a road and parade ground. A more accurate second attack occurred at 18:25hrs when a single Dornier Do17 caused yet more damage with the partial collapse of a barrack building and damage inflicted to the roof of 'E' hangar. Two Wellingtons and a Magister were destroyed. Tragically, a number of personnel were killed and injured while reportedly queuing up for food on the parade ground. Despite the attack and the daily threat of invasion, 311 Squadron's training continued. Those airmen with a degree of flying experience gleaned while serving in the Czechoslovak or French Air Force would join 'A' Flight. They, however, were still required to complete a short refresher course at No.11 Operational Training Unit, the first batch being posted in late August. Those with less experience joined 'B' Flight. Recent events had brought to light at Group H.Q some issues with the vulnerability of R.A.F Honington and the satellite airfield at East Wretham and other of its stations. Despite efforts to camouflage the airfield and disperse the aircraft, it was felt that prominent local landmarks could aid any German attacks. Given this, on the 19th, 311 (Czech) and 9 Squadron pilots carried out flights to identify potential useful landmarks.

*Czechoslovak president Dr. Eduard Beneš captured while inspecting one of 311 Squadron Wellingtons, Honington, 6th August 1940. (Pavel Simet)*

*British liaison officer Squadron Leader P. C. Pickard, DSO, DFC with the first Czechoslovak 311 Squadron Commanding Officer, Wing Commander Karel Mareš-Toman. (Pavel Vančata)*

*Two of the experienced RAF pilots posted to the squadron in the early days. Both had a wealth of operational experience gleaned with No.38 Squadron. Left Squadron Leader Geoffrey Noden Amison MiD and Flight Lieutenant Richard MacFadden DFC. (Authors)*

Three locations were quickly identified for immediate attention. The 127ft tall Elveden War Memorial, the 110ft high Kilverstone Hall Water Tower, and the buildings and ranges at Berners Heath. The bombing war took a different course on the night of the 25th. 3 Group bombed Berlin for the first time in reprisal for the attack on London. R.A.F Honington sat the raid out while Nos. 99 and 149 Squadrons dealt the first blow. The squadrons of R.A.F Honington waited patiently for their turn. On the 29th, three Wellingtons of 311 Squadron took off at 16:37hrs for the short flight to East Wretham to carry out night landing practice. It would turn out to be a busy night.

The first of two incidents occurred at 01:00hrs when Sergeant Čestmír Hanuš misjudged his approach and clipped the upper most branches of some trees. Damage was slight, but it was a close shave, a few feet lower and the crew could have paid a heavy price for a momentary lapse of concentration.

| Date | August 30th 1940 |
|---|---|
| Mark | Mk.Ic |
| Serial | T2577 |
| Code | KX-A |
| Taken On Charge | 09/08/1940 via No.10 MU. |
| Manufacturer | Vickers |
| Contract | B.38600/39 |
| Pilot (s) | Sergeant Václav Korda / Sergeant Čestmír Hanuš |
| Flight | Night Circuit and Landings |
| Time | 01:00hrs |
| Cause | Hit trees on approach to flare path, misjudged height. |

The accidents were not over. Two hours later, Flight Lieutenant Josef Ocelka was instructing Sergeant Arnošt Zábrš, who made a rather heavy landing on the grass runway at 03:00hrs. Thankfully, the sturdy construction of the Wellington prevailed. Both aircraft and crew survived, if a little shaken.

| Date | August 30th 1940 |
|---|---|
| Mark | Vickers Wellington Mk.I |
| Serial | L4332 |
| Code | KX-T |
| Taken On Charge | 03/08/1940 via 214 Squadron |
| Manufacturer | Vickers Armstrong |
| Contract | 54968/36 |
| Pilot (s) | Flight Lieutenant Josef Ocelka / Sergeant Arnošt Zábrš |
| Flight | Night Cross Country |
| Time | 03:00hrs |
| Cause | Heavy Landing |

The training continued despite the mishaps. The Czech crews, especially those with previous operational experience, found this period of inactivity particularly frustrating, such was their desire to get at the enemy. A Bristol Blenheim landed at East Wretham at 00:05hrs on August 31st. Its stay was brief, and it departed at daybreak. R.A.F Honington had three unidentified Hawker Hurricanes land throughout the day. The first, at 11:14hrs, was suffering with an overheating engine. It returned to R.A.F Coltishall after a short stay. Two Hurricanes followed at 13:40hrs, both appeared to be lost. On discovering their location, they immediately took-off.

*A parade of No.311 Squadron members on 6th August 1940 at Honington. President Dr. Edvard Beneš is accompanied by Czech Commanding Officer Acting Wing Commander Karel Mareš-Toman and his British counterpart Wing Commander J. F. Griffiths. (Anna Trehy)*

# September 1940 : Operations

September started with 9 Squadron on operations. They would send seven Wellingtons against the Bitterfeld (Aluminium Works) in Saxony, while three attacked Cologne Marshalling Yards. Once again 311 sat the operation out. A 75(NZ) Wellington struck some trees in harsh weather and crashed after losing sight of the flare path while landing at East Wretham at 04:15hrs. The Wellington, R3159 AA-K on landing immediately burst into flames. The Czechs camped at the airfield, rushed to the crash site and showing commendable courage, entered the blazing Wellington and extracted the injured crew. Unknown to the Czechs, the Wellington was still carrying its full bombload, having been unable to identify the target at Hanover. At 07:58hrs on the morning of September 2nd, R.A.F Honington received a call from Wing Commander Maurice 'Buck' Buckley, commanding officer of 75(NZ) Squadron expressing his thanks for the assistance of the Czechs at East Wretham in saving one of his crews. It was a nice gesture. Night flying continued at East Wretham on the 2nd, unfortunately resulting in yet another accident when Flight Lieutenant Josef Ocelka forgot to lower the flaps when coming into land and overshot the flarepath on return from a night circuit and bumps exercise. Damage was repairable, but the accident was reported as *'carelessness on the part of the pilot'*.

| Date | September 2nd 1940 |
|---|---|
| Mark | Mk.Ic |
| Serial | T2561 |
| Code | N/K |
| Taken On Charge | 09/08/1940 via No.10 MU. |
| Manufacturer | Vickers |
| Contract | B.38600/39 |
| Pilot (s) | Squadron Leader Josef Schejbal / Flight Lieutenant Josef Ocelka |
| Flight | Night Circuit and Landings |
| Time | 04:00hrs |
| Cause | Failed to lower flaps. Overshot flarepath. |

R.A.F Honington was a refuge for eight diverted bombers on the early morning of September 5th. Three Hampdens, one from 49 and two from 83 Squadrons, wasted no time and were the first to depart as dawn broke. September 5th would witness 311 Squadron despatch six Wellingtons at 16:40hrs on a long-range cross-country flight culminating in bombing practice at Towyn Bombing Range on the coast of Cardigan Bay, Wales. The following afternoon, 311 Squadron's Commanding Officer oddly reported that one of the Wellingtons had been engaged by flak and attacked by a fighter over the Bay. Thankfully, the aim of both was off! The squadron flew another night exercise on the 6th, and a returning crew reported the blackout at Salop, Shrewsbury, and Wisbech could have been more effective. R.A.F Honington was visited once again by the AOC 3 Group AVM J.E.A Baldwin CB, OBE, DSO on the 6th. Baldwin was one of Bomber Commands more active Group Commanders, his frequent visits to his stations went a long way to foster an excellent working relationship with his Station and squadron commanders.

Finally, on September 10th, 1940, the squadron was instructed to prepare for operations. A meeting between 9 and 311 Squadron was held at 10:30hrs to discuss the night's activities. Excitement swept the squadron. At last 311 (Czech) Squadron was going to war. However, training still had to be completed, and bombing practice at Berners Heath was laid on. The three crews chosen to open the squadron's bombing war were Squadron Leader J Schejbal, Sergeant F Taiber and Sergeant V Korda. There would have been in all probability a crowd gathered as the three crews taxied towards the grassed runway. Both air and ground crews would have been immensely proud as the Wellingtons and their Czech crews departed. The first away at 19:55hrs was Sergeant F Taiber, followed at 20:05hrs by Squadron Leader J Schejbal and finally at 22:05hrs Sergeant V Korda.

*One of the first group photos of 311 Squadron members, both flying and ground staff, Honington, September 1940. (Pavel Vančata)*

### September 10th 1940 : Z40 – Brussel's Marshalling Yards

| Pilot | 2nd Pilot | Serial | Code | Bomb Load | Result |
|---|---|---|---|---|---|
| S/Ldr J Schejbal | Sgt J Hrnčíř | L7778 | KX-U | 8x250lb | Completed |
| Sgt F Taiber | F/Lt J Ocelka | L7785 | KX-R | 8x250lb | Not Completed |
| Sgt V Korda | Sgt J Bala | P9235 | KX-C | 8x250lb | Completed |

Only Squadron Leader Josef Schejbal and Sergeant Václav Korda reported finding and bombing the target, the former from just 700ft. The night may not have been a total success, but to everyone on the squadron, the night was a triumph that needed to be celebrated. The following day, four crews were placed on standby from 08:30hrs. On the 12th, three crews were again detailed and briefed for operations. The target was the Marshalling Yards at Haren, Brussels. Wing Commander Karel Mareš Toman, chose this night to lead his squadron.. He departed first at 19:40hrs. One crew was over the target area between 21:14hrs and 21:46hrs. Once again, conflicting information confuses the operation. The Squadron Operations Record Book reports all three crews successfully attacked the target. This is, however at odds with other documents which state that Wing Commander Mareš Toman, was unable to locate the primary or secondary target due to cloud and returned with his bombload. Sergeant F Taiber was also unsuccessful, he was unable to bomb due to a faulty bomb release adaptor.[1] Only Sergeant V Korda is known to have attacked from 3,000ft, he reported upon return, *'Fires on arrival. Bursts seen in target, dummy airfield 7 miles NNW of Brussels'*.

### September 12th 1940 : Marshalling Yards at Haren Brussels

| Pilot | 2nd Pilot | Serial | Code | Bomb Load | Result |
|---|---|---|---|---|---|
| W/Cdr K Toman | Sgt J Hrnčíř | L7778 | KX-U | 7x25lb+2x250lb+60x4lb IB | Not Completed |
| F/Lt J Ocelka | Sgt Taiber | L7785 | KX-R | 7x25lb+2x250lb+60x4lb IB | Not Completed |
| Sgt V Korda | Sgt J Bala | P9235 | KX-C | 7x25lb+2x250lb+60x4lb IB | Completed |

The squadron sat out the operation to Calais and Krefeld Sidings on the 15th, flown by 9 Squadron. The Czechs busied themselves preparing for a partial move to East Wretham the following day. Six experienced crews drawn from 'A'

---

[1] *The No.3 Group Results of Operations, and Form E report the crews as returning with bombload.*

Flight plus equipment and ground personnel would be transported to the satellite airfield and commence operations from there.

*An aerial view of the Type C hangars at Honington taken from No. 311 Squadron Wellington in summer 1940. (Jaroslav Popelka)*

The remaining 'A' Flight crews would remain at R.A.F Honington under the capable Flight Lieutenant 'Pick' Pickard DFC. Conditions at East Wretham were archaic. The NCOs would sleep in tents, which were considered far too small, especially for the aircrew with their flying clothing, parachutes and personal belongings. After some clever improvisation, the Czechs quickly sorted out their sleeping arrangements by merging tents. There were similar issues for the officers. They were billeted in a rather dilapidated building nearby. Accommodation was better than a drafty tent, but space was at a premium with 3-4 officers per room. There was no running water, the water pump was located in the courtyard, and there were no washing basins. The six Wellingtons arrived at East Wretham just after 14:00hrs and quickly dispersed, it was a prudent decision. At 15:50hrs, a low-flying Do17 appeared over the airfield, taking everyone by surprise. What airfield defence was available opened fire on the Dornier, which climbed and entered the clouds only to emerge over R.A.F Honington, where it dropped its bombs at 15:53hrs. No serious damage was reported, and it was later established that a total of fifteen 50kg bombs had been found, and none had exploded. The only damage was to the roof of one hangar, where a 50kg bomb had penetrated but failed to explode. The following morning more issues occurred, namely organising breakfast for the NCOs. The in-the-field cooking facilities had not yet been fully set up. There were no such issues at the Officer's site, where an improvised Czech breakfast was prepared, which the R.A.F Officers seemed to enjoy. During the afternoon, Wing Commander Mareš Toman assembled all the Czechoslovakian members of the squadron and explained what he expected from every individual while operating from East Wretham in his usual no-nonsense fashion. Squadron Leader Amison assumed command of the 311 Squadron detachment based at Honington on the 17th. On the 18th, a very grateful bunch of officers were moved to the more comfortable Manor Farm, just 1 mile from the airfield, to ease the cramped conditions.

Four crews were required for operations on the 19th. These were duly flown to R.A.F Honington in the afternoon while the ground crews followed by road. Unfortunately, after a number of delays, the operation was eventually scrubbed. On the 20th, Invasion Alert No.1 found the squadron and its crews waiting for the expected German invasion to finally materialise. Both aircrew and ground crews waited at their dispersals, gas masks at the ready. The tension was eased somewhat when three Wellingtons were given a special night reconnaissance flight. Departing at 19:21hrs, the 'targets' were Ipswich and the ancient market town of Hadleigh, Suffolk. Other than one unrecorded Wellington being fired on by flak, the operation was successful. The invasion ports of Calais and Dunkirk were the target for thirty Wellingtons of 3 Group on the night of September 21st, six of which would be provided by 311 (Czech) Squadron. The crews departed for R.A.F Honington late morning closely followed by their ground crews by road. At Honington, the Wellingtons were fuelled, checked and bombed up by their attentive ground crews. The thirty-six members of crews assembled in the Operations Room at 17:15hrs for briefing.

September 21st 1940 : C37 – Calais Docks

| Pilot | 2nd Pilot | Serial | Code | Bomb Load | Result |
|---|---|---|---|---|---|
| S/Ldr J Schejbal | Sgt J Hrnčíř | L7778 | KX-U | 8x250lb+60x4lb IB | Completed |
| F/Lt J Ocelka | Sgt Taiber | L7785 | KX-R | 8x250lb+60x4lb IB | Completed |
| Sgt V Korda | Sgt J Bala | P9235 | KX-C | 8x250lb+60x4lb IB | Completed |
| P/O B Landa | Sgt K Novotný | P9230 | KX-B | 8x250lb+60x4lb IB | Completed |
| P/O F Janoušek | Sgt K Fák | R3177 | KX-L | 8x250lb+60x4lb IB | Completed |
| P/O K Trojáček | Sgt A Zábrš | L7788 | KX-E | 8x250lb+60x4lb IB | Completed |

Three crews would be undertaking their first operation on this night, Pilot Officers Bohumil Landa, František Janoušek and Karel Josef Trojáček. Take-off commenced at 21:50hrs when Squadron Leader Josef Schejbal trundled down Honington's grass runway, followed in intervals of 5 minutes by the remaining crews. Soon after take-off, Sergeant František Taiber reported engine trouble and returned early with his bombload.[2] A total of 16 Wellingtons were allocated Calais, which was already ablaze when the squadron arrived, courtesy of 9 Squadron. Bombing between 8,000ft and 13,000ft, the crews quickly picked out the main and tidal harbour entrance. Such were the conditions Pilot Officer Bohumil Landa reported three, possibly four large ships of over 100 tons moored in the main harbour entrance. The flak was heavy and accurate, but as the crews turned for home, they had the satisfaction of seeing a large fire, with intense yellow flames taking hold at Bassin de Chasse. On the homeward flight, much to Pilot Officer Bohumil Landa's annoyance, he realised his incendiaries had hung up. The first two crews were back at R.A.F Honington at 23:52hrs, followed within the hour by their four colleagues. No sooner had the first pair landed than they were ordered to return to East Wretham. On reaching the fog-covered satellite airfield, Flight Lieutenant Josef Ocelka overshot his landing, damaging his Wellington in the process. Orders were immediately sent for the remaining Wellingtons to either return to or stay at R.A.F Honington, and their crews return to East Wretham via lorry.

| Date | September 21st 1940 |
|---|---|
| Mark | Mk.Ic |
| Serial | L7785 |
| Code | KX-R |
| Taken On Charge | 11/09/1940 via 9 Squadron |
| Manufacturer | Vickers Armstrong |
| Contract | B.992424/39 |
| Pilot (s) | Flight Lieutenant Josef Ocelka / Sergeant František Taiber |
| Flight | Transit flight Honington – East Wretham |
| Time | 01:29hrs |
| Cause | Overshot flarepath due to local mist. |

It would appear that Wing Commander Griffith's patience had run out with Flight Lieutenant Ocelka recent run of accidents. He reported, *'Flight Lieutenant Ocelka has crashed another Wellington under similar circumstances and his*

---

[2] *This is at odds with Squadron ORB Form 540, the No.3 Group Results of Operations, and Form E report show the crew completing the operation.*

*reactions appear to be somewhat slower than usual for pilots flying at night. I have transferred him to the position of co-pilot until further assessment of its usefulness for night flying can be carried out'.*

There was a real buzz around both R.A.F Honington and East Wretham on the morning of September 23rd. The target for that night was the Nazi capital Berlin. The squadron would provide three Wellington crews. Their primary target was the Wilmersdorf Power Station. The selected crews flew to Honington, where they were briefed and supplied with maps and charts while each Wellington was loaded with eight 250 pounders, two of which were delayed action. It would prove to be a night to remember for all the wrong reasons. Group would despatch a total of 45 Wellingtons to Berlin, while a further ten would keep up the pressure on the invasion barges gathered at Le Havre on the Normandy coast.

### September 23rd 1940 : B60 – Wilmersdorf Electrical Power Station, Berlin

| Pilot | 2nd Pilot | Serial | Code | Bomb Load | Result |
|---|---|---|---|---|---|
| W/Cdr K Toman | Sgt J Hrnčíř | L7778 | KX-U | 6x250lb+2x250lb | Completed |
| P/O B Landa | Sgt K Novotný | P9230 | KX-B | 6x250lb+2x250lb | Secondary Target |
| P/O K Trojáček | Sgt A Zábrš | L7788 | KX-E | 6x250lb+2x250lb | MISSING |

The first crew away was Pilot Officer B Landa at 21:34hrs. Wing Commander Mareš Toman and Pilot Officer K Trojáček quickly followed. Conditions over the target were not ideal. Patchy cloud and considerable ground haze made identifying the Power Station difficult. Arriving over Berlin at 01:20hrs, Wing Commander Mareš Toman, had a particularly tough time. Starboard engine trouble just before the target resulted in the crew losing precious altitude and causing the bomb run to be spoilt. Showing steely determination, he made another run over the capital. It was on this run over Berlin that a night fighter attacked the Wellington. The encounter proved short and inconclusive. With yet another approach to the target spoilt, Wing Commander Toman dropped his bombs from 12,000ft in the general target area, intense flak and searchlights making visual confirmation impossible. It is understood that Pilot Officer K Trojáček managed to bomb the primary, confirmed by a W/T message received back at base. Pilot Officer B Landa could not locate the Power Station due to a faulty engine, so he bombed the secondary target, Berlin's Grunewald Marshalling Yards. The crew made their bomb run from 12,000ft, dropping their bombload in one stick from SW-NE. These were seen to explode across the marshalling yards, producing a small fire. The operation appeared to be a success until a message was received by the crew of Pilot Officer K Trojáček asking for a 'Fix' while flying over the Frisian Islands. Some twelve minutes later, another message was received stating that one of the Wellington's engines damaged by flak over Berlin was showing worrying signs of trouble. There then followed an S.O.S message *'Am landing'*. Then at 04:55hrs, another message was received, leaving yet more doubt about the crew's fate – the aircraft force landed and the crew was safe. It was unclear to those back at Honington if the crew had ditched or crash-landed.

The crew of Wellington L7788 KX-E had made a successful forced landing at Leidschendam in Holland. Having failed to set light to the Wellington, the crew split up in an attempt to escape. Five of the crew were eventually captured. However, one crewmember, Wireless Operator Sergeant Karel Kunka, took his own life when cornered in a granary in Wassenaar by Dutch Police. Using a Verey pistol, he shot himself, believing he was protecting his family back in Czechoslovakia. He succumbed to his injuries in Zuidwal Hospital on September 25th. This brave young Czech was buried on September 27th 1940, at the Gravengage Central Cemetery.

### Vickers Wellington Mk.Ic L7788 – KX-E

| Manufacturer | Vickers Armstrong | |
|---|---|---|
| Contract | B.992424/39 | |
| Taken on Charge | 09/09/1940 via 9 Squadron | |
| Cat E Missing | 23/09/1940 | |
| Struck Of Charge | 24/09/1940 | |
| Total Flying Hours | N/K | |
| Take-Off Time | 21:35hrs | |
| Bomb Load | 6x250lb+2x250lb (Delayed action) | |

|  | CREW |  |
|---|---|---|
| Captain | Pilot Officer Karel **Trojáček** 82580 RAFVR | PoW |
| Second Pilot | Sergeant Arnošt **Zábrš** 787225 RAFVR | PoW |
| Navigator | Pilot Officer Zdeněk **Procházka** 82628 RAFVR | PoW |
| Wireless Operator | Sergeant Karel **Kunka** 787252 RAFVR. Age 27. | Allied Plot. Row 3. Grave 56. |
| Front Gunner | Pilot Officer Václav **Kilián** 82606 RAFVR | PoW |
| Rear Gunner | Sergeant František **Knotek** 787548 RAFVR | PoW |
|  |  |  |
| Posting History | P/O K Trojáček via Cosford Depot, 29/07/1940. Sgt A Zábrš via Cosford Depot, 29/07/1940. |  |
| Operations Flown | P/O K Trojáček : 1 / Sgt A Zábrš : 1 |  |
| Buried | THE HAGUE (WESTDUIN) GENERAL CEMETERY |  |

By late morning of the 24th, after the usual checks had been made at Group H.Q and the Observer Corps it was soon apparent that 311 (Czech) Squadron had suffered its first operational loss. At 06:10hrs, several aircraft from Honington and other 3 Group stations set out on a sea-search for the crew. A report of lifejackets being found by a High Speed Launch had proven to be incorrect. There was a real sense of sadness throughout the squadron. The loss of friends, especially those who had faced the dangers of escaping their homeland and being captured by the Germans, seemed worse when lost so early in their operational tour.

During the previous night local searchlights and Anti-Aircraft Batteries were active and Bury St Edmunds Observer Corps reported an aircraft had crashed at around 04:03hrs at Elveden Estate Suffolk. Reports initially stated that the aircraft was Spitfire L5351 and had burst into flames on impact, and the pilot was dead. At 06:05hrs, a further report was received stating that it was not a Spitfire, and two bodies had been recovered and a third parachute found. At 07:05hrs, a message was sent by Wing Commander Mareš Toman, from East Wretham that he had taken charge of three bodies, which appeared to be Polish. It was later discovered that the Fairey Battle crew from 301 (Polish) Squadron based at R.A.F Swinderby may have been inadvertently attacked by a R.A.F Bristol Blenheim.

On the 25th, the squadron was again on the Battle Order. Operation *'Lucid'* was to take place that night but was cancelled on instructions from the Royal Navy, forcing Bomber Command to switch part of its efforts to Boulogne. R.A.F Honington was ordered to destroy all copies of Form B.282. The two R.A.F Honington-based squadrons were given the invasion barges at Boulogne to attack. Two crews would be provided by 311 (Czech) Squadron, one of which would include South African Squadron Leader Henry Graham, who would replace Sergeant František Taiber, who reported sick.

### September 26th 1940 : C29 – Boulogne-sur-Mer Harbour

| Pilot | 2nd Pilot | Serial | Code | Bomb Load | Result |
|---|---|---|---|---|---|
| S/Ldr H Graham | F/Lt J Ocelka | L7788 | KX-U | 6x250+240lb IB | Completed |
| P/O F Janoušek | Sgt K Fák | P9230 | KX-B | 6x250+240lb IB | Not Completed |

The crew of Pilot Officer F Janoušek returned within 45 minutes with reported engine trouble. They landed back at East Wretham with their bombload.[3] Both 9 and 311 Squadron crews were over the target between 03:05hrs and 04:12hrs. Conditions over the port were largely favourable. Fires had already taken hold when Squadron Leader Graham started his bomb run from 7,000ft. The defences were active, having been stirred up by the preceding bombers. Squadron Leader Graham's bombs were seen to burst across the harbour entrance, producing a number of fires. Despite the optimistic reports by Honington's returning crews, bombing accuracy over Boulogne docks was generally poor. Some 100 bombs and incendiaries had fallen within the Greater Boulogne area, producing superficial damage. The dock entrance watch office was destroyed, as was a light flak post and two harbour tugs. Incendiaries burnt out a motor boat

---

[3] *Other sources report hydraulic failure to the rear turret.*

*Inquisitive German troops are seen here inspecting Vickers Wellington Mk.IA L7788 KX-E. The damage was relatively minor, and once repaired, the Wellington would be test-flown and evaluated by the Luftwaffe.*

*Once repaired, the Wellington was flown to Schiphol airfield. Below : Looking rather battered. Note the application of the Balkenkreuz.*

*Wireless Operator Sergeant Karel Kunka who preferred to shoot himself by the Very pistol rather than be captured, became the first operational victim of 311 Squadron over the enemy territory. (Pavel Vančata)*

moored in Bassin à Flot, plus four commandeered French lorries had also been destroyed. Unfortunately, the bombing spilled over into the residential area. Seventeen houses and shops and a school were destroyed. In rue d'Orléans, the offices of Vidor and Sarraz, a famous fishing boat building firm, were set ablaze for the second time in three days and burnt out. The civilian death toll was thankfully low, with just ten civilians killed.

On the 26th, the fate of the crew of Pilot Officer Karel Trojáček was learnt with the announcement in a report by the news agency Reuters published in the Daily Mirror. The following gave the members of 311 (Czech) Squadron some answers. *'The German-controlled Dutch wireless gave in their news report that somewhere in Holland, a British bomber landed yesterday. The crew were unhurt and disappeared. The only thing the Germans found was the damaged aircraft. A reward of sixty pounds was offered to anyone telling where the crew was'.*

On the 27th, a number of Polish Officers of 301 Squadron arrived at R.A.F Honington to attend the funeral of their dead comrades at Honington (All Saints) Churchyard, killed during the early hours of September 25th. It was a sombre occasion as it was the first Polish Air Force loss in Bomber Command.

*During early September 1940 NCOs of the No.311 Squadron were billeted in tents scattered around the East Wretham airfield. L-R: Air Gunner Sgt Augustin Šesták, Wireless Operators Sgt Josef Cibulka, Sgt Jan Plzák and Sgt Karel Kunka. Only Plzák will celebrate the end of war in May 1945 with 311 Squadron in the rank of F/Lt. Šesták will be lucky enough to return from German captivity in the rank of W/O in the spring of 1945. The other two will perish. (Jaroslav Popelka)*

The mood was lifted slightly with an ENSA party arriving in the early evening of the 27th performing 'Bouquets' in the N.A.A.F.I. It was well attended and appeared to be enjoyed by all. The squadron was stood down on the 28th, unlike 9 Squadron, which briefed seven crews to attack an aluminium works at Hanau, plus a further four on the marshalling yards in Cologne. Perhaps not wishing the crews to become too complacent, a seven-mile cross-country was organised for all ranks of 311 Squadron. More to the crew's liking was the news that the squadron would take delivery of six brand new Wellingtons within days, followed by possibly three more. The night's operations proved costly for 9 Squadron. One crew crashed on returning, killing one crew member and injuring the rest, while another was reported 'missing'. The squadron was not required for the remainder of September, which had been a momentous month, for the squadron and the country. R.A.F Honington welcomed Group Captain S.B. Harris, DFC, AFC on temporary attachment.

Fighter Command had thwarted any thoughts the Luftwaffe had about air superiority over England and stopped any possible invasion. The Luftwaffe unable to bomb during the day shifted almost entirely to night raids on Britain's industrial centres, The 'Blitz' was about to start over Britain.

# October 1940 : Striking Back

October could not have started any worse for the squadron when an Avro Anson on a NAVEX flight crashed at Elton, 7 miles SW of Peterborough. The first hint of trouble arrived at R.A.F Honington at 13:30hrs when news arrived that a 311 Squadron Anson had crashed near Peterborough. It soon became apparent that a number of the occupants had been killed. Shock and disbelief swept the squadron, how could this have possibly happened on what should have been a routine training flight. At 15:25hrs, the crew's names and fate were known. Pilot Officer Jaroslav Skutil, Pilot Officer Josef Slovák, Sergeant František Koukol, Sergeant Oskar Valošek, and R.A.F Wireless Instructor Sergeant George Powis were killed. Two managed to parachute to safety, the captain, Pilot Officer Ludvík Němec and one of two navigators, Pilot Officer Jaroslav Kula.

### Avro Anson Mk.I R9649

| Manufacturer | Avro (Yeadon) | |
|---|---|---|
| Contract | 32842/39 | |
| Taken on Charge | 12/08/1940 via No.38 MU. | |
| Cat E Missing | N/A | |
| Struck Of Charge | 09/04/1941 | |
| Total Flying Hours | N/K | |
| Take-Off Time | N/K | |
| | **CREW** | |
| Captain | Pilot Officer Ludvík **Němec** 82564 RAFVR | Survived |
| Second Pilot | Sergeant Oskar **Valošek** 787224 RAFVR. Age 24. | Div. 4. Block 9. R.C. Joint grave 174. |
| Navigator | Pilot Officer Jaroslav **Kula** 82615 RAFVR | Survived |
| Navigator | Pilot Officer Josef **Slovák** 82638 RAFVR. Age 28. | Div. 4. Block 9. R.C. Joint grave 165 |
| Wireless Operator / Air Gunner | Sergeant František **Koukol** 787615 RAFVR. Age 25. | Div. 4. Block 9. R.C. Joint grave 174 |
| Wireless Operator / Air Gunner | Pilot Officer Jaroslav **Skutil** 82529 RAFVR. Age 30. | Div. 4. Block 9. R.C. Joint grave 165. |
| Wireless Operator – Instructor | Sergeant George Owen **Powis** 627228 RAF. Age 20. (British) | Sec. F. Grave 1428 |
| | | |
| Posting History | P/O L Němec via Cosford Depot, 29/07/1940. Sgt O Valošek NFD. | |
| Operations Flown | Nil | |
| Buried Czechs | PETERBOROUGH (EASTFIELD) CEMETERY | |
| Buried British | DERBY (NOTTINGHAM ROAD) CEMETERY | |

Wing Commander Griffiths DFC and Flight Lieutenant Provazník wasted little time and immediately set off to the crash site. The following details were collated by Wing Commander John Griffiths DFC.

> *I proceed to and examine the scene of the crash within a few hours of the accident. The burnt-out wreckage covered an area approximately 15 yards square. The position of the engines indicated the aircraft had dived vertically into the ground. This fact was subsequently confirmed by a witness who stated that the aircraft approached at a low altitude omitting large quantities of smoke before suddenly bursting into flame and diving into the ground. Prior to examining the wreckage, it was ascertained from one of survivors that the fire had broken out very suddenly somewhere on the floor in the vicinity of the navigators table. This indicated the*

*likelihood of someone having fired the Verey pistol. The pistol was discovered amongst the wreckage and found to contain a spent cartridge.*

*From the statements :*

*The aircraft turned back for base due to bad weather somewhere in the vicinity of Leicester. The navigator, Pilot Officer Kula asked for bearing and the second navigator and captain engaged themselves in attempting to ascertain the aircrafts position by means of map reading. The British wireless operator, Sergeant Powis came forward and handled a bearing to Pilot Officer Kula approximately 30 minutes before the time of crash. About 10 minutes later, wireless operator, Sergeant Koukol came forward with a second bearing which he handled to Kula.*

*Pilot Officer Kula apparently had difficulty in laying down the bearing and some sort of argument ensued between Sergeant Koukol and himself. Sergeant Powis meanwhile had moved forward and positioned himself at the end of navigators table directly beside the Verey Pistol stowage. The position of various members of the crew is show on sketch. No material alteration of position took place prior to the outbreak of fire. The captain of aircraft Pilot Officer Němec was looking out the starboard window when he heard a loud detonation. He turned suddenly and saw a fire on the floor under the navigation table. At this moment members of the crew were engaged as follows:*

*Sergeant Valošek – flying the aircraft.*
*Pilot Officer Slovák – map reading from second pilot seat.*
*Pilot Officer Němec – map reading main spar.*
*Pilot Officer Kula – navigating and laying down bearing.*
*Sergeant Koukol – assisting navigator in laying down bearing.*
*Pilot Officer Skutil – seated in rear turret.*
*Sergeant Powis – doing nothing in particular and standing directly over Verey Pistol stowage.*

*Pilot Officer Němec immediately shouted 'FIRE' and attempted to stamp out the fire. This he was unable to do and for some unknown reason commenced beating the flames with the butt of the Verey Pistol. The aircraft meanwhile was rapidly filling with smoke, and Pilot Officer Němec then attempt to turn off the petrol. He was prevented from doing this by the pilot who knocked the Pilot Officers Němec arm away as he attempted to reach for the petrol cocks. Pilot Officer Němec then gave the order to jump and having fitted his parachute passed down the fuselage and abandoned aircraft. During his passage through fuselage he saw one wireless operator trying to get through the roof exit hatch and another through the side window. He noticed that the altimeter was registering 600-700 feet at this time. The ground of this point is about 200 feet above the level of Honington which would make the absolute height of the aircraft approx. 400 feet. Pilot Officer Kula saw the fire under the table after having heard a loud bang and apparently adorned his parachute and left the aircraft without further ado. I am of the opinion that this accident is attributable to some member of the crew having discharged the Verey pistol through the floor of the aircraft. Three persons only were within reach of the pistol:*

*P/O Kula*
*Sgt Koukol*
*Sgt Powis*

*Of these three, Sergeant Powis was by far the nearest to the stowage and was in fact standing directly over the pistol. The activities of Pilot Officer Kula and Sergeant Koukol are accounted for at the time of outbreak of fire. Sergeant Powis was apparently standing by in the role of spectator. I consider it is highly probable that he absentmindedly commenced fingering the pistol in the stowage with the result that he eventually pulled the trigger and discharged the Verey light through the floor of the cockpit. Both Pilot Officer's Němec and*

> Kula report a detonation immediately before the outbreak of fire. This combined with the discovery of pistol housing spent cartridge appears to me as a clear indication of the means by which the fire was started.
>
> In view of the above mentioned I am forced to the following conclusion with regards of the cause of accident:
>
> 1) The aircraft was set on fire by a Verey light fired through the fuselage floor.
> 2) The evidence points toward Sgt Powis as having been responsible for the discharge of the pistol.
>
> **Griffiths W/C Commanding 311 Squadron.**

*One of No.311 Squadron's Avro Ansons. These forgiving twin-engine aircraft were pivotal in ensuring squadron training kept pace with operational losses. Pilot Sergeant Josef Kalenský (left) and Air Gunner Sergeant Václav Bozděch (right), posing by Anson Mk.I R9600. (Zdeněk Hurt)*

311 (Czech) Squadron sat out the operations planned for that night, unlike 9 Squadron who would detail and brief 11 crews against three targets. Three crews would visit Berlin and Hanau while another five would attack Gelsenkirchen. The raid on Berlin would claim another 9 Squadron crew. The following morning at 09:20hrs, two 311 Squadron Wellingtons took off on a sea search for the crew of Flight Lieutenant Cox, who ditched off Lowestoft at around 02:46hrs. Sadly, despite an extensive search lasting most of the day, no sign of the crew was ever found. Flight Lieutenant 'Pick' Pickard DFC was obliged to make a forced landing with a faulty engine at R.A.F Tangmere at 13:45hrs. The reason behind the flight to the south coast is unclear. On the 3rd, the AOC AVM J.E.A Baldwin CB, OBE,

DSO visited R.A.F Honington and lunched with the Commanding Officer in the Mess. On the 5th, the squadron had the sad task of burying its former colleagues. Wing Commander Mareš Toman, and Pilot Officer Zdeněk Pekárek, the Equipment Officer, flew to Peterborough for the funeral from East Wretham. Also attending from R.A.F Honington were Wing Commander John Griffiths DFC, Squadron Leader Geoffrey Amison MiD and the two survivors, Pilot Officers Ludvík Němec and Jaroslav Kula.

Frustrated at the lack of operation, 311 Squadron's C/O rang 3 Group H.Q on the morning of October 7th, stating that they hoped to be included in the operation planned for that night. It was an audacious call but typical of Wing Commander Mareš Toman. Within five minutes of the call, 3 Group replied that the squadron would almost certainly be required the following night and, as such, would not be required that evening. Satisfied with the reply, the squadron started preparations almost immediately. A Vickers Wellington of 149 Squadron crashed at R.A.F Honington at 20:03hrs after a reported encounter with a German night fighter. The damaged Wellington, OJ-G struck and wrecked the Chance Light on landing before careering into and destroying a lorry and utility van, tragically killing four Chance Light and Flare Path crew members.

Just after lunch on the 8th, orders were received from Group H.Q for the squadron's possible involvement in Operation *'LUCID'* as prepared in Form B.298, within the hour further instructions arrived cancelling any involvement. Finally, after nearly two weeks of inactivity, 311 (Czech) Squadron was ordered to prepare six crews for operations on the night of October 8th. Their target was the Deutsche Schiff und Maschinenbau AG shipyards at Bremen.

### October 8th 1940: D1 – Bremen Shipyards

| Pilot | 2nd Pilot | Serial | Code | Bomb Load | Result |
|---|---|---|---|---|---|
| S/Ldr J Schejbal | Sgt J Hrnčíř | R1021 | KX-W | 3x500lb+1x250lb(D)+60x4lb IB | Completed |
| Sgt F Taiber | F/Lt J Ocelka | N2772 | KX-J | 3x500lb+1x250lb(D)+60x4lb IB | Completed |
| Sgt J Bala | P/O J Breitcetl | N2773 | KX-K | 3x500lb+1x250lb(D)+60x4lb IB | Completed |
| S/Ldr J Veselý | Sgt F Zapletal | N2771 | KX-H | 3x500lb+1x250lb(D)+60x4lb IB | Completed |
| P/O F Janoušek | Sgt K Fák | R3177 | KX-L | 3x500lb+1x250lb(D)+60x4lb IB | Completed |
| P/O B Landa | Sgt K Novotný | L7844 | KX-T | 3x500lb+1x250lb(D)+60x4lb IB | Not Complete |

Three pilots would be making their operational debut on this night. Squadron Leader Jan Veselý, Pilot Officer Jindřich Breitcetl and Sergeant František Zapletal, while Sergeant Jaroslav Bala would captain a crew for the first time. The first crew away at 18:36hrs was Sergeant Bala, followed roughly every 10 to 20 minutes by the remaining crews. No sooner had the last Wellington departed than Pilot Officer B Landa requested permission to land after reporting an encounter with an enemy fighter, this was later amended by the crew to an overheating engine. The remaining crews crossed the Dutch coast via corridor 'B' to make landfall in the region of Alkmaar. Flak and searchlights made their presence felt almost the entire route to Bremen. The Czechs would not be the only squadron over the target. 9 Squadron had detailed nine crews. The first bombs started falling at 20:18hrs. Opposition over a cloud-free Bremen was fierce, with barrage flak especially effective. Attacking from between 11,000ft and 13,000ft, the squadron produced some accurate bombing. Numerous explosions were observed in the target area and near a factory complex. The returning crews were enthusiastic, reporting the target was hit repeatedly, producing many explosions with red, orange and yellow flames. These were visible from over 40 miles. A bombload appeared to hit a large factory complex, producing a huge fire that quickly started belching dense smoke which was followed by a colossal explosion that threw debris high into the air showering the surrounding area with masonry and roof tiles. R.A.F Honington reported two Wellingtons, one each from 75(NZ) and 37 Squadron landing just after midnight having mistaken the airfield for their home base at R.A.F Feltwell!

The squadron was stood down on the 9th, giving the ground crews a few welcome hours to work on the Wellingtons and make them secure before the forecast gales arrived. On the 10th, 3 Group detailed forty-eight Wellingtons for operations over northern Germany and the Ruhr. 311 Squadron briefed three crews for an attack on the Rhenania Ossage Mineralolwerk AG in Hamburg. The squadron would be joined over the target on this occasion by 9 and 214 Squadrons.

### October 10th 1940: A10 – Oil Refinery Hamburg

| Pilot | 2nd Pilot | Serial | Code | Bomb Load | Result |
|---|---|---|---|---|---|
| P/O B Landa | Sgt K Novotný | R1021 | KX-W | 1x500lb+52x50lb+60x4lb IB | Completed |
| P/O F Janoušek | Sgt K Fák | R3177 | KX-L | 1x500lb+52x50lb+60x4lb IB | Completed |
| F/Lt J Šnajdr | Sgt L Anderle[4] | N2772 | KX-J | 1x500lb+52x50lb+60x4lb IB | Completed |

The three Wellingtons were away between 22:32hrs and 23:50hrs and headed out over the Suffolk coast. Identifying Hamburg was simplified by the fires left by the Wellingtons of 214 Squadron, who had bombed an hour before. Arriving over Hamburg between 00:50hrs and 01:30hrs, smoke and haze made pin-pointing the refinery difficult. However, Pilot Officer B Landa reported upon return, *'burst seen with large fire burning like a geyser. Red and yellow fire seen as far as Bremen'*. Pilot Officer F Janoušek bombed from 12,000ft, reporting, *'Bombs landed 100 yards North of target'*. The fires created at Hamburg were still visible from Emden on the return trip. The operation was marred on return when Sergeant Leo Anderle undershot his approach to R.A.F East Wretham, damaging the Wellington in the process. Fortunately, none of the crew were injured. The unfamiliar sound of Rolls Royce engines reverberated over R.A.F East Wretham in the early hours of the 11th. A Whitley of 58 Squadron landed at 02:42hrs with engine trouble.

| Date | October 11th 1940 |
|---|---|
| Mark | Mk.Ic |
| Serial | N2772 |
| Code | KX-J |
| Taken On Charge | 28/09/1940 via No.18 MU. |
| Manufacturer | Vickers Armstrong |
| Contract | B.992424/39 |
| Pilot (s) | Flight Lieutenant Josef Šnajdr / Sergeant Leo Anderle |
| Flight | Operations |
| Time | 06:05 |
| Cause | Undershot, stalled on approach. |

The raid appeared to be a success, and apart from the mishap on landing, the participating crews were confident that the oil refinery had been hit. Bomber Command's reprisal raids on Berlin continued on the 12th when four crews were chosen. The selected crews were briefed to attack the Reichsluftfahrtministerium in Berlin's Wilhelmstrasse, the centre of German government.

### October 12th 1940: Area 'T' – Berlin

| Pilot | 2nd Pilot | Serial | Code | Bomb Load | Result |
|---|---|---|---|---|---|
| S/Ldr J Schejbal | Sgt J Hrnčíř | R1021 | KX-W | 3x500lb+1x250lb+60x4lb IB | Completed |
| Sgt F Taiber | F/Lt J Ocelka | L7841 | KX-S | 3x500lb+1x250lb+60x4lb IB | Completed |
| Sgt J Bala | F/Lt J Šnajdr | N2773 | KX-K | 3x500lb+1x250lb+60x4lb IB | Completed |
| S/Ldr J Veselý | Sgt F Zapletal | N2771 | KX-H | 3x500lb+1x250lb+60x4lb IB | Completed |

3 H.Q Group allocated just seven crews to attack the Reichsluftfahrtministerium. R.A.F Honington's 9 Squadron would provide the additional three crews. Conditions en route were less than favourable, with low cloud and haze making navigation difficult. Over Berlin between 22:15hrs and 22:58hrs, 311 Squadron ran the gauntlet of heavy flak and searchlights. Bombing heights varied dramatically between the crews. Sergeant František Taiber opted for 8,000ft, while Sergeant Jaroslav Bala decided height was safer, his crew bombed from 15,000ft. Squadron Leader J Schejbal could not locate the primary due to 10/10th cloud. In its place, he selected the Tegel Gas Works, producing a reddish blaze on departure. The remaining crews were more successful. Each claimed to have attacked the primary. Sergeant J Bala reported his entire bombload landed 1000 yards SW of the ministry, creating a sizable fire. Squadron Leader J Veselý

---

[4] *The ORB contradicts itself with this crew. The Form 540 claimed the aircraft was flown by Sergeant Leo Anderle while the Form 541 records F/Lt J Šnajdr.*

and Sergeant F Taiber each claimed their bombs fell within 500 yards of the aiming point. On leaving the target, several fires had taken hold. One large fire was visible from 40 miles away.

*Two squadron Wellingtons are photographed over the flat Fens of East Anglia. Wellington KX-M R1410 would complete 13 operations, while KX-K R1378 would fly just five. (Authors collection via John Costin)*

On the 13th, the four crews flew the short distance back to East Wretham thrilled at the previous night's success. At lunch in the Dining Hall the raid was celebrated by the participating squadron crews. The joy was somewhat dampened by the news that the highly regarded Squadron Leader Henry Graham was to be posted back to Honington, his time with the Czechs almost over. There would however be one final trip together. Fourteen officers departed East Wretham at 11:00hrs leaving for Norwich for lunch, the guest of honour was the departing Squadron Leader Graham. His loss was keenly felt, the ORB records, *'Squadron Leader Graham, O/C Op's Flight. He was very popular, and it is to his credit that the squadron started operations so soon'.*[5]

On the 14th Bomber Command switched its attention to Le Havre with a low-key raid carried out by just 14 bombers, ten of which were supplied by 3 Group. 311 (Czech) Squadron and the New Zealanders of 75(NZ) Squadron would each supply three crews, while 37 Squadron provided four.

October 14th 1940: CC.24 – Le Havre Docks

| Pilot | 2nd Pilot | Serial | Code | Bomb Load | Result |
|---|---|---|---|---|---|
| W/Cdr K Toman | Sgt K Novotný | L7844 | KX-T | 8x250lb | Completed |
| Sgt J Bala | Sgt L Anderle | N2773 | KX-K | 10x250lb | Completed |
| P/O F Janoušek | Sgt K Fák | R3177 | KX-L | 8x250lb | Completed |

There were some concerns about the weather conditions before take-off. H.Q had already changed the primary target, and now the squadron's participation in the night's activity hung in the balance yet again. A 'Red' Warning alert was received at 19:37hrs, after hostile aircraft were reported in the area. Despite the danger, Sergeant J Bala and crew were the first to depart at 19:42hrs. Given the threat, the station lights were extinguished in between take-offs in fear of an attack. Finally, by 20:42hrs, R.A.F Honington's Wellingtons were safely airborne. The crews crossed the south coast

---

[5] *Henry Graham would take command of the Short Stirling equipped No.7 Squadron in 1941. Survived the war awarded a DSO and DFC.*

and headed almost directly to the target. Conditions en route and over the docks were better than forecast. A large fire aided the later crews to identify the dock complex, which was visible from 40 miles away. Flying over the target between 12,000ft and 16,000ft in the clear conditions gave the crews an opportunity for some accurate bombing. A total of 75 250-pounders and 180 4lb incendiaries were dropped between 22:02hrs and 00:50hrs, inflicting Le Havre's worst raid since early September. The returning crews reported they had hit most of the docks, including a lock gate, hangars along the Bassin Bellot, a tug plus a small ship in one of the Bassin de Citadelle dry docks. There was some wayward bombing with bombs falling well to the northwest of the port area. The fires in the docks were so fierce that every local fireman was needed to tackle the inferno. The only incident in an otherwise quiet operation involved Sergeant J Bala and crew who had an inconclusive encounter with a German night fighter over Le Havre.

Bomber Command switched its attention to northern Germany on the 16th. Thirty-eight Wellingtons of 3 Group were given orders to attack the *'Scharnhorst'* and *'Gneisenau'* located in Kiel while 311 (Czech) Squadron was allocated Bremen and the Deutsche Schiff- und Maschinenbau AG. The night would prove to be a tragic and costly one for the squadron.

### October 16th 1940: D1 – Bremen – Deutsche Schiff- und Maschinenbau AG

| Pilot | 2nd Pilot | Serial | Code | Bomb Load | Result |
|---|---|---|---|---|---|
| P/O F Janoušek | Sgt K Fák | R1021 | KX-W | 5x500lb SAP | Completed |
| F/Lt J Šnajdr | Sgt L Anderle | N2773 | KX-K | 1x500lb+5x250lb+60x4lb | Secondary Target |
| S/Ldr J Veselý | Sgt F Zapletal | N2771 | KX-H | 1x500lb+5x250lb+60x4lb | Early return Crashed |
| P/O B Landa | Sgt K Novotný | L7844 | KX-T | 1x500lb+5x250lb+60x4lb | MISSING |

The first crew away from R.A.F Honington[6] was Flight Lieutenant Josef Šnajdr at 18:30hrs at the controls of Wellington N2773 KX-K. No sooner had the last Wellington departed, reports began to filter through to R.A.F Honington that flares and incendiaries had been dropped a few miles east of East Wretham. Hostiles were known to be in the region. As a precaution, Honington was placed on 'Red' Air Raid Warning at 20:31hrs. The first of the night's tragedies took place at 21:45hrs[7]. Squadron Leader J Veselý and crew encountered freezing conditions while en route to Bremen, the Wellington becoming severely iced-up. A message was sent at 20:05hrs stating the crew were turning for home. Base continuously tried to contact the crew between 20:05hrs and 20:50hrs, without success.

Presumably experiencing W/T failure, the crew were unable to get a fix and found themselves heading dangerously towards the outer flak defences of London. Perhaps unsure of its position, the Wellington was heard to be circling above Fighter Command H.Q for approximately 15 minutes before its eventual crash. The exact cause of the crash of Wellington N2771 is unclear. Various theories have been put forward. Flak damage, a collision with a balloon cable or engine malfunction due to icing have been mentioned. Whatever the reason, five young Czechs were killed when their Wellington crashed and burst into flames near the tennis courts at the Headquarters of Fighter Command at Bentley Priory. The bodies of the crew were recovered and taken to the Harrow Hill Mortuary. The first realisation of trouble was received at R.A.F Honington at 22:40hrs when a message arrived stating; *'A/C 'H' of 311 had crashed onto the tennis courts of FC HQ with pilot badly burnt, who spoke little English'*.

### Vickers Wellington Mk.IC N2771 KX-H

| Manufacturer | Vickers Armstrong | |
|---|---|---|
| Contract | B.992424/39 | |
| Taken on Charge | 28/09/1940 via No.18 MU. | |
| Cat E Missing | N/A | |
| Struck Of Charge | 16/10/1940 | |

---

[6] *Conflicting information regarding where the crews took off. Air 81 reports East Wretham.*
[7] *Strangely the squadron ORB states 23:59hrs, and crew attacked target, this is at odds with RAF Honington Station Records Book which records the first report of a tragedy at 22:40hrs.*

| | | |
|---|---|---|
| Total Flying Hours | N/K | |
| Take-Off Time | 18:45hrs | |
| Bomb Load | 1x500lb+5x250lb+60x4lb | |
| | | |
| | **CREW** | |
| Captain | Squadron Leader Jan **Veselý** 82582 RAFVR. Age 34. | Sec. G.4. Coll. grave 17 |
| Second Pilot | Sergeant František **Zapletal** 787242 RAFVR. Age 30. | Sec. G.4. Coll. grave 17 |
| Navigator | Pilot Officer Jaroslav **Slabý** 82637 RAFVR. Age 25. | Sec. G.4. Coll. grave 17 |
| Wireless Operator | Pilot Officer Jaroslav **Matoušek** 82524 RAFVR. Age 25. | Sec. G.4. Coll. grave 17 |
| Front Gunner | Pilot Officer František **Truhlář** 82643 RAFVR | Survived |
| Rear Gunner[8] | Pilot Officer Josef **Albrecht** 787410 RAFVR. Age 25. | Sec. G.4. Coll. grave 17 |
| | | |
| Posting History | S/Ldr J Veselý via Cosford Depot, 29/07/1940. Sgt F Zapletal NFD. | |
| Operations Flown | S/Ldr J Veselý : 2 / Sgt F Zapletal : 2 | |
| Buried | PINNER CEMETERY | |

Squadron Leader Jan Veselý was born on May 17th, 1906 in Přehořov. On completion of his studies as a teacher he spent two years at the Military Academy in Hranice, where he finished as an infantry lieutenant in July 1928. In October 1931, he finally got his wish when he was accepted into the 18th Observer School of the Military Aviation School in Prostějov. During the Czech mobilisation in 1938, he served as the first officer on two squadrons in 5th Air Regiment in Brno. Like many young Czechs, he fled Czechoslovakia via Poland. He and many others sailed to France with the condition he served five years' service in the Foreign Legion. In France, his flying experience would see him train to be a bomber pilot with the Armée de l'Air, and it is understood that he saw action during the Battle of France. With the fall of France, he and many other brave Czechs were forced to flee, this time to North Africa, from where Jan Veselý finally arrived in Liverpool. A wonderful memorial was erected in his honour in his home town in 2018. The sole survivor, Pilot Officer František Truhlář, was badly burnt, but alive, the rear section of the Wellington having broken-off on impact. He was initially taken to the Brockley Hill Orthopaedic Hospital with burns to his face, hands and foot. He would spend the next year having treatment becoming one of the first Czechoslovakian 'Guinea Pig's' at East Grinstead under the pioneering care of surgeon Sir Archibald McIndoe and his wonderful team of doctors and nurses. František courageously returned to operations in 1943 as a Spitfire pilot with 312 (Czech) Squadron. In 1944, he was involved in another serious flying accident and once again suffered severe burns. He survived the war only to be killed while flying a Spitfire over his hometown in Czechoslovakia on 3rd December 1946. Another tragedy was unfolding over Holland. Pilot Officer B Landa and crew had the misfortune of being intercepted by Lt Ludwig Becker of 4./NJG1 at 21:25hrs while en route to Bremen. The encounter was reported to have been the first RAF bomber to have been intercepted and shot down by a night fighter assisted by the ground control radar system, *Freya*.

<u>Vickers Wellington Mk.IC L7844 KX-T</u>

| | | |
|---|---|---|
| Manufacturer | Vickers Armstrong | |
| Contract | 992424/39 | |
| Taken on Charge | 28/09/1940 via No.22 MU. | |
| Cat E Missing | 16/10/1940 | |
| Struck Of Charge | N/K | |
| Total Flying Hours | N/K | |
| Take-Off Time | 18:50hrs | |
| Bomb Load | 1x500lb+5x250lb+60x4lb IB | |
| | | |

---

[8] *Compared to the ORB entry, the Air Gunners switched places in fact. Sgt Albrecht occupied the front turret and P/O Truhlář was manning the rear turret.*

|  | **CREW** |  |
|---|---|---|
| Captain | Pilot Officer Bohumil **Landa** 82557 RAFVR. Age 43. | Plot 1. Row 2. Grave 61 |
| Second Pilot | Sergeant Emanuel **Novotný** 787250 RAFVR. | PoW |
| Navigator | Pilot Officer Hubert **Jarošek** 82605 RAFVR. Age 30. | Plot 2. Row 2. Grave 62 |
| Wireless Operator | Sergeant Karel **Klimt** 787547 RAFVR. Age 28. | Plot 1. Row 2. Grave 59 |
| Front Gunner | Sergeant Otto **Jirsák** 787141 RAFVR. Age 33. | Plot 2. Row 2. Grave 60 |
| Rear Gunner | Sergeant Augustin **Šesták** 787153 RAFVR | PoW |
|  |  |  |
| Posting History | P/O B Landa via Cosford Depot, 29/07/1940. Sgt E Novotný NFD. |  |
| Operations Flown | P/O B Landa : 4 / Sgt E Novotný : 5 |  |
| Buried | OOSTERWOLDE GENERAL CEMETERY, OLDEBROEK |  |

Having been grievously damaged and set on fire, Pilot Officer Landa turned the Wellington back towards the relative safety of the North Sea. Unfortunately, the damage was just too extensive and the Wellington crashed. The crash site was subsequently investigated by the Germans who recovered four bombs. Two more, both 250-pounders were recovered in 1981. The encounter was described by the victor, Lt Becker.

*'I was controlled very well by Leutnant Diehl to the correct height 3,300 metres in a curve-of-pursuit. Approach to the enemy machine from starboard and astern with frequent course correction until, in the moonlight and about 100 metres to port and above I saw an aircraft. As I came closer, I recognized it as a Wellington. I gradually positioned myself astern of it and manoeuvred very close aiming at the fuselage and wing-roots. I gave it a burst of about five or six seconds. The starboard motor caught fire at once, I pulled my machine above the bomber and away. For a short period the British flew on losing height, then the fire went out and I saw the enemy aircraft dive to earth in a typical spinning motion. There was a fire where he hit the ground.'*

*Wireless Operator Sergeant Karel Klimt posing by Wellington Mk.IC L7844/KX-T. On the night of 16th/17th October 1940 the failed to return from the raid on Bremen with the crew of P/O Bohumil Landa including Sgt Klimt. (Pavel Vančata)*

Wellington L7844 KX-T crashed near Oosterwolde (Gelderland) 17 miles NNW of Apeldoorn, killing four of the crew, including the pilot, Pilot Officer Bohumil Landa. The crew were initially buried at Osterwyk.[9] The survivors, oblivious to the fate of their friends, eventually arrived over the cloud covered docks of Bremen. Weather conditions prevented both crews from attacking their primary objective. Flight Lieutenant J Šnajdr bombed CC.20 in Bremen at 21:40hrs from 15,000ft. Pilot Officer F Janoušek headed towards Kiel and the Krupp-Germania shipyard, which they bombed from 10,000ft. On return to base, they reported *'two bursts on Docks, five bursts near ships, two near Gneisenau'*.

While on the return journey, Flight Lieutenant J Šnajdr experienced trouble with the Wellington's wireless set. With quickly deteriorating weather, the crew had difficulty establishing their position, and without the wireless equipment, they groped their way back towards England. Over England, fog, rain and reported icing made any attempt to establish

---

[9] *Air 81.*

*Survivor, front gunner Pilot Officer František Truhlář. Photographed in his Czechoslovak Air Force uniform. (John Costin)*

their position almost impossible. Finally, with his fuel tanks registering almost empty, Flight Lieutenant J Šnajdr had no option but to order his crew to their parachutes. The first three to leave were Pilot Officer J Fürbach, Sergeants O Langer, and L Anderle. All three landed without mishap. Pilot Officers M Vejražka and J Richter quickly followed. The last to leave was the pilot, Flight Lieutenant J Šnajdr. The time was 03:22hrs, and they had been airborne for 6 hours 10 minutes. Tragically, Pilot Officer Miroslav Vejražka's parachute failed to deploy, and he was killed on impact near East Farm, West Willoughby, near Grantham.[10] Some of the parachute was tangled around his lower body while parachute cords were tangled around his left hand. Wellington N2773 KX-K crashed near Blidworth, 10 miles NNE of Nottingham and burnt out.

<u>Vickers Wellington Mk.IC N2773 KX-K</u>

| Manufacturer | Vickers Armstrong | |
|---|---|---|
| Contract | 992424/39 | |
| Taken on Charge | 28/09/1940 via No.18 MU. | |
| Cat E Missing | 16/10/1940 | |
| Struck Of Charge | 25/10/1940 | |
| Total Flying Hours | N/K | |
| Take-Off Time | 18:30hrs | |
| Bomb Load | 1x500lb+5x250lb+60x4lb IB | |
| | **CREW** | |
| Captain | Flight Lieutenant Josef **Šnajdr** 82575 RAFVR | Survived |
| Second Pilot | Sergeant Leo **Anderle** 787563 RAFVR | Survived |
| Navigator | Pilot Officer Josef **Richter** 82629 RAFVR | Survived |
| Wireless Operator | Pilot Officer Miroslav **Vejražka** 82533 RAFVR. Age 29. | Row C. Grave 5. |
| Front Gunner | Pilot Officer Jan **Fürbach** 82599 RAFVR | Survived |
| Rear Gunner | Sergeant Oldřich **Langer** 787196 RAFVR | Survived |
| | | |
| Posting History | F/Lt J Šnajdr via Cosford Depot, 29/07/1940. | |
| Operations Flown | F/Lt J Šnajdr : 2 / Sgt L Anderle : 2 | |
| Buried | HONINGTON (ALL SAINTS) CHURCHYARD, | |

The tragic events of the night were by and large unknown to the squadron until messages were received at R.A.F Honington which were then passed onto East Wretham. It is interesting to see the sequence of events at R.A.F Honington on what would turn out to be a busy night. I have excluded the landing times of 9 Squadron.

22:40hrs : No.3 Group informed that M.L.O reported that aircraft 'H' of **311** Squadron had crashed into the tennis courts at FCH.Q.
23:41hrs : Wellington 'W' **311** Squadron landed.
01:45hrs : Wellington 'P' of 214 Squadron landed.
01:45hrs : X2987 of 61 Squadron landed East Wretham.
01:45hrs : A message from Wells that Sergeant Anderle of No.34 Squadron (**311**) had landed by parachute at Brayfield Park Farm, 2 miles from Wells.
02:00hrs : Two Wellingtons of 214 Squadron landed.
02:35hrs : Aircraft 'K of 37 Squadron landed.
03:10hrs : Aircraft 'M' of 144 Squadron landed.
03:55hrs : Message from RAF Digby. 4 members of **311** Squadron aircraft bailed out near Newark. Various messages received. No news of aircraft 'T' 311 Squadron.

---

[10] *There are unconfirmed reports that he was shot and killed by an overzealous Home Guard member.*

*Left; Original war time cross with inscription of Imperial War Graves Commission (IWGC) renamed to the present Commonwealth War Graves Commission (CWGC) in 1960. Pilot Officer Miroslav Vejražka was one of the first three 311 Squadron members buried at All Saints Churchyard at Honington on 21st October 1940. (Pavel Vančata & Authors collection)*

There was a bazaar and strange twist to the operation. At 08:50hrs on the morning of the 17th, a message was received sent by an aircraft identifying itself as 'K' of 311 Squadron asking permission to land at R.A.F East Wretham. Nothing more was heard from this aircraft, which did not land. No explanation could be found for the message, or who could have sent it!

The realisation that the previous night's raid had resulted in one missing crew, the majority of another all dead, and the third crew obliged to take to their parachute was hard to comprehend. Shock and sorrow swept the squadron. Six friends and colleagues dead, and another six missing hit the surviving squadron members hard. There was little, if anything, the survivors could do but continue with their day-to-day activities. A crew were airborne late morning on local flying practice, it did little to lift the gloom. Conditions on take-off could have been better. A blanket of low cloud down to 300ft in places was reported. However, the Deputy Squadron Commander allowed the training flight to go ahead. At 12:10hrs, Bury Observer Corp reported that a 'Wellington' had crashed in flames between the village of Greeting St Mary and Needham Market. There was some relief when a follow-up message from Group H.Q reported the aircraft was a Hawker Hurricane. At 12:26hrs, a more ominous message arrived, the aircraft was, in fact, a Wellington, its identity unknown at that time. Another message arrived at 13:00hrs, and to the utter disbelief of everyone, it confirmed that the Wellington was L7786 of 311 Squadron. Both the pilot, Sergeant Oldřich Tošovský and his second pilot, Sergeant Karel Lang, had been killed when their low-flying Wellington hit power cables and crashed at Pipps Farm, Coddenham near Needham Market at 11:20hrs. During the afternoon, the squadron commanding officer visited the crash site, and immediately upon his return, reported to the Station Commander, Group Captain Harrison DFC, AFC that it was his opinion that the crew had flown into a balloon cable. The news of yet another crash was not taken well by the Czechs, yet more friends needlessly killed. Anger and frustration had started to creep into the various Messes. Aware of the irritation, Wing Commander Mareš Toman requested that his personnel not discuss the incident as a certain amount of discontent and criticism was being voiced. With the death of more colleagues, the obviously upset crews posed the question of the squadron's possible deployment to North Africa, or the Colonies.[11] All were fully aware of what awaited them, and their families back in Czechoslovakia if shot down and captured by the Germans. Brutal interrogation and possible execution for them and perhaps their love-ones. It is hard to imagine the strain this must have placed on the aircrew, who bravely climbed aboard their Wellingtons almost nightly. Given the circumstances surrounding the deaths of Sergeants Tošovský and Lang, the Station Commander ordered an inquest be immediately carried out by the Coroner of Stowmarket, the inquest would take place at R.A.F Wattisham.

Flight Lieutenant Richard MacFadden DFC was ordered to fly the squadron Avro Anson to R.A.F Digby to collect the survivors of N2773 during the afternoon. Unbeknown to the squadron, three were already returning to East Wretham via road and rail. At 17:57hrs, Anson R9600 landed back at base with Flight Lieutenant Josef Šnajdr and Pilot Officer Joseph Richter. Unaware, they were briefed on the tragic events that had unfolded over the previous 24 hours. Safety concerns raised at the crash site of N2773 by the recovery teams were dispelled when the interrogated crew confirmed all the bombs had been dropped.

<u>Vickers Wellington Mk.IA L7786 KX-?</u>

| Manufacturer | Vickers Armstrong | |
| --- | --- | --- |
| Contract | 992424/39 | |
| Taken on Charge | 11/09/1940 via 9 Squadron. | |
| Cat E Missing | N/A | |
| Struck Of Charge | N/K | |
| Total Flying Hours | N/K | |
| Take-Off Time | ?? | |
| Bomb Load | N/A | |
| | | |
| | **CREW** | |
| Captain | Sergeant Oldřich **Tošovský** 787237 RAFVR. Age 23. | Row C. Grave 6. |
| Second Pilot | Sergeant Karel **Lang** 787416 RAFVR. Age 24. | Row C. Grave 7. |
| Navigator | - | |
| Wireless Operator | - | |
| Front Gunner | - | |
| Rear Gunner | - | |
| | | |
| Posting History | Sgt O Tošovský NFD.<br>Sgt K Lang NFD. | |
| Operations Flown | Nil | |
| Buried | HONINGTON (ALL SAINTS) CHURCHYARD | |

Group H.Q obviously had their doubts about Wing Commander Mareš Toman judgement of the cause of the crash of L7786. At 19:41hrs, Group H.Q stated that if 311 Squadron wished to support the claim that a balloon cable caused the Wellingtons crash, they would need to provide evidence to support the claim. The subsequent investigation found that high tension cable had wound around the port propeller and engine nacelle. Wing Commander Griffiths DFC closed the accident investigation on October 21$^{st}$ with an unequivocal verdict: *'The root cause of the accident was the disobedience of an order by a pilot who was conducting illegal flying at low altitude. A secondary cause of the crash – an impact with high-voltage wires that prevented the aircraft from flying over a tall oak.'*

---

[11] *Air 27/1686 - 311 Operations Record Book.*

A thick blanket of fog settled over Honington and East Wretham on the evening of the 17th, ending a tense and emotional day. The 54-year-old Czech Foreign Minister, Mr Jan Masaryk, arrived at the squadron on the 18th accompanied by Group Captain Frank Beaumont from the Air Ministry. All the Czechs were paraded. For over two hours, the Foreign Minister spoke with, and listened to the assembled crews, the recent tragic events being foremost in the discussion. A drop in confidence on the squadron may have resulted in Wing Commander Mareš Toman taking the unprecedented step of informing 3 Group HQ on the morning of the 18th that given the recent spate of accidents it was his opinion that 311 Squadron should be withdrawn from operations that night. This prompted a visit by Group Captain Barrett and Mr Masaryk to East Wretham. Just before lunch, Group H.Q requested that Wing Commander Toman Mareš plus an interpreter should travel to Group H.Q for a meeting to be held 16:15hrs to *'discuss weather conditions over Czechoslovakia'*. This was perhaps a veiled reference to the issue of squadron morale. The outcome of this meeting is unknown.

On the 21st, the funeral of Squadron Leader Jan Veselý and crew took place at Pinner Cemetery. Flight Lieutenant Provazník and fellow officers attended. Strangely, no Czech officials were in attendance, which left the crews slightly bewildered. The same day Sergeants Tošovský and Lang were buried at Honington (All Saints) Churchyard. That night two of the most experienced crews were detailed and briefed for a raid on Hamburg. They would on this occasion take-off from R.A.F East Wretham.[12]

<u>October 21st 1940: D2 – Hamburg 'Bismarck' Blohm & Voss Shipyards</u>

| Pilot | 2nd Pilot | Serial | Code | Bomb Load | Result |
|---|---|---|---|---|---|
| S/Ldr J Schejbal | Sgt J Hrnčíř | R1021 | KX-W | 4x500lb SAP | Completed |
| Sgt F Taiber | F/Lt J Ocelka | L7841 | KX-S | 3x500lb SAP | Completed |

The Group would detail 23 Wellington crews against Hamburg, drawn from 9, 37, 75(NZ) and 311 Squadrons. The first crew away was Squadron Leader J Schejbal at 18:15hrs, followed two minutes later by Sergeant F Taiber. There would have been some tension at East Wretham, considering the run of bad luck the squadron had suffered of late. Conditions over Hamburg were terrible, with thick ground mist making visual identification of the target impossible. The small force was over the target area between 20:19hrs and 21:09hrs. Ten of the 23 Wellingtons deployed would bomb last resort targets due to the conditions. However, both 311 Squadron crews were more successful. Squadron Leader J Schejbal bombed from 11,000ft reporting, *'Bombs dropped in target area. Many fires, some visible from 20 miles'*. Sergeant F Taiber dropped his all HE load from 12,000ft, experiencing heavy flak all the while. On return to base, he reported, *'Numerous explosions observed mostly white colour. Many fires seen'*. Both crews landed at R.A.F Honington on return, as did five diverted Wellingtons of 37 and 75(NZ) Squadrons, their home airfield being fog-bound. The raid may not have been a complete success, but the safe return of both crews was just the tonic the squadron needed.

The squadron was visited by Group Captain Beaumont, Wing Commanders Carter and McKee and Air Commodore RNDr Karel Janoušek, the Inspector of Czechoslovak Air Force in the R.A.F, on the 23rd. The purpose of the visit was to discuss the underlying issues on the Squadron. On conclusion of the visit Air Commodore Janoušek had a meeting with 3 Group AOC, AVM Baldwin CB, DSO, OBE. There was obviously a genuine problem on 311 Squadron which was affecting morale and relations between the Czechs and certain senior R.A.F officers. Air Commodore Janoušek's own investigation highlighted a number of issues: *'1) Lack of cooperation between Czech and British personnel was the result of an inappropriate choice of the British squadron leader. Czech personnel were also partly to blame, the Czech squadron leader tended to downplay the problems instead of informing the base commander about them. 2) Crews are declared operational before they have been properly trained. 3) Although there have been cases where individuals had to be sent from the squadron due to poor morale, suspicions of subversive activities have not been confirmed.'*

---

[12] *Honington Station Operations Record Book.*

It was an honest report on the state of the squadron. The main issue appeared to be Wing Commander John Griffiths DFC style of command. Born in 1905 in Stamford City, Ontario, his military career started aged 19 when he attended the Royal Military College between 1924 and 1926. That same year, he obtained his flying badge. He accepted a commission in the R.A.F and served with a number of squadrons during the 20s and 30s, including 13, 28 and 31 Squadrons. In 1932, he undertook a Russian language course at Kings College. After serving in Estonia, Ethiopia and Malta, John Griffiths returned to Great Britain and took command of 99 Squadron just before the outbreak of war. He was a typical example of a pre-war R.A.F officer, a well-groomed, tidy man with a well-manicured moustache. He epitomised the service requirements of the pre-war R.A.F. His courage was not in doubt. However, the task of commanding and teaching the brave and superstitious Czechs was something completely different. An empathic disposition was required to inspire confidence and get the best out of the squadron. It would appear that Wing Commander Griffiths DFC did not possess this most delicate of traits. His job may have been made more challenging by the overprotective nature of Wing Commander Mareš Toman, who, like a proud parent, did all he could to protect his men even when circumstances required a firm hand. Mareš Toman was highly respected by all, but like his opposite number, he had been trained in the stale and authoritarian ways of a peacetime Air Force whose standards and rules were now out of place in a war of survival. That same day the squadron Operations Record Book mentions the squadron was taken off operations. R.A.F Honington's Station Records Book reports that the Commanding Officer of 311 Squadron stated that his squadron would be unavailable until further notice.

There was much to consider at Group H.Q. Everything had to be handled carefully and tactfully, Czech national pride was at stake as was the reputation of 3 Group and R.A.F Bomber Command. Diplomacy was key, it was important that no one was blamed or considered at fought. Finally, to try and resolve the delicate situation, AVM Baldwin CB, DSO, OBE had the following measures implemented: *'(1) The R.A.F commanding officer would be replaced. (2) Transfer the entire squadron to R.A.F East Wretham as soon as possible and give the squadron a sense of autonomy by having the entire base to itself.'* The third and final recommendation would have been a difficult one given the war situation and the need for every possible squadron: *'(3) Suspend operations for two or three months and use East Wretham as an operational training unit to train new crews.'*

The job of training would stay with the capable pair and good friends, Squadron Leader J Schejbal and Flight Lieutenant 'Pick' Pickard DFC. Together they planned and organised an intensive training programme. On the 25th and 27th, firm action was taken by Wing Commander Mareš Toman when two Czech airmen chose to disobey orders. Both were sent to R.A.F Cosford. An unidentified Bristol Blenheim crew chose R.A.F East Wretham as a makeshift target range during the late afternoon of the 25th. The aircraft made a number of simulated dive-bombing attacks on the airfield, much to the annoyance of everyone on the ground. A strongly worded message was passed on to 2 Group H.Q. On the 27th, R.A.F Honington was attacked by the Luftwaffe. A total of thirty-six HE bombs were dropped, destroying one Wellington and damaging a further two. Superficial damage was reported, and sadly, three airmen were killed and another injured in the attack.

Given the ill-feeling and issues on the squadron over the previous two weeks, the Czechs decided to arrange a celebratory day in the N.A.A.F.I at R.A.F Honington on October 28th, the occasion, Czech Independence Day. Considerable care and attention went into decorating the N.A.A.F.I, adorned with traditional Czech decorations. A large Czechoslovakian flag took centre spot on the stage flanked on either side by photographs of King George IV and Czechoslovakian President Edvard Beneš. The squadron invited 3 Group AOC, AVM Baldwin CB, DSO, OBE, Honington's Station Commander, Group Captain Richard Harrison DFC, AFC[13], 9 Squadron commanding officer, Wing Commander A.E. Healy, Wing Commander J Griffiths DFC and all the R.A.F servicemen on 311 Squadron. The celebration started at the relatively early hour of 10 a.m. In a typical Czech fashion, the entire day was minutely planned, with Czech songs, poems, music and a choir interspersed with fairy tales to entertain their guests. If anything, it was an opportunity to display and celebrate Czechoslovakian culture. A football match was also organised, officers versus a combined sergeant & airmen XI. The final score was 2-2. It was all very new to the R.A.F, who were more accustomed

---

[13] *Would in February 1943 take command of No.3 Group.*

*Working out at a distant dispersal in the summer could be pleasant. However, with the onset of winter, conditions quickly deteriorated, and for the poor groundcrews, it became a battle with the elements. In this photograph, the ground staff inspects a Bristol Pegasus engine of one of 311 Squadron Wellingtons. (Pavel Vančata)*

to a riotous piss-up to let off steam. Other than the late arrival of the AOC and an Air Raid Warning the entire day and evening was enjoyed by all and did much to heal the obvious rift between the Czechs and senior R.A.F officers.

The following evening a solitary Do17 was reported by the personnel manning the 'K' Site to be heading towards R.A.F Honington. The alarm was sounded and at 18:08hrs explosions were heard, but the station was spared. The only damage, a piece of shrapnel had hit the engine cowling of a dispersed 9 Squadron Wellington. The following day the station was visited by H.R.H The Duke of Kent, who along with AVM Baldwin CB, DSO, OBE visited the Operations Room and watched as eleven crews of 9 Squadron were briefed for operations that night. With the temporary withdrawal from operations the Czechs now had an ideal opportunity to implement a more structured and unhurried training programme. It was a welcomed reprieve allowing some of the recent arrivals, and those on the Training Flight the prospect of some additional and needed training.

*Some of the individuals who set the standards and guided the squadron in the early days. Standing outside Wretham Hall are L-R, Wing Commander J G Griffiths DFC, Flight Lieutenant M J Earle, Squadron Leader J Schejbal, Wing Commander K toman-Mareš, Flight Lieutenant T Kirbt-Green, Squadron Leader M Provaznik, Group Captain J Beronounský and Flight Lieutenant P Pickard.*

# November 1940 : Back to Basics

With much of the pressure lifted, the aircrews could now concentrate solely on their training. Despite the squadron having a handful of experienced aircrews within its ranks, it is apparent that its operational debut was premature. This, in part, was due to a few high-ranking and influential Czech military officials insisting that the Czech airmen were ready to commence operations when they were clearly not. The opportunity for refresher training at Operational Training Units was rebuffed and deemed by those high-ranking officials unnecessary. National pride and political motives may have been at work. Sadly, the only ones to pay were the brave young Czech aircrew. Clearly, the whole training regime needed to be overhauled, with particular attention on night-time navigation, which in pre-war Czechoslovakia was almost unheard of. The events of October 16th also highlighted an urgent need for more intensive training for the squadron wireless operators on the R1082 receiver and the T1083 transmitter. It was a vital role that seemed insignificant to some on the squadron. The first week of November found the squadron flying almost daily despite the weather. When the weather was too bad for flying, it was to the classroom for lectures and talks. On the morning of November 5th, two crews and passengers were ordered to R.A.F Aldergrove, Northern Ireland to collect five Wellingtons. At around 18:37hrs the following evening, R.A.F Honington was informed that of the five Wellingtons collected, three had landed at R.A.F Abington, one in 'Cambridge' and worryingly, one was unaccounted for. Three minutes later, Honington contacted Group H.Q for information on the missing Wellington. At 20:45hrs, Regional Control was instructed to put the procedure for overdue aircraft into operation. For the next two hours, the squadron anxiously waited for news. Finally, at 23:05hrs, news arrived that the Wellington and its crew had landed at R.A.F Silloth, Cumbria, but no attempt to inform either Aldergrove or Honington had been made. The collection and delivery service continued on the 8th when crews were ordered to R.A.F Aldergrove, R.A.F Edzell in Scotland and Hawarden in Chester to collect and deliver urgently needed Wellingtons. Although unglamorous, it gave the observers much-needed experience in navigation. The squadron reported an accident involving Pilot Officer J Breitcetl on the 7th. While on his fifth landing attempt, he inadvertently tried to land across East Wretham's flarepath. Realising his mistake, he opened up to go around again but stalled the Wellington, which hit the ground from 20ft. The crew were shaken but unharmed. The Wellington, which had been on the squadron less than a month, would need specialist repair and would not see further service on the squadron.

| Date | November 7th 1940 |
|---|---|
| Mark | Mk.Ic |
| Serial | T2467 |
| Code | KX-? |
| Taken On Charge | 20/10/1940 via No.22 MU. |
| Manufacturer | Vickers |
| Contract | 38600/39 |
| Pilot (s) | Flight Lieutenant Josef Ocelka / Pilot Officer Jindřich Breitcetl |
| Flight | Night Circuit and Landings |
| Time | 20:45hrs |
| Cause | Approached flare path out of wind, opened up and stalled. |

Whitleys of 10 and 78 Squadron departed R.A.F Honington at 17:55hrs on the early evening of the 8th for operations having used Honington as a forward airfield. While taking off, a 'Blenheim' was observed circling the station, no alarm had been raised and there seemed no apparent danger. At 18:10hrs, the first bombs exploded near the flarepath, the 'Blenheim' was in fact a Junkers Ju88 (V4+ER) of 7./KG1. The airfield defences opened up and Army Lewis gunner, Gunner Tom Sudbury opened fire with his Lewis gun hitting the Ju88 which crashed near 'D' hangar, killing the pilot Lt Peter Ungerer and his crew. On the 11th, two Wellingtons, held back at R.A.F Edzell due to weather conditions, were raising concerns at Honington. Neither had made contact, and their actual whereabouts were unknown. The training continued. Weather conditions and frequent warnings of 'hostile' aircraft apart, the crews appeared to appreciate and benefit from the new training regime. The 'missing' Wellingtons and their crews were finally found when they delivered their aircraft to R.A.F Oakington on the 13th. Carelessness on the part of Sergeant Hugo Dostál would see him severely reprimanded on the 14th when he struck a hangar while taxiing too fast at East Wretham. It was a silly mistake and one that would see Wellington Mk.I L4338 require the attention of Vickers Armstrong. The aircraft would not return to squadron service.

| Date | November 14th 1940 |
|---|---|
| Mark | Mk.I |
| Serial | L4338 |
| Code | KX-R |
| Taken On Charge | 04/08/1940 via 214 Squadron. |
| Manufacturer | Vickers Armstrong |
| Contract | 38600/39 |
| Pilot (s) | Sergeant Hugo Dostál |
| Flight | Not Known |
| Time | Not Known |
| Cause | Taxiing from dispersal too fast struck hangar |

R.A.F Honington was inspected on the 14th, by the Group AOC and Bomber Command's Commander-in-Chief. On completion, they travelled to East Wretham where in torrential rain they carried out a brief inspection of the rather sparse facilities. At 00:25hrs on the 15th, an S.O.S was received at R.A.F Honington from Wellington 'P' that it was in trouble and preparing to ditch in the North Sea. This was followed by another S.O.S from the Wellington at 01:10hrs, followed 13 minutes later by a faint signal, *'Abandoning aircraft, raft'*. At 03:45hrs, 311 Squadron was requested to have three crews available at first light for a sea search. Locating the crew quickly was essential. The North Sea was no place to be in mid-November.

Airborne as ordered, they headed out towards the last known position of Wellington 'P'. At 11:00hrs, Sergeant Jan Hrnčíř, flying Wellington R1021 KX-W located a dinghy. He immediately contacted base with *'We have dinghy'* and started to circle the crew huddled below and requested a fix. The delighted Czechs were relieved at 12:45hrs by another squadron crew. All but the second pilot were eventually picked up by boat. Wet, cold and suffering from exposure, they had been extremely lucky. Vickers Wellington T2509 KO-W had been badly damaged by flak near Hamburg. That evening, a message was received from the commanding officer of 115 Squadron, Wing Commander H.I. Dabinett, thanking everyone involved in the search.

It was on this day that 311 Squadron parted ways with Wing Commander Griffiths DFC on posting to H.Q Bomber Command. His departure was in response to the investigation in October. Griffiths was a first-rate and courageous officer but not best suited to the diplomatic juggling act of commanding a squadron manned by non-R.A.F personnel. His replacement was Acting Wing Commander William Stephen Pomeroy Simmons MiD. Born in London in 1906, he sailed to New Zealand in 1928

*Air Vice Marshal J.E Baldwin K.B.E., C.B., D.S.O. is seen here with Air Chief Marshal Edgar Ludlow Hewitt, Inspector General of the R.A.F. Hewitt had been unceremoniously replaced in April 1940 as A.O.C. R.A.F Bomber Command by Air Chief Marshal Portal due to his insistence on forming Operational Training Units, which many senior staff officers believed deprived men and aircraft from front-line squadrons.*

to try his hand at farming. His time there was brief as he had returned to England by 1931 and was granted a Short Service Commission as a Pilot Officer on probation from December that year. He spent some time serving in both India and the Middle East. By September 1939, with the rank of Squadron Leader he was officer-in-charge of training with 52 Squadron.

The ongoing issue of discipline was again raised on the 15th with the posting to R.A.F Cosford of brothers Pilot Officers Ludvík and Pavel Kozák, both of whom refused to fly. It was an unpleasant business, but the squadron was slowly getting rid of its disruptive individuals. On the 16th, another sea search was organised, and again, three crews were airborne just after lunch. Weather conditions were not ideal, and despite flying below the clouds and murk for nearly 4 ½ hours nothing was found. Training continued unabated. Six squadron Wellingtons were airborne just after mid-day on the 24th on a formation flying exercise, followed that night, by two crews on a night training flight. The following day, the Bombing and Gunnery Range at Berners Heath was allocated exclusively to 311 Squadron for practice. More cross-country training flights were flown on the morning of the 26th, while Wing Commander Toman Mareš organised a long-range sweep over the Irish Sea for the following day. The sweep was cancelled, and in its place the squadron was allotted ferrying duties. Three Wellingtons left for R.A.F Edzell at 10:40hrs, followed by five Wellingtons at 11:50hrs, two destined for R.A.F Edzell, while three were on their way to R.A.F Hullavington. At 16:55hrs news arrived that R1036 'F' flown by Pilot Officer František Cigoš had forced landed at Bradwell Common damaging the propellers and wing tip due to engine failure.

*William Stephen Pomeroy Simmons MiD. His posting was anything but straightforward. Diplomacy, tact and a firm but fair hand were needed to ensure confidence from the disgruntled Czechs. It would be a troubled alliance.*

| Date | November 27th 1940 |
| --- | --- |
| Mark | Mk.Ic |
| Serial | R1036 |
| Code | KX-F |
| Taken On Charge | 30/10/1940 via No.48 MU. |
| Manufacturer | Vickers Armstrong |
| Contract | 992424/39 |
| Pilot (s) | Pilot Officer František Cigoš |
| Flight | Ferrying Flight |
| Time | 16:05hrs |
| Cause | Undercarriage collapsed. |

None of the crew were hurt, but an almost new Wellington was again destined for repair. With an intensification of training there was a resurgence in confidence amongst the aircrew. The troublemakers, or more vocal individuals, had been sent to Cosford, and a feeling of camaraderie had begun to develop amongst the Czechs and R.A.F. The last few days of November would see the squadron successfully complete a number of night training sorties and cross-country exercises, a challenging thing to achieve given the constant warnings of intruders and abysmal weather. November had been a difficult month for the whole squadron. A new commanding officer, the removal of disruptive individuals plus the added opportunity to concentrate on their training needs had witnessed a shift in temperament throughout the squadron. The remaining Czech aircrew were equal to the challenge and proved themselves highly capable and willing.

*A lovely photograph of Vickers Wellington Mk.IC T2469 KX-H. This Wellington saw only limited service with 311 Squadron from September 20th 1940. It would eventually be transferred to 214 Squadron. (Zdeněk Hurt)*

# December 1940 : Back on Operations

The month started with another avoidable flying accident. On the afternoon of the 2nd, Pilot Officer Josef Šejbl was airborne practising single engine flying and landings in Wellington P9212. During one practice landing, Pilot Officer Šejbl misjudged his speed, and the Wellington landed heavily tail first, causing structural damage. The crew emerged unhurt, apart from a few cuts and bruises. The cause of the accident was attributed to carelessness on the part of Pilot Officer Šejbl, who was rather harshly reprimanded. The Wellington came to a standstill 200 yards right of R.A.F Honington's floodlights and would prove a problem moving. A planned night exercise was cancelled as fog was expected, and as a precaution, the squadron was warned to have Gooseneck and Money Flares ready.

| Date | 02/12/1940 |
|---|---|
| Mark | Mk.IA |
| Serial | P9212 |
| Code | KX-J |
| Taken On Charge | 11/10/1940 via NZ Flight R.A.F Stradishall |
| Manufacturer | Vickers Armstrong |
| Contract | 549268/36 |
| Pilot (s) | Pilot Officer Josef Šejbl / Sgt Josef Kalenský |
| Flight | Training |
| Time | 15:45hrs |
| Cause | Pilot Error |

The squadron welcomed Czech-born Pilot Officer Jan Gellner RCAF on the 3rd. He had fled to America pre-war, joined the RCAF, and would serve with some distinction on the squadron as a navigation instructor, then Navigation Leader. Four crews were airborne on the evening of the 4th despite the threat of fog, which would play havoc with training over the next week. R.A.F Honington welcomed a new station commander on the 5th, Group Captain John Astley Gray DFC, fresh from a stint at No.1 Group H.Q. He would replace the outgoing Group Captain Harrison DFC, AFC, who would replace him at No.1 Group H.Q.

December 6th was a milestone in the squadron's training. A message was sent to Group H.Q stating that from Monday, December 9th, 311 Squadron would be ready to operate four crews. In fact, the squadron was ordered to prepare two crews on the 8th. However, this operation was cancelled at 16:17hrs due to conditions over the continent. In its place, the frustrated crews had to settle for night flying training. Finally, on December 9th, 1940 and after all the hard work, 311 Squadron was ready to resume operations. That morning, the squadron informed Group H.Q that it could provide four crews for that night, but as the day drew-on, this was reduced to two.

### December 9th 1940: C.29 – Boulogne Docks

| Pilot | 2nd Pilot | Serial | Code | Bomb Load | Result |
|---|---|---|---|---|---|
| S/Ldr P Pickard DFC | Sgt J Křivda | R1021 | KX-W | 4x500lb+180x4lb IB | Completed |
| Sgt J Bala | Sgt P Uruba | L7842 | KX-T | 4x500lb+180x4lb IB | Not completed |

Operating that night was Squadron Leader 'Pick' Pickard DFC. He and his fellow instructors, both R.A.F and Czech had worked tirelessly to get the squadron ready. Joining him over Boulogne would be one of the squadron's most experienced Czech captains, Sergeant Jaroslav Bala. The first away from R.A.F Honington was Sergeant Bala and crew. Thirty minutes later Squadron Leader Pickard DFC was airborne. The main target for 3 Group on this night was Lorient. 14 Wellingtons drawn from 9 and 75(NZ) Squadrons would be given the eastern bank area of Port Militaire. The two Wellingtons of 311 Squadron would be the only bombers over Boulogne-sur-Mer. Conditions over Boulogne prevented

*A wonderful photograph of Jan Gellner. A Czech lawyer of Jewish descent who emigrated to the United States before the war. In 1940 he joined the RCAF and was trained in Canada as an Observer. He joined the 311 Squadron as an Astronavigation Instructor, but he insisted that he should also take part in operational flights. He got the permission and carried out 37 sorties. As he was one of the very few navigators in the squadron who had a good command of English, he was also appointed to the functions of Squadron Navigation Officer and Bombing Leader. (Pavel Vančata)*

Sergeant Bala from bombing. He jettisoned his entire bombload due to severe icing. Squadron Leader Pickard DFC was more fortunate. He arrived over the docks at 19:15hrs depositing his entire load from 11,000ft. On his return he reported, *'Buildings near dock bombed. Fires started near first building'*.

Another unpleasant incident blurred the recent good work when Sergeant Karel Fák refused an order to fly. He had previously completed six operations as a second pilot but pleaded for a transfer to fighters. His requests were rejected, and he was transferred to the Army after being sent to Honington. Group H.Q was informed on the morning of the 10th that the squadron could provide three crews for that night if required, in the end they were not needed. A Bristol Beaufort of 22 Squadron landed at East Wretham on the 11th due to poor weather conditions over it home base. It was the only excitement on a day of constant rain and sleet.

Another cancelled operation followed on the 12th due to weather. Despite the constant rain, sleet and clouds, the squadron was keen to press on with training. An early morning cross-country flight on the 13th would see a crew off to Arbroath, Scotland, followed almost immediately by another crew on a cross country to Silloth, Cumbria, both were completed without mishap. One of the squadrons R.A.F pilots, Flight Lieutenant Thomas Kirby-Green appears to have been on ferrying duty on the 13th. While collecting Wellington T2846 from 22 Maintenance Unit, Birmingham, for delivery to 149 Squadron the Wellington's wheels sank into soft ground, the abrupt stop threw the tail upwards, and the nose pitched downwards into the ground. The crew were uninjured, but the tail assembly of the Wellington collapsed. Flight Lieutenant Kirby-Green had been on the squadron since his posting from 9 Squadron mid-September.

The 14th would be a busy day for postings. The hugely popular and respected Flight Lieutenant Michael Earle departed, his time on 311 Squadron over. This quietly spoken Kiwi had served as both Navigation Instructor and Squadron Adjutant. The squadron welcomed Squadron Leader Frederick James Powell MC, a Great War veteran who would take command of the Headquarters Flight Honington. Also arriving was Pilot Officer Angus MacNicol, who would assume the role of Squadron Adjutant. The squadron logged another forced landing on the 15th when an unrecorded Czech crew but believed to be skippered by Sgt Jaroslav Hájek were on a routine ferrying flight from Silloth to Honington aboard Wellington Mk.IA N3007. Thick fog and a defective wireless set resulted in the crew making a unplanned landing in a field at Great Limber, Lincolnshire at 14:30hrs. The crew were ordered to arrange guard for the aircraft until recovered.

Both R.A.F Honington and East Wretham were a scene of intense activity throughout the 16th. Bomber Command H.Q had planned something big. In retaliation for the devastating attacks on London and particularly Coventry on November 14th, Bomber Command wanted to show Germany that it, too, could deliver a devastating blow. The target chosen was Mannheim. Weather conditions over the bomber bases restricted the number of bombers to be used. However 134 aircraft, sixty-three of them provided by 3 Group would be detailed. This was the largest force to date sent to a single target and it would be the first recorded 'area' attack. The operation code named '*Abigail Rachel*' was intended to cause as much destruction and terror as possible. As far as Bomber Command was concerned the gloves were now off. Four crews would be offered by 311 Squadron for the operation.

<u>December 16th 1940: D.55 – Mannheim</u>

| Pilot | 2nd Pilot | Serial | Code | Bomb Load | Result |
|---|---|---|---|---|---|
| S/Ldr P Pickard DFC | Sgt B Baumruk | T2519 | KX-Y | 3x500lb+3x250lb | Completed |
| Sgt J Hrnčíř | W/Cdr Toman | R1021 | KX-W | 3x500lb+180x4lb | Completed |
| Sgt F Taiber | Sgt V Bufka | R1022 | KX-K | 3x500lb+180x4lb | Completed |
| Sgt J Křivda | Sgt J Pavelka | T2577 | KX-G | 3x500lb+180x4lb I | Crashed |

Squadron Leader Pickard was once again on the Battle order. He would be joined by Sergeant Bohuslav Baumruk on his first operation with the squadron. Two experienced pilots would be making their operational debut as captains on this raid, Sergeants Jan Hrnčíř and Jan Křivda. Sadly, for one, it would be his last. The first away was Squadron Leader Pickard DFC, followed closely by Sergeant F Taiber. Wellington T2577 KX-G followed and safely cleared the airfield, only to turn back flying at around 100ft. To those on the ground, it appeared that the pilot was attempting to land. The struggling Wellington circled the station, then struck the topmost branches of some trees before crashing on the road between East Wretham and Wretham Hall and was immediately engulfed in flames. Navigator, Pilot Officer Nedvěd, was by a miracle uninjured in the crash and managed to escape via the Astro hatch. On exiting the Wellington he found the semi-conscious and badly injured 2nd pilot, Sergeant Josef Pavelka a few yards from the wreckage. Nedvěd dragged his pilot clear and left him a safe distance from the now blazing aircraft. With no thought for his own safety, he courageously raced to the rear of the Wellington and desperately tried to pull Pilot Officer Jaromír Toul clear from his shattered rear turret. It was then a 500-pounder exploded, the explosion tearing the aircraft in two. Finally, after a few agonising minutes the fire tender arrived along with the crash crew. Both of whom immediately went to work dousing the fire and locating and extracting the trapped crew.

## Vickers Wellington Mk.IC T2577 KX-G

| Manufacturer | Vickers Armstrong | |
|---|---|---|
| Contract | 38600/39 | |
| Taken on Charge | 09/08/1940 via No.10 MU. | |
| Cat E Missing | N/A | |
| Struck Of Charge | 18/12/1940 | |
| Total Flying Hours | N/K | |
| Take-Off Time | 18:10 | |
| Bomb Load | 3x500lb+180x4lb IB | |
| | **CREW** | |
| Captain | Sergeant Jan **Křivda** 787209 RAFVR. Age 27. | Row C. Grave 10. |
| Second Pilot | Sergeant Josef **Pavelka** 787230 RAFVR | Seriously injured |
| Navigator | Pilot Officer Vladimír **Nedvěd** 82624 RAFVR | Not injured |
| Wireless Operator | Pilot Officer Josef **Doubrava** 82595 RAFVR | Minor Injuries |
| Front Gunner | Sergeant Jiří **Janoušek** 787545 RAFVR. Age 23. | Row C. Grave 9 |
| Rear Gunner | Pilot Officer Jaromír **Toul** 82642 RAFVR. Age 23. | Row C. Grave 8 |
| | | |
| Posting History | Sgts J Křivda and Sgt J Pavelka NFD. | |
| Operations Flown | Sgts J Křivda : 1 / Sgt J Pavelka : 0 | |
| Buried | HONINGTON (ALL SAINTS) CHURCHYARD | |

Pilot Officer Toul was reported to have died on his way to West Suffolk General Hospital, where the injured crew members were taken for treatment. The following was written post-war by Pilot Officer Vladimír Nedvěd.

*'Started with a full tank of gasoline and bombs was a normal load, the aircraft climbed to 30 meters, but no more. The speed reached 95 knots, which is about 170 km/h, but then began to slowly fall to 85-80 or less. I sat at the table with navigation speedometer and altimeter ahead. When the rate began to slowly decline below 80 knots. I said to myself, how is it possible that they still hold in the air? I quickly got up from the navigation table and went to the back of the plane - the fuselage was very spacious – there was an astrodome in the middle of the aircraft from which you could see in all directions around and is also the focus of a sextant for astro navigation. I saw clearly that both engines that still ran flawlessly on full power. Peaks of some trees were already higher than our plane and I saw clearly before us light our airfield runway about 1.5 km away. Pilot Jan Křivda tried a slight bend in the track and return to land. We were only about 1 km from the runway threshold when the left half of the wing caught a tree, the machine turned left, drove into the ground and immediately caught fire. Everything was extremely hot. Shortly after that, I managed to open the astro hatch and jumped onto the wing, left engine and fuselage were already in flames. When I finally stood on solid ground, I saw under the right engine Sergeant Josef Pavelka. I immediately dragged him about 50 meters from the plane and there he hid in a ditch safety. On the way back to the plane, I met Josef Doubrava, who was not injured and rested in a nearby forest. Ammunition in front of the turret began to explode from the heat. When I came back to the turret, the door was closed. I opened it and found Jaromír Toul unconscious or shocked. I immediately woke him up and said, 'the plane is burning, you have to get out quickly.' But he could not lift himself as his legs were wedged under the seat, into which he was crushed on impact. Bombs began exploding about four meters from us, and the flames spread to us. Jaromir asked me: 'please do not leave me', I told him, 'do not worry I'll stay with you'. (When I write these words again, I have tears in my eyes... that decision and the situation is just incredible, I know that only God gave me the courage, strength and protection ...) I tried in vain to get Jaromir removed from the turret. He was trapped there and it was not in my human power to free him. The flames were approaching*

*and more bombs exploded. While this drama was going on, the rescue service from the airfield was waiting at a safe distance, till all six bombs blows, each with 120 kg of explosives and guns ammunition would stop shooting. Jaromir and I, we were in the middle of all of that. Still fully conscious and quite uninjured (except for legs trapped under the seat) he told me: 'I have a revolver in my overalls pocket, pull it out and shoot me.' He just didn't want to burn alive, that's what we airmen were always afraid of. The moment of decision came: what should I do? Should I shoot the friend I want to save - me? Fate took its course when the last and nearest 250lb bomb exploded. The pressure wave threw me about five metres through the air and I landed on all fours. Miraculously, I was still unharmed - but I knew it was physically impossible to get Toul out of the plane and decided not to return to the machine. Once all the explosions ended, rescue crews came to the aircraft, extinguished the fire, and when about 20 men lifted the rear of Wellington – Jaromír was pulled from the turret. With the explosion of the last bomb, which threw me aside, Jaromír had to hit the rear of the machine guns head-on and suffered a fatal wound. He died during the ambulance to the hospital, and that was the end of all his*

*Air Gunner Pilot Officer Jaromír Toul who died from injuries sustained in the tragic crash of Wellington Mk.IC T2577/KX-G on 16th December 1940. (Pavel Vančata). Headstone, Honington Churchyard. (Authors Collection)*

*suffering.'*

The following investigation into the crash reported that Sergeant Jan Křivda put the flaps up after take-off. He returned the flap handle to neutral but in doing so, moved it too far, which resulted in the flaps lowering. Loss of speed and insufficient height meant that the Wellington struck the trees. The crew's navigator, Vladimír Nedvěd was born in Brno, Czechoslovakia, on March 27th, 1917 and educated at Kyjov High School. He joined the Czech Air Force in October 1936 and trained as a navigator, graduating from the Military College in 1938 as a Lieutenant. Like many, following the German occupation in 1939, he fled Czechoslovakia, travelling by train and on foot through the Balkans. In Lebanon, he boarded a ship for France, arriving in early 1940. With the fall of France in June, he escaped to England, where Vladimír joined the R.A.F Volunteer Reserve and undertook yet more training as a navigator before joining 311 Sqn.

*Four days after the crash of Vickers Wellington Mk.IC T2577/KX-G on 16th December 1940, the funeral of three airmen, Pilot Officer Jaromír Toul, Sergeant Jan Křivda and Sergeant Jiří Janoušek, took place at All Saints Churchyard at Honington. (Milan Šindler). Below, All Saints Churchyard, Honington. (Authors Collection)*

Nedvěd was rightly recommended for the George Cross for his heroic actions on this night but was subsequently appointed a Member of the Order of the British Empire for gallantry.

The remaining three crews forged on encountering less than favourable weather. Of the 63 Wellingtons of 3 Group detailed, 61 took off, of which 47 reached Mannheim. For those that persevered, they found Mannheim largely free of cloud. The raid was divided into two phases. The first, *'Abigail'* would comprise of the Group's most experienced crews, who would open proceedings by dropping incendiaries over the city centre. These would provide the following phase, *'Rachel'* with an easily identifiable aiming point. The three Czech crews arrived over the target between 21:16hrs and 22:20hrs. Both Squadron Leader Pickard DFC and Sergeant J Hrnčíř dropped their bombs from 15,000ft. On return, Squadron Leader Pickard reported, *'Bombs seen to burst and fires started in the eastern part of town north of 'Lindenhof'*. Sergeant J Hrnčíř dropped his bombs east of 'Lindenhof' producing a small fire. Sergeant F Taiber and crew were over Mannheim at 12,000ft. Back at Honington the crew reported, *'All bombs dropped in target area. One fire started. Small fires seen on arrival about ½ mile south of target'*. The devastating blow that H.Q Bomber Command had hoped to deliver was despite the initial euphoria, a disappointment. The primary marking was woefully inaccurate. The lead incendiary force had missed the city centre almost entirely, and as such, the bombing was scattered. This was confirmed by post-raid reconnaissance. Some useful damage had been inflicted, over 240 buildings had been destroyed or seriously damaged, and more than a thousand people bombed out.

There was no time to lament the previous night's tragic events. The following morning, Wing Commander Mareš Toman informed Group H.Q that the squadron's 'Freshmen' crews were available for operations that night if required. As the day progressed, it was apparent the worsening weather would either postpone or cancel the operation. It was not until 18:00hrs that Group H.Q informed Honington, Wyton and Stradishall that the planned operation was cancelled due to the possibility of fog. On the 18$^{th}$, 311 Squadron was again on the battle order, only for this operation to be cancelled. The following morning, Form B.371 arrived at East Wretham. The squadron would provide four 'Freshmen' crews that night. The Wellingtons were fuelled and bombed up and the crews brief, only for yet another cancel message to arrive at 22:10hrs. There would have been genuine disbelief and frustration for all concerned. These cancellations did little to help the already fragile nerves of some of the aircrews.

The now standard message was sent to Group H.Q on the morning of 20$^{th}$ informing them of the availability of crews on 311 Squadron. The preliminary details of the raid for that night arrived before lunch and the squadron once again set about preparing four Wellingtons. News arrived at East Wretham during the afternoon of a change in take-off time, but most prepared themselves for the inevitable cancellation signal. The Met officer had forecast a heavy frost which meant that the four Wellingtons standing at their exposed dispersals were covered in heavy canvas tarpaulins. It was not until the early hours of December 21$^{st}$ that the crews climbed aboard their freezing Wellingtons.

### December 21$^{st}$ 1940: CC.13 – Oostende

| Pilot | 2$^{nd}$ Pilot | Serial | Code | Bomb Load | Result |
|---|---|---|---|---|---|
| F/Lt J Ocelka | Sgt A Jedounek | R1021 | KX-W | 8x250lb+120x4lb | Completed |
| P/O F Cigoš | Sgt P Uruba | L7842 | KX-T | 8x250lb+120x4lb | Completed |
| P/O A Kubizňák | Sgt B Baumruk | R1022 | KX-K | 8x250lb+120x4lb | Completed |
| Sgt L Anderle | Sgt J Filler | L7841 | KX-S | 6x250lb+120x4lb | Completed |

Three of the pilots sitting in the freezing cockpits would be making their operational debut, Arnošt Jedounek, Antonín Kubizňák and Josef Filler. Sergeant Leo Anderle would be starting his first as captain and first operation since his crash on October 16$^{th}$. The four Wellingtons were all safely airborne between 04:30hrs and 04:38hrs from a frost covered East Wretham. Only seven Wellingtons of 3 Group were given Oostende, and they would be joined by Wellingtons of 1 Group plus a handful of Whitleys of 4 Group. Conditions over the port area were clear of clouds. Arriving over the already alerted target between 05:42hrs and 06:40hrs, the four crews encountered intense and accurate flak which made the run over the target particularly uncomfortable. Pilot Officer F Cigoš had the unnerving experience of being attacked by an enemy fighter *'fitted with a searchlight'*. Vigorous evasive action saved the crew on this occasion. Conditions were such that moored ships of various tonnage and the extensive dock area were easily observed, as were numerous explosions. On leaving the target, returning crews reported fires were easily visible. The jubilant crews landed back at East Wretham between 07:20hrs and 07:55hrs. It was an excellent performance. There was more good news, Pilot Officer Josef Doubrava had been discharged from hospital and was back on the squadron. Two incidents not mentioned in the Squadron ORB are reported in the 3 Group Records Book for the 21st.[14] The squadron's Avro Anson R9600 hit a post while taxiing causing minor damage, while an unrecorded squadron Wellington made a forced landing due to fog. There was no damage or injuries reported. The following night the squadron found itself once again on the Battle Order. Three crews would be detailed and briefed to attack Flushing along with Wellingtons drawn from Marham's 115 Squadron and Feltwell's 75(NZ) Squadron. The Group's main effort was a continuation of the raids on Mannheim.

December 22nd 1940: CC.7 – Flushing

| Pilot | 2nd Pilot | Serial | Code | Bomb Load | Result |
|---|---|---|---|---|---|
| P/O F Cigoš | Sgt P Uruba | L7842 | KX-T | 4x500lb+120x4lb | Completed |
| P/O A Kubizňák | Sgt B Baumruk | R1022 | KX-K | 4x500lb+120x4lb | Completed |
| Sgt L Anderle | Sgt J Filler | L7841 | KX-S | 3x500lb+120x4lb | Completed |

The three crews chosen had flown on the successful Oostende operation only 24 hours before. Arriving over the port facility between 18:33hrs and 19:52hrs, the small force of Wellingtons was subjected to a vicious barrage of flak and searchlights. Undeterred, the Czechs pressed home their attacks. Pilot Officer F Cigoš bombed from 12,000ft, running over the target from west to east. His bombs landed accurately on the target, starting four small fires and one large fire. The crew's rear gunner, Pilot Officer Karel Křížek, gleefully reported that the fires were still visible ten minutes after leaving the target. Sergeant L Anderle attacked the oil storage tanks, leaving a number ablaze. The fires at the oil facility attracted the crew of Pilot Officer A Kubizňák, who dropped his high explosives amongst them, while dropping their incendiaries between docks 4 & 5. It was another excellent effort by the crews, who were all safely back at East Wretham by 20:17hrs.

On the 23rd, the squadron was informed that it would once again be required for operations that night. The order arrived late afternoon but the operation was subsequently cancelled. Christmas Eve 1940, and both 9 and 311 (Czech) Squadrons were ordered at 09:30hrs to prepare for operations that night. To the delight of all, a message was received at 10:20hrs, that the AOC had cancelled the operation. However, a number of aircrews, maintenance crews and armament staff had to remain on stand-by if called-up.

The second wartime Christmas was celebrated in traditional style across 3 Group. The Czechs had their own way of celebrating. In place of turkey or pork, fried fish, usually carp, is served. The Christmas tree gets decorated on Christmas Eve. Traditionally, the tree gets decorated with apples, sweets, and traditional ornaments. There was, however, one common theme, lots of drinking, singing songs and games. A Christmas Party was organised for the 24th at Wretham Hall. Wing Commanders Mareš Toman and Bill Simmons MiD attended, as did many high-ranking visitors. It was, by all accounts, an outstanding success. Christmas Day was celebrated in a '*Jolly manner*', with the R.A.F tradition of the officers and Senior NCOs serving the dinners to the lower ranks. A concert was held at R.A.F Honington during the evening. The celebrations and various parties were just what the squadron needed.

A preliminary order for operations was received on the 26th, which was quickly countermanded. Both Lorient and Le Havre were to be visited by the squadron on the 27th. Four crews were required. However this was reduced to three.

---

[14] *Air 25/51*

### December 27th 1940: CC.26 – Lorient

| Pilot | 2nd Pilot | Serial | Code | Bomb Load | Result |
|---|---|---|---|---|---|
| Sgt J Hrnčíř | Sgt B Blatný | R1021 | KX-W | 3x500lb+120x4lb | Completed |
| Sgt J Bala | Sgt A Rozum | T2553 | KX-B | 3x500lb+120x4lb | Early Return |

Twenty-eight-year-old Sergeant Alois Rozum joined the experienced crew of Sergeant Jaroslav Bala for his first operation. Born in Plzeň, western Bohemia, he had already accumulated a considerable number of flying hours pre-war back in Czechoslovakia. Sergeant J Bala was the first away at 16:35hrs, followed ten minutes later by Sergeant J Hrnčíř. It was not long before Sergeant Bala was back in the circuit with a defective engine. He made an excellent landing under challenging circumstances at 17:06hrs, receiving the congratulations and praise of the squadron commander. Sergeant J Hrnčíř and crew benefited from being one of the last over Lorient. Fires had already taken hold in the Smithery, barracks and power station as well as the dock area and submarine pens. On his return, Sergeant J Hrnčíř reported, *'About 30 heavy explosions seen over the Smithery of which a fire extended'*.

### December 27th 1940: C.24 – Le Havre

| Pilot | 2nd Pilot | Serial | Code | Bomb Load | Result |
|---|---|---|---|---|---|
| F/Lt J Šnajdr | P/O A Kubizňák | R1022 | KX-K | 4x500lb+120x4lb | Completed |

Only six 3 Group Wellingtons were ordered to attack shipping and barge concentrations at Le Havre. Over the target at 20:10hrs, the crew dropped their mixed bombload on what was believed to be the glow of fires seen below the solid clouds blanketing the target. Dunkirk was the intended target for the squadron on the 28th, and three crews were available if required. This operation was cancelled due to unfavourable weather late afternoon. The cancellations continued the following day, again due to adverse weather. Wing Commander Simmons MiD and Wing Commander McKee attended a meeting with R.A.F Honington Station Commander on the 29th. This may have been in response to a document drawn up by the recently posted Flight Lieutenant Michael Earle. The full contents are unknown, but the 311 (Czech) Squadron ORB reports that it was Flight Lieutenant Earle's opinion that the squadron should *'cease to operate, and the older trained crews to instruct new crews'*. A withdrawal from operations was something the squadron was keen not to happen given the recent success.

Operations were planned for the 30th, but as expected by almost everyone, the cancel signal from Group H.Q was received late afternoon. You did not have to be a meteorologist to know the weather was terrible. An Air Raid Warning was issued during the day. A reported Dornier Do215 was causing some anxiety in the general area. East Wretham was spared, but Honington's defences had an opportunity to open fire on the Dornier without success. The 'All Clear' was sounded only for the Do215 to reappear 20 minutes later! There was no reported damage or injuries at East Wretham, everyone was just amazed that the Germans were flying in such filthy weather. The bad weather continued the following day, keeping the squadron grounded. Drizzle and low cloud made any chance of flying impossible.

# January 1941 : New Year, Fresh Start

*Winters grip. Vickers Wellington Mk.IA P9230 KX-X photographed on a freezing morning at East Wretham during the winter of 1940–1941. (Václav Kolesa)*

The recent series of loss-free operations had boosted confidence and lifted morale. Everyone on the squadron hoped the new year would continue in the same way. The squadron sat out the operation on the night of January 1st. It was a brief reprieve. The following morning, Group H.Q requested the availability of 311 Squadron. The reply was quick, four crews were available if required.

Four crews were detailed and briefed that evening. However, Flight Lieutenant J Ocelka was withdrawn just before take-off with intercom failure. Two crews would attack Emden while Sergeant Vilém Bufka and crew would bomb Bremen, Germany's second-largest seaport.

<u>January 2$^{nd}$ 1941: D.184 – Emden</u>

| Pilot | 2$^{nd}$ Pilot | Serial | Code | Bomb Load | Result |
|---|---|---|---|---|---|
| P/O F Cigoš | Sgt P Uruba | L7842 | KX-T | 4x500lb+120x4lb | Alternative Target |
| P/O A Kubizňák | Sgt B Baumruk | R1022 | KX-K | 4x500lb+120x4lb | Alternative Target |

Both crews were away by 18:00hrs. Snow showers were encountered at the Dutch coast, and solid cloud over the primary meant neither crew were able to bomb. Not wishing to return with his bombload, Sergeant F Cigoš daringly, given the appalling conditions, decided to continue onto Bremen, where the crew bombed a railway junction from 17,000ft. Bombs were seen to burst 500 yards from a railway junction, producing four fires and an explosion visible from 20 miles. Pilot Officer A Kubizňák encountered the same conditions. He, too, decided to press on to Bremen, where he started a group of fires south of the railway station. Bremen's flak and searchlight defences were particularly active. It had been heavily bombed the previous night and suffered significant damage. Luck was on the side of Pilot Officer A Kubizňák. His Wellington was holed by flak twelve times but thankfully nothing vital was hit and the crew were all surprisingly uninjured.

January 2$^{nd}$ 1941: N.35 – Bremen

| Pilot | 2$^{nd}$ Pilot | Serial | Code | Bomb Load | Result |
|---|---|---|---|---|---|
| Sgt V Bufka | P/O J Breitcetl | T2553 | KX-B | 2x500lb+360x4lb | Completed |

Sergeant Vilém Bufka would captain a crew for the first time while Pilot Officer Jindřich Breitcetl was flying his first raid since October 8$^{th}$. Like their colleagues, they encountered snow showers and icing, which was especially troublesome in the clouds. The crew arrived over the target around 20:20hrs and bombed from 14,000ft. They were on the receiving end of Bremen's accurate barrage flak, which came close to wounding the front gunner with shrapnel. Despite the opposition, their bombs were seen to burst 200 yards south of a road and railway junction, producing four good fires, which were observed by the tough rear gunner, Sergeant Vilém Jakš as the Wellington turned for home. Thirty-year-old Vilém Jakš was the pre-war middleweight boxing champion of Czechoslovakia and had fought the world middleweight champion, Frenchman Marcel Thil, in 1935 for the world title. Jakš lost after 11 gruelling rounds in front of 20,000 spectators. All three crews were safely back by 23:30hrs. It had been an excellent night's work, and Pilot Officers Cigos and Kubizňák's decision to bomb Bremen was an indication of the squadron's growing confidence.

Both 9 and 311 Squadrons were warned the following morning that they would be operating again that night, once again visiting Bremen. For some unknown reason, 311 Squadron was withdrawn soon after lunch. The waiting continued the following night when the squadron sat out attacks on Duisburg and Brest. On the 5$^{th}$, the now almost daily warnings of 'hostile' aircraft in the area reverberated around Honington. Preventative action was implemented, but on this occasion, despite the warning, the station was caught unawares. At 13:36hrs, the alarm sounded, and almost immediately, a German Do17Z with bomb doors open was over the airfield. Eight 100kg bombs and two 50kg bombs[15] were dropped. Nine landed harmlessly on the airfield, while one landed close to the equipment store. Sadly AC2 Jindřich Liebold was too late to reach the shelter and was struck and killed by a bomb splinter. The all-clear sounded at 13:43hrs. Two crews were required for the 6$^{th}$. Preparations were completed, and the Wellingtons and their crews waited. At 15:40hrs, the crews were warned to stand down until midnight, followed almost an hour later by the cancellation message. For the poor ground crews, this meant a late-night working in freezing conditions.

The waiting continued on the 7$^{th}$. That day, a Do215[16] made a surprise appearance over a snow covered Honington at 12:59hrs dropping a number of bombs which thankfully on this occasion resulted in no injuries. It would be a busy day at Honington, no sooner had the Do215 departed than the AOC No.3 Group AVM J.E.A Baldwin CB, OBE, DSO arrived for a meeting with the Station Commander. On the 9$^{th}$, both of Honington's squadrons were once again informed that they would be required for an operation that night. The enforced lay-off and cancellations were sapping the spirits of the Czechs eager to operate, too much waiting around was not good for morale. The groundcrews out at the distant snow covered dispersals bombed up and prepared five Wellingtons for the night's activities.

---

[15] *Honington Station Records Book. The Luftwaffe never used, or dropped 100kg bombs, probably all 50kg.*
[16] *Honington Station Records Book. Aircraft was a Dornier Do17Z.*

On the 9th, 3 Group turned its attention to oil targets. Two locations would be attacked. The Groups 'Freshmen' crews would bomb the petroleum harbour at Rotterdam, while the more experienced crews would be visiting the Ruhr and the Nordstern Synthetic-Oil Plant in Gelsenkirchen.

### January 9th 1941: Z.3 – Rotterdam Petroleum sheds and storage facility

| Pilot | 2nd Pilot | Serial | Code | Bomb Load | Result |
|---|---|---|---|---|---|
| Sgt A Šedivý | Sgt J Čapka | T2561 | KX-A | 4x500lb +120x4lb | Completed |

Sergeant Šedivý departed at 18:03hrs, joining a small force of novice crews over Rotterdam. Conditions over the target were ideal. No cloud or ground haze resulted in some accurate bombing. Sergeant Šedivý and crew were over the oil plant at 20:08hrs, dropping their entire bombload from 20,000ft on an already blazing target. As the bomber crews turned for home, numerous fires were observed. A few of the large oil tanks were seen on fire, producing dense smoke which billowed into the night sky.

### January 9th 1941: A.71 – Gelsenkirchen

| Pilot | 2nd Pilot | Serial | Code | Bomb Load | Result |
|---|---|---|---|---|---|
| F/Lt J Ocelka | Sgt A Jedounek | R1021 | KX-W | 5x500lb | Completed |
| S/Ldr P Pickard DFC | F/Lt J Šnajdr | R1022 | KX-K | 5x500lb | Completed |
| P/O F Cigoš | Sgt P Uruba | L7842 | KX-T | 5x500lb | Alternative Target |
| Sgt L Anderle | Sgt J Filler | T2553 | KX-B | 5x500lb | Completed |

As Sergeant Šedivý and crew were landing from their 1st operation, the remaining crews were climbing for altitude and passing over the English coast. Pilot Officer Cigoš was unable to identify the target, a combination of map reading problems, clouds and Germany blanketed in snow meant that after a 35- minute search they opted to bomb Duisburg, creating a number of fires between buildings.

The other crews were more fortunate. They located the target and bombed from between 16,000ft and 18,000ft. Clouds over Gelsenkirchen were encountered, but vertical visibility between them was excellent. Squadron Leader Pickard DFC bombed at 23:30hrs, his bombs bursting between factory buildings. Sergeant Anderle and crew reported on return, *'Fires started south of target, explosions seen. After leaving target, a big dull fire immediately SW of target'*. The crew had shown real pluck, the heating system aboard their Wellington had failed soon after take-off, but despite the sub-zero conditions, they opted to continue on. With all the crews safely home, it was off to de-briefing and a hot cup of cocoa before catching up with some well-earned sleep. The squadron had dropped a total of 12,480lb of bombs. It was a good night's effort.

There was a real buzz around R.A.F Honington on the morning of January 11th. The target for that night was the Fiat Works in Turin. The order to fit long-range fuel tanks to 9 Squadron's Wellingtons had already been received, and preparations were quickly made. Not surprisingly, 311 (Czech) Squadron were not required. It was clear at Group H.Q that the squadron needed more time to be ready for operations of this complexity and range. For the Czechs, it was a bitter disappointment, but their time would come.

On the morning of January 13th, R.A.F Honington signalled Group H.Q they could provide nine crews from 9 Squadron and three crews from 311 Squadron for that night. Northern Italy was once again the destination, however, this was later cancelled. The cancellations continued on the night of the 15th. A raid on Bordeaux was planned only for this to be cancelled due to weather conditions. Despite the freezing conditions throughout the 16th, five squadron Wellingtons were prepared for operations that night. The target for 9 and 311 Squadrons was Wilhelmshaven and the German heavy Cruisers. Being a superstitious bunch, the Czechs would have no doubt hoped that the 16th would not be so costly for them as it had been in October and December.

## January 16th 1941: D.197 – Wilhelmshaven Docks

| Pilot | 2nd Pilot | Serial | Code | Bomb Load | Result |
|---|---|---|---|---|---|
| S/Ldr J Ocelka | Sgt A Jedounek | R1021 | KX-W | 4x500lb+120x4lb | Early Return |
| F/Lt J Šnajdr | Sgt B Blatný | R1022 | KX-K | 4x500lb+120x4lb | Completed |
| Sgt A Šedivý | Sgt J Čapka | T2561 | KX-A | 4x500lb+120x4lb | Completed |
| Sgt V Bufka | P/O J Breitcetl | T2553 | KX-B | 4x500lb+120x4lb | Alternative Target |
| P/O A Kubizňák | Sgt B Baumruk | T2519 | KX-Y | 4x500lb+120x4lb | MISSING |

### Vickers Wellington Mk.IC T2519 KX-K

| | | |
|---|---|---|
| Manufacturer | Vickers | |
| Contract | 38600/39 | |
| Taken on Charge | 12/11/1940 via No.8 MU. | |
| Cat E Missing | 16/11/1940 | |
| Struck Of Charge | 16/01/1941 | |
| Total Flying Hours | N/K | |
| Take-Off Time | 17:48hrs | |
| Bomb Load | 4x500lb+120x4lb | |
| | **CREW** | |
| Captain | Flying Officer Antonín **Kubizňák** 82556 RAFVR. Age 29. | Runnymede Memorial 30 |
| Second Pilot | Sergeant Bohuslav **Baumruk** 787186 RAFVR. Age 24. | Runnymede Memorial 39 |
| Navigator | Flying Officer Josef **Hudec** 82604 RAFVR. Age 28. | Runnymede Memorial 30 |
| Wireless Operator | Pilot Officer Jindřich **Leskauer** 82617 RAFVR. Age 23. | Runnymede Memorial 33 |
| Front Gunner | Sergeant Rudolf **Bolfík** 787583 RAFVR. Age 27. | Runnymede Memorial 39 |
| Rear Gunner | Pilot Officer Jaromír Oldřich **Král** 82612 RAFVR. Age 27. | Runnymede Memorial 33 |
| | | |
| Posting History | F/O A Kubizňák posted via Cosford Depot, 29/07/1940. Sgt B Baumruk NFD. | |
| Operations Flown | F/O Kubizňák : 4 / Sgt Baumruk : 4 | |
| Remembered | RUNNYMEDE MEMORIAL | |

The crew of Antonín Kubizňák are known to have reached Wilhelmshaven, a NGZ message having been received back at Honington at 20:30hrs. An hour later, the first sign of trouble was received back at base when a message was received stating, *'port motor trouble'*. This was followed almost immediately by *'port motor not going'*. At 21:42hrs, the crew asked if Honington was clear to land, the initial concern eased by the apparent confidence of the crew in getting home despite the engine failure. Five minutes later, the crew asked Honington if they were overhead. They were informed they were not and given a QDM 283. Something had obviously gone wrong with the navigation, as several further requests were made for a homing bearing. Finally, at 22:18hrs, an S.O.S was received at Honington. The last known fix timed at 22.18hrs places Wellington T2519 midway between England and the Dutch coast. It was the last contact with the crew. The following morning, four Wellingtons from 311 and six from 9 Squadron were airborne to search for the missing crew from 09:10hrs until 16:05hrs. Nothing was found.

The loss of yet another crew was the catalyst for another round of verbal confrontation on the squadron. The arrival of His Majesty King George VI and the Queen at R.A.F East Wretham on the 18th was a welcome interlude in the growing resentment and anger by certain individuals. Just before 09:00hrs, the squadron, who had been ordered to parade well before their arrival, stood smartly in freezing conditions. Eventually, the Royal Party arrived. Accompanying their Majesties was C-in-C R.A.F Bomber Command, Air Marshal Sir Richard Peirse on his first visit to the squadron, and 3 Group's AOC Air Vice Marshal John Baldwin. The visit went smoothly, and as usual, the Queen was at her regal best, talking at length with various individuals. Although there appeared to be harmony within the squadron, individual airmen were unhappy and willing to voice their grievances publicly. Sadly, this included a few senior airmen who created difficulties. A veteran navigator reported sick, and another refused to fly when his request for pilot training was refused. A commissioned air gunner then requested to be transferred to the Army. These refusals to fly or transfer requests were a continual sap on morale, especially among novice crews. Lessons had been learnt, and action on the part of the squadron commanders was relatively swift with the removal of the airmen. Thankfully, 311 (Czech) Squadron had amongst its ranks some intensely loyal and professional airmen, who were the backbone of the squadron, and their steadying presence prevented the spread of any discontent.

*A snow-covered and freezing East Wretham. An informal photograph of the King and Queen was taken during their visit on January 18th, 1941. L-R: ?, Acting Wing Commander Mareš-Toman (by back), A glimpse of Bomber Commands new C-in-C Air Marshal Sir Richard Peirse, Queen Elizabeth, King George VI, AVM Baldwin CB, DSO, OBE, ? Acting Group Captain Berounský (partially). (Marie Lambourne)*

Help was at hand with the arrival of the Training Flight from R.A.F Honington, commanded by Flight Lieutenant Kirby-Green. This would form 'B' Flight but would, for the time being, remain effectively non-operational. The handful of crews at East Wretham had done an excellent job thus far, but fatigue, illness and reliance on the same crews nightly was putting them under unnecessary strain. The arrival of the Training Flight would reduce the pressure and, at the same time, allow the novice crews an opportunity to operate alongside their more experienced colleagues. The wintery weather resulted in both East Wretham and Honington informing Group H.Q on the morning of the 21st that the airfields were unserviceable for operational flying. On January 23rd, the London Gazette published the award of the Czech Military Cross of 1939 to the following R.A.F members of the squadron, Wing Commander John Griffiths DFC, Squadron Leader Henry Graham, Navigational Instructor and Flight Lieutenant Anthony Brown, Gunnery Leader. The remainder of the month was a sequence of almost daily operational cancellations. To keep the crews active, ground training in Astro navigation was organised for the observers while the wireless operators were given intense training. This training would result in almost daily tests for aptitude and performance. The squadron notched up an avoidable accident on the 29th due to pilot error. Sergeant Rozum managed to strike a civilian vehicle with the tail of his Wellington while taxiing. Repair to the starboard tailplane and elevator would require replacing.

| Date | 29/01/1941 |
|---|---|
| Mark | Mk.IC |
| Serial | T2561 |
| Code | KX-W |
| Taken On Charge | 09/08/1940 via No.10 MU. |
| Manufacturer | Vickers Armstrong |
| Contract | 38600/39 |
| Pilot (s) | Sergeant Alois Rozum |
| Flight | Training |
| Time | 12:10hrs |
| Cause | Pilot Error |

On the 30th, the Luftwaffe made an unwelcome appearance. At 12:37hrs, the alarm sounded, and 'hostile' aircraft were reported overhead. The airfield appeared to have escaped when the 'All Clear' sounded three minutes later. Those who took cover gingerly left the shelters and slit trenches and resumed their duties. At 12:49hrs, the alarm rang out again, and 'hostiles' were again reported in the vicinity of the airfield. It was at 13:23hrs that a low-flying Dornier Do17Z appeared machine gunning the dispersal areas. No damage or injuries were reported. Again, an 'All Clear' was sounded, only to be raised almost immediately with 'hostiles' reported overhead. This would continue on and off until 15:53hrs, when the final all-clear was sounded. These warnings continued throughout the following day, thankfully without any appearance of the Luftwaffe. Fortunately during these warnings the squadron was not required for operations.

*One of the crews who were undergoing intensive training in the winter of 1940–1941. The photo was probably taken during the Royal visit. L-R: Pilot Officer František Cigoš, Sergeant Petr Uruba, Pilot Officer Jaroslav Partyk, Pilot Officer Arnošt Valenta, Pilot Officer Eduard Šimon, Sergeant Gustav Kopal. The crew started the tour of operations on 2nd January 1941 and failed to return from their third sortie (with P/O Emil Bušina who replaced Pilot Officer Partyk due to illness) on 6th February 1941. All but one survived the war in the German captivity. P/O Valenta was amongst those 50 Allied prisoners of war who were executed after the escape from Stalag Luft III in March 1944. (Milan Šindler)*

# February 1941 : Operations Intensify

February 1st found East Wretham on almost constant alerts, German bombers were again over the region. The planned operation that night was cancelled, weather being the issue. The following day, the 'Training Flight' made use in the lull in air raid warnings and were up carrying out navigational flights and circuit and bumps. A 'Freshmen' operation was planned for that night but as expected cancelled by Group H.Q in the late afternoon.

Given the recent lull, Group H.Q was keen to resume operations and requested fourteen crews from Honington for a planned raid on the 3rd. Preparations were in hand when at 09:50hrs the East Wretham Air-Raid alarm sounded. At 10:08hrs around twenty bombs exploded on the north-west corner of the airfield causing minor damage to a dispersed Wellington. Nine Junkers Ju88s of III/KG.30 were creating problems over 3 Group's airfields carrying out random attacks against East Wretham, Feltwell, Honington, Newmarket Heath and Waterbeach.

Squadron Leader J Ocelka almost wrote-off Wellington N3010 and his co-pilot on the night of February 3rd. An issue with the undercarriage when coming into land resulted in the crew aborting the landing and carrying out a second circuit of the airfield in an attempt to rectify the problem. The cause of the problem was believed to have been resolved but on landing the undercarriage immediately collapsed.

| Date | 03/02/1941 |
|---|---|
| Mark | Mk.IC |
| Serial | N3010 |
| Code | KX-L |
| Taken On Charge | 20/09/1940 via 214 Squadron |
| Manufacturer | Vickers Armstrong |
| Contract | 549268/36 |
| Pilot (s) | Squadron Leader Josef Ocelka / Sergeant Arnošt Jedounek |
| Flight | Local Flight |
| Time | 21:50hrs |
| Cause | Pilot error |

The damage was extensive. Both airscrews were wrecked. Bomb doors and the bomb beam were badly bent and buckled, and the undercarriage twisted. To add to the problems, the crumpled Wellington was creating an obstruction, and salvage assistance was requested. It would be over two months before N3010 returned to service on the squadron. Once again a certain amount of criticism was aimed at Squadron Leader Ocelka for his handling of the incident. This unfortunate incident closed a series of mishaps by Squadron Leader Ocelka, it would be the last for a long while. Another cancelled operation for the more experienced crews on the 5th resulted in a call for the 'Freshmen' crews to be allowed to test their skills on a French target. Predictably, however, it was left to the more experienced crews to fill the void.

### February 6th 1941: CC.29 – Boulogne Docks

| Pilot | 2nd Pilot | Serial | Code | Bomb Load | Result |
|---|---|---|---|---|---|
| S/Ldr J Ocelka | Sgt A Jedounek | T2971 | KX-H | 8x250lb+120x4lb | Completed |
| F/Lt J Šnajdr | Sgt B Blatný | R1022 | KX-K | 8x250lb+120x4lb | Completed |
| Sgt L Anderle | Sgt J Filler | T2553 | KX-B | 8x250lb+120x4lb | Completed |
| P/O F Cigoš | Sgt P Uruba | L7842 | KX-T | 8x250lb+120x4lb | MISSING |
| Sgt A Šedivý | Sgt J Čapka | T2561 | KX-A | 8x250lb+120x4lb | Completed |
| Sgt V Korda | Sgt A Rozum | R1371 | KX-F | 8x250lb+120x4lb | Completed |

*Vickers Wellington Mk.IC L7842 KX-T photographed the day after its unscheduled landing at Saint-Paul Airport near the French city of Flers. Below : In Luftwaffe hands! This was the last of two 311 Squadron Wellingtons which fell into German hands. (Pavel Vančata)*

There were promising signs of a productive night as the crews began to climb away from a snow-covered R.A.F East Wretham around 18:00hrs. There was little, if any cloud over East Anglia, but the night was freezingly cold. Group H.Q committed only three squadrons against Boulogne-sur-Mer, the Czechs, plus nine Wellingtons from both 9 and 214 Squadrons. While en route, conditions gradually worsened and the initial hopes of a clear target quickly began to evaporate the closer they flew towards the French coast. Conditions over the target were far from ideal as the first Wellingtons began their attacks at 19:20hrs. Almost solid cloud cover obscured the primary target, but undeterred, the squadron pressed home their attacks. Sergeant V Korda bombed from 13,000ft, dropping his entire load in one stick. He reported upon return, *'Bombs from other aircraft seen to explode across basin 4 & 6'*. Flight Lieutenant Šnajdr and Sergeant Anderle dropped in the general target area but were unable to positively observe results. Only Squadron Leader Josef Ocelka was confident about his bombing. Back at base, he reported. *'Bursts of own bombs observed followed by five big fires and seven small fires, followed by explosions.'* The last of the bombers departed at 21:00hrs. Group H.Q were somewhat subdued in their assessment of the raid, stating, *'Fires and explosions were seen but could not be confirmed'*.

Stronger than forecast winds, clouds and poor navigation resulted in the crew of Pilot Officer František Cigoš failing to return from this routine operation. A contributing factor to this situation was that the crew's usual navigator, Pilot Officer Jaroslav Partyk, was ill. He was hastily replaced by Pilot Officer Emil Bušina, who had not yet completed full training and this was his first operation. The hopelessly lost crew fumbled around, their last message being received at 21:10hrs. They eventually forced landed short of fuel at the former French airfield at Flers south of Caen, believing they were in England. The crew were all taken prisoner. Vickers Wellington L7842 was quickly repaired and later flown by the Luftwaffe Experimental & test Facility at Rechlin.

### Vickers Wellington Mk.IC L7842 KX-T

| | | |
|---|---|---|
| Manufacturer | Vickers Armstrong | |
| Contract | 992424/39 | |
| Taken on Charge | 20/10/1940 via No.22 MU. | |
| Cat E Missing | 06/02/1941 | |
| Struck Of Charge | 06/02/1941 | |
| Total Flying Hours | N/K | |
| Take-Off Time | 18:06hrs | |
| Bomb Load | 8x50lb+120x4lb IB | |
| | | |
| | **CREW** | |
| Captain | Pilot Officer František **Cigoš** 82541 RAFVR | PoW |
| Second Pilot | Sergeant Petr **Uruba** 787198 RAFVR | PoW |
| Navigator | Pilot Officer Emil **Bušina** 82588 RAFVR | PoW |
| Wireless Operator | Pilot Officer Arnošt **Valenta** 82532 RAFVR. | PoW, Coll. grave 9. A.[17] |
| Front Gunner | Sergeant Gustav **Kopal** 787232 RAFVR | PoW |
| Rear Gunner | Pilot Officer Karel **Křížek** 82903 RAFVR | PoW |
| | | |
| Posting History | P/O F Cigoš posted via Cosford Depot, 29/07/1940. Sgt P Uruba NFD. | |
| Operations Flown | P/O F Cigoš : 4 / Sgt P Uruba : 5 | |
| Buried | POZNAN OLD GARRISON CEMETERY | |

Pilot Officer František Cigoš reported on his release from PoW Camp in 1945:

---

[17] *Murdered by Gestapo late March 1944, following mass escape from Stalag Luft III. Reported he was amongst early pairs of escapers from the escape tunnel. Recaptured in Gorlitz area & last seen alive on 31st March 1944*

*'We took off from East Wretham in the Wellington at 18:00hrs on February 6th 1941 on a bombing mission to Boulogne. After we had bombed the target and on the return journey to base, the navigator lost consciousness. The Wireless Operator was able to obtain only one Q.D.M. I flew on this course for about an hour over the sea. I then saw land and owing to bad weather I decided to force-land. I landed the aircraft in a field about 23:00hrs and it was surrounded by Germans immediately. I detonated the secret equipment. The whole crew were captured and I then heard we had landed at Flers. We were taken by lorry to Flers and interrogated. We remained there that night and the following day were taken by train to Dulag Luft Oberursel.'*[18]

Freezing conditions, light rain and drizzle, added to poor visibility, meant that the squadron would focus its energy on training on the 8th and 9th. Operations against Germany and French ports had effectively been cancelled. On the 9th, the squadron was visited by 48-year-old Air Vice Marshal Karel Janoušek of the Czechoslovak Inspectorate General (C.I.G). An extremely capable and experienced individual, he worked tirelessly to ensure that the Czechoslovakian airmen were given the necessary help they needed during the early days of their arrival. He was a keen advocate of Czech officers commanding Czech squadrons. One of his many roles was to inspect the R.A.F's Czechoslovak squadrons, organise their training and operations, also to ensure sufficient reserves would be available and listen to any issues. The squadron notched up another incident on the 9th. Sergeants Jaroslav Hájek and Karel Mazurek were airborne on a training flight late morning. All appeared to be going well until the crew attempted to land at R.A.F Honington. Failing to notice the wind speed and direction, the Wellington undershot its approach, resulting in a collapsed undercarriage. Damage was extensive, the port mainplane, port airscrew and port undercarriage all needing extensive repair.

| Date | 09/02/1941 |
|---|---|
| Mark | Mk.IA |
| Serial | L7785 |
| Code | KX-R |
| Taken On Charge | 11/09/1940 via 9 Squadron. |
| Manufacturer | Vickers Armstrong |
| Contract | B.992424/39 |
| Pilot (s) | Sergeant Jaroslav Hájek / Sergeant Karel Mazurek |
| Flight | Training |
| Time | 12:00hrs |
| Cause | Pilot inexperience |

Damage was such that the Wellington needed the specialist intervention of 54 MU. KX-R never returned to service with the squadron. Sergeant Jaroslav Hájek was held responsible for the accident and severely reprimanded. It was on the 10th that Air Marshal Sir Richard Peirse C-in-C Bomber Command implemented his now customary once a month all-out assault on a German town or city. The target chosen was Hanover. The city was positioned at the confluence of the River Leine and its tributary the Ihme, in the North German Plain. It was an important crossing point for the German railway system with lines connecting from east to west, from Berlin to the Ruhr, as well as Düsseldorf and Cologne and north to south, Hamburg, Frankfurt, Stuttgart and Munich. It also had a number of prominent war producing factories, including the tank factory M.N.H. Maschinenfabrik Niedersachsen (Badenstedt). Bomber Command assembled 222 aircraft for the night, as usual the bulk of the numbers and bombs dropped would be provided by 3 Group. Every available squadron within the group would be involved supplying an impressive 101 bombers for Hanover, plus 17 'Freshmen' crews for an attack on Rotterdam's oil storage tanks. It was a magnificent effort by the whole group. The Czechs at R.A.F East Wretham were pushed to supply just four crews for the assault. One crew was sick, and another on leave exposing the squadron's chronic lack of operational crews.

---

[18] *WO208/3331*

## February 10th 1941: Hanover

| Pilot | 2nd Pilot | Serial | Code | Bomb Load | Result |
|---|---|---|---|---|---|
| S/Ldr J Ocelka | Sgt A Jedounek | T2971 | KX-H | 3x500lb + 120x4lb | Completed |
| Sgt A Šedivý | Sgt J Čapka | T2561 | KX-A | 3x500lb + 120x4lb | Completed |
| Sgt L Anderle | Sgt J Filler | T2553 | KX-B | 3x500lb + 120x4lb | Alternative target |
| S/Ldr P Pickard DFC | Sgt J Bernát | R1371 | KX-F | 3x500lb + 120x4lb | Completed |

Flying alongside Squadron Leader 'Pick' Pickard DFC was 26-year-old Sergeant Josef Bernát on his first operation. A graduate of the Military Aviation School in Prostějov, he had served in the Czechoslovak Air Force pre-war. Like so many brave young Czechs, soon after the occupation of Czechoslovakia, he made his way through Poland to France, where he enlisted in the Foreign Legion. After the capitulation of France, he was evacuated to Great Britain. Over the coming months, he would prove himself a fearless and skilful pilot. The four crews had safely departed by 18:52hrs and began the long flight over the North Sea, making landfall north of Rotterdam. Weather conditions over the German frontier were ideal. Bright moonlight, broken clouds and clear skies were just what the navigators had hoped for. Regrettably, the weather quickly deteriorated. The early arrivals encountered almost solid cloud over Hanover, and consequently, the bombing was somewhat scattered, falling in the eastern and western suburbs of the city. Squadron Leader J Ocelka and crew bombed on ETA, dropping their entire load in one stick from 14,000ft, while Sergeant A Šedivý made his bomb run from west-to-south over Hanover. Braunschweig was on the receiving end of Sergeant L Anderle's bombs. Unable to locate the primary, they dropped from 15,000ft and were able to observe their bombs burst, producing six greenish explosions and seven fires. Squadron Leader 'Pick' Pickard DFC dropped his load at Maschee, situated just south of the city centre. The majority of the force bombed in the general target area, producing a number of large explosions and numerous fires. Considering the number involved, the results were considered disappointing.

The following morning, the squadron provided four crews for a sea search over the North Sea for a ditched crew. Unbeknown at East Wretham, Group H.Q had set in action plans for an attack on Bremen that night and another crack at Hanover. This put the squadron in a particularly awkward position regarding the availability of crews and aircraft. Two of the sea search Wellingtons were immediately recalled, and R.A.F East Wrethams groundcrews quickly started preparing what Wellingtons were available for that night.

## February 11th 1941: Bremen

| Pilot | 2nd Pilot | Serial | Code | Bomb Load | Result |
|---|---|---|---|---|---|
| Sgt A Šedivý | Sgt J Čapka | R1022 | KX-K | 3x500lb + 120x4lb | Completed |
| Sgt V Korda | Sgt J Filler | T2972 | KX-G | 3x500lb + 120x4lb | Early Return |
| Sgt L Anderle | Sgt F Radina | T2971 | KX-H | 3x500lb + 120x4lb | Completed |
| S/Ldr P Pickard DFC | Sgt F Fencl | R1021 | KX-W | 3x500lb + 120x4lb | Completed |
| F/O J Breitcetl | Sgt J Bernát | R1371 | KX-F | 3x500lb + 120x4lb | Completed |

Six of the ten pilots chosen had flown the previous night. Two second pilots, Sergeants František Radina and František Fencl, would be making their operational debut, while Flying Officer Jindřich Breitcetl would be flying his first operation as captain. It had been a desperately close call providing the required numbers demanded by Group H.Q. The crews slowly climbed away from East Wretham in the last of the daylight and entered a band of medium cloud, which was not considered too much of a problem at the time. One crew was forced to return early. Sergeant V Korda, at the controls of Wellington T2972, began to experience problems with the cockpit instruments as they passed over a cloud-covered Great Yarmouth. The crew initially decided to press on, but without an airspeed indicator, altimeter, rate of climb indicator and a starboard rpm gauge flickering wildly, the crew realised that to continue was reckless. By the time the crew were back over East Wretham, it was blanketed in thick fog sandwiched between the ground and low clouds. Sergeant Václav Korda, an experienced pre-war commercial pilot, put his years of experience to effective use and was responsible for the crew's safe return in difficult conditions. Others would not be so fortunate. The thirty-six Wellingtons of 3 Group found Bremen completely cloud covered, making locating the Focke-Wulf factory almost impossible. To add to the problems, flak over Bremen was intense. Flying Officer J Breitcetl dropped his entire load south of Bremen, his incendiaries producing small fires. Squadron Leader Pickard DFC bombed from the giddy height of 18,000ft and reported his bombs exploding in Bremen's dock area producing a series of fires.

*A classic photograph depicting a trio of 311 Squadron Wellingtons. The furthest aircraft, Wellington T2561 KX-A would complete 32 operations with the squadron.*

The remaining squadron crews unable to identify the primary bombed in the general target area. The disappointing results were made worse on return to England. Widespread and unexpected fog had descended over much of eastern England. East Anglian squadrons were particularly hard hit with tired crews desperate to locate their position and find a fog free airfield. Squadron Leader Pickard DFC and Sergeant L Anderle managed to land at R.A.F Newmarket, while Flying Officer J Breitcetl got down at R.A.F Wyton. The crew of Sergeant A Šedivý were short of fuel and found themselves over Swinderby, Lincolnshire where at 01:05hrs they crash landed with the undercarriage retracted at Wheatly Farm, Fosse Way. The crew were shaken, but uninjured. However damage to the Wellington meant that it would have to be dismantled and transported before requiring the expertise of No.43 Group. Wellington R1022 would not return to squadron service.

| Date | 12/02/1941 |
| --- | --- |
| Mark | Mk.Ic |
| Serial | R1022 |
| Code | KX-K |
| Taken On Charge | 20/10/1940 via No.22 MU. |
| Manufacturer | Vickers Armstrong |

| Contract | 992424/39 |
|---|---|
| Pilot (s) | Sergeant Alois Šedivý / Sergeant Josef Čapka |
| Flight | Operations |
| Time | 01:05hrs |
| Cause | Fuel shortage |

The night was a disaster. The fog, which the met officers had failed to forecast, resulted in twenty-two aircraft either crash landing or being abandoned, eleven of which were from 3 Group. 218 and 115 Squadrons were particularly unfortunate. The only good news on an otherwise disappointing night was recorded in the 3 Group Headquarters Records Book. 311 (Czech) Squadron was for once singled out for praise. The Record Book stating; *'Sorties which managed to get down, put up a first-class effort, the Czechs especially doing well'*.

The squadron spent the 13th and 14th training, the current issues with available crews highlighting the urgent need for additional fully trained crews on the squadron. Vickers Wellington Mk.IA P9226 was damaged by a tractor at R.A.F Honington on the 13th. Unauthorised to drive the tractor, the driver was severely reprimanded.

| Date | 13/02/1941 |
|---|---|
| Mark | Mk.IA |
| Serial | P9226 |
| Code | KX-Z |
| Taken On Charge | 02/11/1940 via No.22 MU. |
| Manufacturer | Vickers Armstrong |
| Contract | 549268/36 |
| Pilot (s) | - |
| Flight | At dispersal |
| Time | - |
| Cause | Hit by tractor |

The squadrons of 3 Group were given two targets on the 15th. 311 (Czech) Squadron would provide just two crews on the raid directed at the Sterkrade-Holten - Ruhrchemie (Ruhr Benzin AG) Synthetic-Oil Plant while three crews would re-visit the docks at Boulogne. The now familiar captain's names appeared on the Battle Order, but with two new additions. Sergeants Hugo Dostál and Jaroslav Doktor would be making their operational debut.

February 15th 1941: A.79 Sterkrade-Holten – Ruhrchemie Oil Plant

| Pilot | 2nd Pilot | Serial | Code | Bomb Load | Result |
|---|---|---|---|---|---|
| Sgt V Korda | Sgt H Dostál | T2972 | KX-G | 5x500lb | Completed |
| Sgt L Anderle | Sgt J Filler | T2553 | KX-B | 5x500lb | Completed |

A total of forty-five 3 Group Wellingtons were given Sterkrade to attack. It was one of the Ruhr's more accessible targets thanks to its position on the western edge of the industrial Ruhr valley, it was also a significant producer of oil and coal. The night was particularly dark, and to make things more challenging, ground haze and an abnormal number of searchlights made accurate pinpointing of the target extremely difficult for the crews. The force was over Sterkrade between 20:10hrs and 21:45hrs, with many crews orbiting the target in the hope of a fleeting glimpse of the oil plant. Sergeant V Korda dropped his entire load in one stick in the general target area. The results were unobserved. He reported upon his return many fires between Oberhausen and Sterkrade. The crew of Sergeant L Anderle bombed from 15,000ft, confident their bombs fell in the target area. Like his colleague, Leo Anderle reported numerous fires. However, disappointingly, these were over Duisburg, 8 miles south of the intended target. The bomber force dropped 138 500-pounders, seven 250-pounders, and 660 4lb incendiaries, plus seven 1000 pounders. Despite the amount of ordinance dropped there was no reported damage to the oil plant.

### February 15th 1941: C.29 – Boulogne Docks

| Pilot | 2nd Pilot | Serial | Code | Bomb Load | Result |
|---|---|---|---|---|---|
| F/O J Breitcetl | Sgt J Bernát | R1371 | KX-F | 3x500lb + 120x4lb | Completed |
| S/Ldr J Ocelka | Sgt F Fencl | T2971 | KX-H | 3x500lb + 120x4lb | Completed |
| S/Ldr Pickard DFC | Sgt J Doktor | R1015 | KX-L | 3x500lb + 120x4lb | Completed |

Group only provided 16 crews for the raid on Boulogne-sur-Mer. Conditions over the docks afforded the participating crews a chance for some accurate bombing. Over the target between 14,000ft and 16,000ft, the defences below put up a withering amount of flak that equalled any target in Germany. All three crews bombed Basin No.4. Squadron Leader J Ocelka started a sizeable red fire and a smaller intense yellow blaze. His counterpart, Squadron Leader 'Pick' Pickard DFC, reported, *'Bombed Southeast side of Basin No.4. Twelve large fires seen, visible from 30 miles from target area'*. Flying Officer J Breitcetl reported that he attacked, *'shipping and docks, two bursts seen followed by two fires'*.

For the next two days and in between some filthy weather, the squadron was up training day and night. The recent spell of good results had been a tremendous morale boost, aided by an excellent party at Wretham Hall. On the 18th, the Luftwaffe made an unwelcome visit when a Heinkel He111 of 4./KG53 flew low over the airfield at 07:49hrs. The German gunners sprayed everything in sight, damaging the port engine of a dispersed Wellington while putting a few holes in the roof of a hut. The explosions from their bombs smashed a few windows of the Sergeant's Mess. It had been a lucky escape. The He111 was later shot down near Ovington, a few miles from Watton. It was Honington's turn to be bombed at 15:15hrs when Dornier Do17s carried out an accurate attack on the airfield. Nine bombs were reportedly dropped. Six landed between 'A' hangar and the

*The harsh winter of 1940/41 is clearly illustrated in this unusual photograph of Air Gunner Sergeant Vladimír Cupák in flying gear during a snowstorm at East Wretham during the winter of 1940-1941. He completed 36 sorties between 9th January and 16th July 1941. Later he was retrained as a Radar Operator and between 22nd February 1942 and 22nd June 1944 he flew Bristol Beaufighter night fighters with 68 Squadron. (Pavel Vančata)*

bomb dump. One exploded on the roof of 'A' hangar while two fell near 'E' hangar. The craters on the aerodrome were sufficient to render the landing area unserviceable for around five hours. Mercifully, there were no casualties, and only two 9 Squadron Wellingtons were slightly damaged. Rain, then snow fell on the 19th effectively keeping the squadron grounded. This continued on the 20th. It was not until the 21st that news arrived informing the squadron that Berlin was the target for that night with Dusseldorf as an alternative. The squadron had not visited Berlin since October 12th, and it was a target everyone wanted to bomb. During the course of the day H.Q 3 Group had second thoughts, primarily due to weather concerns. The target was changed to Wilhelmshaven with Emden as an alternative. The numbers deployed were also reduced from 57 to 34. The Czechs of 311 Squadron would provide five crews.

### February 21st 1941: Town 7 – Wilhelmshaven

| Pilot | 2nd Pilot | Serial | Code | Bomb Load | Result |
|---|---|---|---|---|---|
| S/Ldr J Ocelka | Sgt A Jedounek | T2971 | KX-H | 4x500lb+120x4lb | Completed |
| Sgt V Korda | W/Cdr K Toman | T2972 | KX-G | 4x500lb+120x4lb | Completed |
| Sgt L Anderle | Sgt J Filler | T2553 | KX-B | 4x500lb+120x4lb | Early Return |
| F/O J Breitcetl | Sgt J Bernát | R1371 | KX-F | 3x500lb+1x250lb+180x4lb | Early Return |
| F/Lt J Šnajdr | Sgt B Blatný | T2561 | KX-A | 3x500lb+1x250lb+180x4lb | Completed |

Joining the crew of Sergeant Václav Korda was the squadron commander, Wing Commander Karel Mareš Toman, who unlike his R.A.F counterpart believed in leading by example. Light snow had begun to fall as the first of the Wellingtons took off. Once airborne, the Wellingtons gain altitude and set off for northern Germany above a layer of solid cloud. The squadron notched up two early returns on this night. Sergeant L Anderle and crew experienced starboard engine trouble aboard Wellington T2553 KX-B. Realising that to continue was pointless they dropped their bombs on searchlight and anti-aircraft positions at Den Helder producing small fires from their incendiaries[19]. Flying Officer J Breitcetl was obliged to bomb the airfield at De Kooy when he suffered engine failure and icing issues. The solid cloud continued almost to the target area where a few breaks in the clouds gave the crews some chance to bomb. The defences of Wilhelmshaven were particularly active, intense flak and searchlights adding to the problems. Both Flight Lieutenant J Šnajdr and Squadron Leader J Ocelka claimed to have bombed, Squadron Leader Ocelka having 20 incendiaries hang-up. Neither observed the results. Sergeant V Korda and crew were more fortunate, a gap in the cloud gave them the opportunity they needed to ensure the target was bombed accurately. Observation of the results were however spoilt by the glare from the numerous searchlights. Only twenty returning crews would claim to have bombed the general area of the target, observing some bursts across the docks and a few isolated fires. Flight Lieutenant J Šnajdr and crew landed at R.A.F Bircham Newton at 23:02hrs having received an incorrect Q.D.M from R.A.F Manston of 354, instead of 254. They eventually landed back at base at 02:00hrs. The squadron was informed early on the morning of February 22nd that it was to provide two 'Freshmen' crews for that night. This order was later cancelled and the squadron sat out an attack by twenty-nine 3 Group Wellingtons briefed to attack the Hipper-class cruiser at anchor in Brest Harbour.

It was around this time that Wing Commander William Simmons MiD was posted to R.A.F Driffield, Yorkshire. Although he was a good organiser and administrator, the Czechs did not particularly welcome his command style. Since his arrival in November, he had not flown a single operation which had not gone unnoticed by the crews. Throughout his time on the squadron, he did little to inspire the Czechs. Command passed over to Wing Commander Toman Mareš on February 21st.

Bomber Command H.Q had intended to attack Wilhemshaven on the 23rd after the recent disappointing results. However, due to adverse weather over most of northern Europe they wisely changed the target to the docks at Boulogne-sur-Mer. The squadron would offer six crews.

---

[19] *Another report states that they bombed De-Kooy airfield.*

## February 23rd 1941: C.29 – Boulogne Docks

| Pilot | 2nd Pilot | Serial | Code | Bomb Load | Result |
|---|---|---|---|---|---|
| Sgt L Anderle | Sgt J Filler | T2971 | KX-H | 5x500lb + 120x4lb | Completed |
| Sgt V Korda | Sgt A Jedounek | T2972 | KX-G | 5x500lb+2x250lb+120x4lb | Completed |
| F/Lt J Šnajdr | Sgt B Blatný | T2561 | KX-A | 5x500lb+2x250lb+120x4lb | Completed |
| F/O J Breitcetl | Sgt J Bernát | R1371 | KX-F | 5x500lb+2x250lb+120x4lb | Completed |
| Sgt J Doktor | Sgt H Dostál | R1015 | KX-L | 5x500lb + 120x4lb | Completed |
| Sgt F Fencl | Sgt F Radina | R3234 | KX-N | 5x500lb + 120x4lb | Completed |

*Thirty-year-old Air Gunner Vilém Jakš, the pre-war middleweight boxing champion of Czechoslovakia. He is photographed with Pilot Officer Korda DFC.*

Two pilots would be making their operational debut as captains, Sergeants Jaroslav Doktor and František Fencl. Both were experienced pilots who had previously served in the Czechoslovak Air Force pre-war. František Fencl was born on February 6th 1915 in Dominikální Paseky. After completing his apprenticeship as a locksmith, František moved to Prostějov, where he graduated from the School of Aviation Cadets in 1936. After graduating, he became a member of the 1st Air Regiment 'T. G. Masaryk' in Prague, and in 1937, he was appointed a field pilot. A year later, he completed training in night and instrument flying. With the Czech occupation, he crossed the border into Poland, from where he sailed from Gdynia to the French port of Cherbourg. With the capitulation of France, František arrived in England, and his previous flying experience saw him assigned to a particular group focused on selecting and training pilots for bombers, whether they were pilots, navigators, radio operators or gunners. Fellow pilot Jaroslav Doktor was older than his colleague, born on 20th November 1909 in Německý Brod. He began his military service in 1931. Somewhat of a maverick and often in trouble with authority, he spent a lot of time in military prison. His last sentence resulted in the end of his Czechoslovak Air Force career. He was demoted and finally dismissed in 1933. He then served in the Army before returning to flying in 1935 as a civil flying instructor. There then followed a period serving as a commercial pilot with the Czechoslovakian Airline from 1938. He left the port of Gdynia in August 1939 aboard the ship *Kastelholm*, arriving at Boulogne-sur-Mere a week later. On eventually reaching England, he joined the Royal Air Force Volunteer Reserve at the Czechoslovak Air Force Depot at Cosford and was promoted to the rank of Sergeant and selected for training as a bomber pilot, being posted to No.11 O.T.U at R.A.F Bassingbourn. A wiry individual with a pinched expression but a big smile, he soon settled into squadron life.

All six crews had safely departed by 18:58hrs, and unlike the previous operation, there were no serviceability issues. The first bombers were over the target at 19:51hrs. Conditions initially made identifying the target difficult. Sergeant Sgt V Korda and crew arrived over an eerily quiet Boulogne, with no fires reported and little activity. They bombed from 12,700ft in two sticks in the port area. The later arrivals found a very different target. Many fires were taking hold, especially around Basin 4 & 5, and luckily, a few isolated cloud breaks began to appear. It also coincided with an increase in flak and searchlight activity. Flight Lieutenant J Šnajdr reported Basin 4 was illuminated by fires. His first stick of bombs fell on the pier between Basins 4 & 5, starting a red fire which produced dark black smoke. His second stick was dropped in the same area. The fires around Basin 4 & 6 were the target for the first stick of bombs dropped by Flying Officer J Breitcetl while his second stick landed just south of Basin 7. His rear gunner, Sergeant Jindřich Beneš, reported four new fires started. The two 'Freshmen' crews were in the thick of the action. Sergeant F Fencl reported that dense black smoke was rising north of Basin 6, while an intense whitish fire was seen blazing between Basin 6 and 4. All the while an equally vivid red coloured fire was seen between Basin 7 and 4. Sergeant J Doktor dumped his entire bombload on Basin 4 from 13,000ft, creating a dense column of yellowish smoke. As the squadrons crews turned for home, Boulogne docks appeared to have been clobbered, with fires visible from 30 miles. It had been an excellent effort by the crews who had pressed home their attacks despite the intense flak. The jubilant crews started landing back at East Wretham at 21:16hrs, having dropped 18,720lb of bombs.

Yet more intense training was carried out on the 24th and 25th, the Training Flight making full use of the break in the weather. The squadron sat out an attack on Dusseldorf on the 25th by the squadrons of R.A.F Marham, Stradishall and Wyton. It was the turn of Mildenhall, Feltwell and Honington on the 26th when 47 Wellingtons and two of the new Short Stirlings would begin a brief campaign directed against the Rhineland Capital, Cologne. Bomber Command's intention was simple: *'Cause as much damage as possible'*. Once again, the squadron detailed and briefed six crews for the operation.

<u>February 26th 1941: Town 2 – Cologne</u>

| Pilot | 2nd Pilot | Serial | Code | Bomb Load | Result |
|---|---|---|---|---|---|
| S/Ldr J Ocelka | Sgt F Kráčmer | T2971 | KX-H | 4x500lb+1x250lb+120x4lb | Completed |
| Sgt A Šedivý | Sgt J Čapka | T2561 | KX-A | 4x500lb+1x250lb+120x4lb | Completed |
| F/O J Breitcetl | Sgt J Bernát | R1371 | KX-F | 4x500lb+1x250lb+120x4lb | Completed |
| Sgt F Fencl | Sgt F Radina | R3234 | KX-N | 4x500lb+1x250lb+120x4lb | Completed |
| Sgt L Anderle | Sgt J Filler | R1378 | KX-K | 4x500lb+1x250lb+120x4lb | Early Return |
| Sgt J Doktor | Sgt H Dostál | R1015 | KX-L | 4x500lb+1x250lb+120x4lb | Early Return |

One new name was chalked up on the Operations Board, Sergeant František Kráčmer, a 25-year-old former Czechoslovakian Air Force pilot. He would occupy the right-hand seat next to Squadron Leader J Ocelka aboard Wellington T2971 KX-H. Two crews were obliged to bomb alternative targets soon after crossing the Dutch coast. Sergeant L Anderle experienced electrical trouble aboard Wellington R1378. They bombed the docks and installations at Vlissingen from 14,000ft at 20:20hrs, creating a big, reddish-coloured fire. Within the hour, the crew of Sergeant Doktor, who had suffered engine issues, were also over Vlissingen, adding their bombload to the fires in and around the docks. The remaining crews headed further into occupied territory via corridor 'G' and encountered flak and searchlights from the Scheldt estuary to the target area. The night was incredibly dark, and the industrial haze made identifying the target, even Cologne's size, difficult. The crew of Sergeant A Šedivý were early over the target area at around 21:00hrs. On arrival, only a few scattered fires were visible, and due to the darkness and haze, the crew were initially unsure of their position. Eventually, with the aid of some flares dropped by other equally doubtful crews, they finally identified the target and dropped their bombs at 21:35hrs from 14,000ft. The three remaining crews arrived over Cologne to find half a dozen sizable fires already taking hold. Squadron Leader J Ocelka reported three large fires producing sheets of white flames. They dropped their bombs in the target area and reported that *'Fires were visible from 50 miles away'*.

Returning crews earnestly reported explosions and large fires in the target area, and all indications pointed to a successful raid. Headquarters 3 Group at Harraton House, Exning, reported, *'The results appear to be very satisfactory, and nearly all the crews reported large fires'*. The optimism was misplaced. Local reports recorded just ten high explosives and a paltry 90 incendiaries falling on the outer fringes of Cologne.

The only incident during a routine night occurred back at R.A.F Wretham. Pilot Officer John Gellner mentions an extraordinary incident in his book *Moon Flyer*.[20] Vickers Wellington T2620 of 9 Squadron, skippered by Sergeant Terrence Donnelly, landed at East Wretham at 23:10hrs, having been damaged by flak. On landing, Donnelly queried where he was, and on being told, he requested permission to take-off and return to his home base at R.A.F Honington. He was ordered to wait as the squadrons Wellingtons were landing. Impatient to do so, he nearly collided with Squadron Leader Ocelka and then became entangled with barbed-wire fencing. If this was not enough for one night, the obviously flustered pilot then hit a dispersed Bristol Blenheim of 21 Squadron. The 9 Squadron ORB does not mention the drama one of its crews had caused.

There was another unusual incident on the 26th involving the Honington's Training Flight's Avro Anson R9648 flown by Flight Lieutenant Michael Earle, which thankfully had a happy ending. While en route to Sealand, Flintshire, to collect an aircraft, the cockpit roof escape hatch suddenly blew off with a tremendous whack. Pilot Officer Jiří Engel, the navigator, promptly took to his parachute landing at Aldford, Cheshire, at 15:15hrs. The remaining crew, Sergeant Adolf Musálek, Pilot Officer Otakar Černý and Pilot Officer Jaroslav Zafouk, continued on to make a safe landing at Sealand. The unfortunate Engel was slightly injured on landing. His troubles did not end there. A local farmer armed with a shotgun stopped and detained him. His poor English was considered a good reason to hold him until his identity could be confirmed.

On the 27th, Luftwaffe bombers carried out an audacious series of attacks on 3 Group airfields. R.A.F Oakington, Mildenhall, Honington and Feltwell were all attacked between 11:04hrs and 14:14hrs. The Czechs at East Wretham got off lightly, a low flying Dornier Do17Z arrived over the airfield at 12:43hrs. No bombs were dropped but the airfield was machine-gunned. It was left to the squadrons of R.A.F Marham, Stradishall and Wyton to finish off the month with a low-key operation on the 28th against Wilhelmshaven and the Tirpitz.

---

[20] *Moonlight Flyer.*

*Squadron Leader Josef Schejbal.*

# March 1941 : A Loss Free Month

The month started with a bang for the whole squadron. For the first time since the squadron's formation it was able to detail and brief eight crews for operations. It was a tremendous morale booster, made possible by the wonderful work of Squadron Leader Percy 'Pick' Pickard DFC and every member of the Training Flight. Bomber Command H.Q had called for a maximum effort for a return raid on Cologne. The squadrons of 3 Group would provide 75 crews, but this number was reduced to 57 in the afternoon due to concerns about the weather. Seven crews of 311 (Czech) Squadron would attack Cologne, while a single 'Freshmen' crew would bomb the docks Boulogne-sur-Mer.

### March 1st 1941: CC.29 – Boulogne

| Pilot | 2nd Pilot | Serial | Code | Bomb Load | Result |
|---|---|---|---|---|---|
| Sgt O Helma | Sgt F Kráčmer | T2553 | KX-B | 4x500lb+120x4lb | Completed |

Sergeant Oldřich Helma and Sergeant František Kráčmer would be operating for the first time together. They were relatively young compared to some of the pilots on the squadron, both in their mid-twenties, but were not lacking in skill and courage. The crew departed at 19:00hrs. It would be a long, lonely journey to Boulogne. Group H.Q had allocated two crews to attack, the other provided by the R.A.F Feltwell-based 57 Squadron. Sergeant Helma arrived over the target between 20:31hrs and 20:45hrs and bore the brunt of a vicious flak barrage. All but one bomb was dropped from 15,000ft in the target area, one 500-pounder having hung up. They returned safely back at East Wretham, the operation lasting just 3 hours and 3 minutes.

### March 1st 1941: Town 2 – Cologne

| Pilot | 2nd Pilot | Serial | Code | Bomb Load | Result |
|---|---|---|---|---|---|
| S/Ldr J Ocelka | Sgt A Jedounek | T2971 | KX-H | 4x500lb+1x250lb+120x4lb | Completed |
| F/Lt J Šnajdr | Sgt B Blatný | R1021 | KX-W | 4x500lb+1x250lb+120x4lb | Completed |
| Sgt L Anderle | Sgt J Filler | R1378 | KX-K | 4x500lb+1x250lb+120x4lb | Completed |
| Sgt A Šedivý | Sgt J Čapka | T2561 | KX-A | 4x500lb+1x250lb+120x4lb | Completed |
| F/O J Breitcetl | Sgt J Bernát | R1371 | KX-F | 4x500lb+1x250lb+120x4lb | Completed |
| Sgt F Fencl | Sgt F Radina | R3234 | KX-N | 4x500lb+1x250lb+120x4lb | Completed |
| Sgt J Doktor | Sgt H Dostál | R1015 | KX-L | 4x500lb+1x250lb+120x4lb | Completed |

The familiar route was flown over the North Sea via corridor 'G' with landfall over the Scheldt estuary in conditions of ten-tenths cloud. To the credit of the ground crews, the squadron suffered no early returns, which arrived over an almost cloud-free Cologne. The squadron arrived over the city during the latter half of the raid, which began at 21:22hrs. Squadron Leader J Ocelka and crew reported four substantial fires on arrival. In comparison, Sergeant L Anderle was slightly more cautious, reporting eight scattered fires, with two larger fires near the centre of Cologne. All the crews bombed the target between 14,000ft and 14,500ft in the face of a tremendous flak barrage and frantic searchlight activity. Flight Lieutenant J Šnajdr's crew nearly fell victim to Cologne's flak defences. While on their bomb run, Wellington R1021 KX-W shuddered under the impact of two flak bursts that showered the Wellington with shrapnel. The forward section of the fuselage was severely punctured with numerous holes as the jagged pieces tore through the fabric and geodetics, damaging the oxygen system, port engine, propeller, dinghy, oil tank and wiring. The navigator's table and instruments seemed to have absorbed most of the damage. Thankfully, the navigator, Pilot Officer Karel Bečvář was hunched over the bombsight at the time. Miraculously, none of the crew were injured, but the rear gunner, Sergeant František Chmura, was reported to have been *'Shaken'*. The last of 3 Group's crews departed at 23:58hrs. Nearly all those who claimed to have bombed Cologne reported large fires burning on leaving, Sergeant J Doktor reporting they were visible from 75 miles! A Whitley Mk.V of 51 Squadron landed at East Wretham just after midnight due to poor weather over Yorkshire, while another Whitley from 10 Squadron found sanctuary at R.A.F Honington. Whitley Mk.V

*There was always time for training, even for the officers. A group of 311 Squadron officers during practical training with Vickers Gas Operated (VGO) machine guns. Standing at the work-bench L-R: Pilot Officers Josef Doubek, Jan Parolek, Karel Hančil, Pavel Simet, Ondřej Kacíř, Otakar Černý (standing in the middle of window)? Pilot Officer Miroslav Vild, ?. Behind them is standing one of the British instructors. Czechoslovak field post stamp with the date September 28, 1940 seeped through from the reverse, dating the picture to August–September 1940. (Pavel Simet)*

T4235 flown by Sergeant W.C Marshal of 78 Squadron, was reported to have crashed near the 'B' site at Honington at 04:25hrs. It had previously been damaged by flak over Germany and suffered a port engine issue.

The only activity on the 3$^{rd}$ was flown by a 'Freshmen' crew. The target, unsurprisingly, was Boulogne-sur-Mer. Sergeant Oldřich Helma was once again the captain who, on this occasion, was joined by 26-year-old Sergeant Karel Schoř, making his operational debut.

<u>March 3$^{rd}$ 1941: CC.29 – Boulogne</u>

| Pilot | 2$^{nd}$ Pilot | Serial | Code | Bomb Load | Result |
|---|---|---|---|---|---|
| Sgt O Helma | Sgt K Schoř | T2972 | KX-G | 4x500lb+120x4lb | Completed |

Eight 'Freshmen' crews would be given the docks at Boulogne while 32 bombers of 3 Group would keep the civilian population of Cologne in their air-raid shelters for another night. Proceedings started at 20:23hrs. Sergeant O Helma was one of the first over the target. The crew reported no fires on arrival but pinpointed the docks and installations through the broken cloud. The bombs were dropped in two sticks from 15,000ft. The crew's rear gunner, Sergeant

Ladislav Kadlec, reported the bombs landed in the target area, starting two fires. However, the navigator, Pilot Officer Josef Motyčka, was confident his bombs landed between docks 4 and 7.

While Sergeant Helma and crew were returning from Boulogne, the crew of Flying Officer Josef Šejbl almost managed to write off Wellington P9235 when they overshot their landing at East Wretham at 20:50hrs.

| Date | 03/03/1941 |
| --- | --- |
| Mark | Mk.IA |
| Serial | P9235 |
| Code | KX-C |
| Taken On Charge | 09/09/1940 via 115 Squadron. |
| Manufacturer | Vickers Armstrong |
| Contract | 549268/36 |
| Pilot (s) | Flying Officer Josef Šejbl |
| Flight | Training Flight |
| Time | 20:50hrs |
| Cause | Inexperience |

Only Sergeant Rostislav Kaňovský, the Wireless Operator was injured, suffering a broken knee on landing. He was taken to Bury St Edmunds Hospital. The Wellington had endured a catalogue of damage when the undercarriage collapsed. It would not operate with the squadron again on being repaired. The night's activities were not quite over. While coming in to land at 21:55hrs, Wellington T2972 was reported to have sustained damage to the 'fabric' when followed by an enemy fighter. That same night, a reported ten bombs fell on East Wretham. One landed on a dispersal point while the remainder fell across the airfield. There is no report of injuries.[21]

A period of bad weather followed which would result in planned operations on the 5th, 7th, 8th, 9th and 10th being cancelled. Thankfully, the weather, although unpleasant at times, did not interfere drastically with the Training Flight's activities. The lousy weather, however, did not interfere with German attacks on 3 Group stations. On the 6th, R.A.F Feltwell was bombed by a lone raider just after lunch. The following day, Feltwell was again attacked by a solitary Dornier Do17. R.A.F Alconbury was attacked twice on the 8th, and some damage was inflicted, and a Wellington damaged. In the evening, the Q site at R.A.F Marham was bombed. On the 8th, a big party was held in the Officer's Mess. The occasion was the announcement of the DSO to Squadron Leader Percy Pickard DFC. The citation refers to his service to 311 Squadron and particularly his role in the training of crews. Pickard was mild-mannered, approachable and humorous, but he was a tremendous taskmaster who expected and demanded 100% from all those who served under him. That drive and dedication ensured that the Training Flight, despite numerous setbacks, produced some exceptional crews.

> *Since joining the squadron in July 1940, this officer has invariably taken out the new Czech crews on their initial operation or first long-distance mission. On such occasions, he has been the only British member amongst the crews who have been inspired by his splendid leadership and example. On one occasion it was undoubtedly due to his determined efforts that one Czech crew were rescued after being adrift in the North Sea for over 13 hours. On another occasion when a crew was forced down in the North Sea, his persistence and good airmanship in failing light, and his sound use of recognition signals enabled surface craft to affect a rescue. His complete disregard for danger was particularly shown on one occasion when a fully loaded bomber crashed and caught fire. He led the rescue party and personally extricated two members of the crew and succeeded in eventually conveying them to safety, although compelled to remain prone in the danger area during the explosions of some bombs. He has displayed coolness and courage of a high of order and his magnificent work has contributed largely to the present efficiency of the squadron.*

---

[21] *Headquarter No.3 Group Records Book.*

Pickard and his wife lived off station at East Harling, a sleepy little village a few miles southeast of the airfield. Wherever Squadron Leader Pickard went, so did his sheepdog 'Ming', whom the Czechs were wary of. Around this time, Percy Pickard was chosen to 'star' in a propaganda/factual film depicting a bomber crew on a raid on Germany. Given the recent Luftwaffe Blitz on London, Birmingham, Coventry, Southampton, Bristol, Hull and Swansea, plus a dozen more cities and towns, the British Government was keen to give the public something to cheer about. The filmmakers believed that in the tall, blonde, well-spoken, pipe-smoking Percy Pickard, they had the ideal man, but Pickard was not convinced. The film was made at Blackheath Studios and would see Squadron Leader Pickard DSO, DFC, away for much of March and early April. The producers of the film, titled *'Target for Tonight'* came up with a novel idea, in place of actors, actual aircrew would be used. Percy Pickard would play Squadron Leader Dickson, pilot of 'F-Freddie'. His squadron commander was played by the flamboyant Group Captain John 'Speedy' Powell, one of five brothers, four of which served in the RAF. While the Station Commander was played by Group Captain 'Bull' Staton, a no-nonsense Nazis hating bear of a man. The Crown Film Unit produced the film which became an instant success, winning an Academy Award in 1942. With the film's success came the public adulation to its hero pilot, something that Percy Pickard felt rather uncomfortable with.

*A wonderful photograph of members of 311 Squadron enjoying the spring sun of 1941 on the stairs to the Wretham Hall. L-R: Air Gunner Pilot Officer Eduard Šimon, Acting Squadron Leader P. C. Pickard, DSO, DFC with his fateful dog Ming, Navigator Pilot Officer Karel Bečvář and Acting Squadron Leader Josef Ocelka who would be the third squadrons Commanding Officer. (Pavel Vančata)*

After seven nights during which operations were impossible due to the weather conditions, 3 Group resumed operations on March 11th. The target was Kiel. Disappointingly, 311 Squadron was obliged to sit the raid out. With the prospect of clear weather and an almost full moon, a maximum effort was called for the following night. The Group would be out in force, and attack three separate targets. Fifty-five bombers would bomb the Focke-Wulf factory at Bremen, twenty-eight would visit Berlin, while four 'Freshmen' crews would bomb the docks at Boulogne-sur-Mer. The raid would also witness the squadron operating against three separate targets for the first time. That night would see the start of a new campaign. By the beginning of 1941, the country was gripped with U-boat fever. The German success in the North Atlantic had Britain and its Royal Navy almost beaten. The North Atlantic convoys had lost nearly 3 million tons of urgently needed supplies since the fall of France the previous summer. Despite all its posturing, the Royal Navy could not oppose the threat as it was woefully under strength and under equipped. Coastal Command was still flying obsolete aircraft, and those more modern types in service did not have the range to make any worthwhile contribution. To add to the U-Boat menace, the German Battle-cruisers *Scharnhorst* and *Gneisenau* were on the prowl in the Atlantic, creating mayhem. If this was not enough, the mighty Battleship *Bismarck* was being prepared and made ready to set sail. It was not just at sea the Germans appeared to have the upper hand. It was also in the air. The Luftwaffe had the long range Focke-Wulf Fw200 Condor. With a range of nearly 2,000 miles, this formidable aircraft was causing havoc over the Atlantic. Armed with 2000lb of bombs, the Condors repeatedly attacked and sank merchant vessels. More worryingly, they reported the positions of the convoys to the U-boat packs. Winston Churchill called these four-engine giants *'The scourge of the Atlantic'*. By March 1941, Britain faced the alarming prospect of its vital ocean links to Canada and America being severed. Germany was winning the war in the Atlantic. Prime Minister Winston Churchill was forced into giving a simple instruction – *'Bomber Command will direct all its efforts against the targets that housed or sourced the threat to British shipping'*. On March 9th 1941, Air Chief Marshal Sir Wilfred Freeman (Vice Chief of the Air Staff) informed Air Marshall Sir Richard Peirse of a new directive. Peirse was to *'dedicate his energies to defeating the attempt of the enemy to strangle our food supplies and our connection with the United States'*. The directive could not be more explicit, *'We must take the offensive against the U-Boat and the Focke-Wulf wherever and whenever we can'*. A list of twelve targets was drawn up. Each was divided into individual yards, plants, factories and assembly buildings. The targets chosen were, **Germany** : Kiel, Bremen, Vegesack, Hamburg, Augsburg, Mannheim, Dessau, **France** : Lorient, St.Nazaire, Bordeaux, Bordeaux-Merignac and **Norway**, Stavanger. The C-in-C Bomber Command Air Marshall, Sir Richard Peirse, was not overly happy about the policy change. Peirse felt that his command was on the verge of success with his attacks on Germany, especially the attacks on Germany's oil targets. The Air Ministry, more in hope than judgement believed the accuracy of reports filtering in from H.Q Bomber Command and allowed Sir Richard Peirse to devote a proportion of his efforts to continue the attacks directed against oil targets. It would be 3 Group, the most potent Group in Bomber Command at the time, which would bear the brunt of the forthcoming campaign. The timing of the new directive coincided with 311 (Czech) Squadron finally being supplied with accomplished and motivated crews on a regular basis from its diligent Training Flight. These new crews would be tested over the coming weeks and months. Training took priority over the coming days when the English weather allowed.

<u>March 12th 1941: Town 13 – Berlin</u>

| Pilot | 2nd Pilot | Serial | Code | Bomb Load | Result |
|---|---|---|---|---|---|
| F/Lt J Šnajdr | Sgt B Blatný | R1410 | KX-M | 1x500lb+1x250lb+300x4lb | Complete |
| Sgt V Korda | Sgt F Kráčmer | T2972 | KX-G | 4x500lb | Complete |

Finally, on the 12th, word arrived on the squadron that Berlin would be visited that night. The Nazi capital would be attacked by 72 bombers, 28 provided by the Wellington squadrons of 3 Group. The group would also send its crews to the docks at Bremen. The Czechs of 311 Squadron would be involved in both raids. The first crews away were those given Berlin. Flight Lieutenant J Šnajdr lifted off East Wretham's grass runway at 19:45hrs, followed three minutes later by Sergeant V Korda. Conditions were ideal. The North Sea was crossed in the light of a full moon in an almost cloud-free sky. The Czechs made landfall over the Dutch coast slightly north of Ijmuiden, keeping well clear of the flak defences of Amsterdam. The route took the crews past Osnabruck, Minden and south of Hanover before running into the target over Brandenburg and Potsdam. Berlin's defences were already fully alert. Searchlights seemed to encircle Berlin while a wall of heavy flak burst menacingly between 12,000ft and 16,000ft. Fires, mostly scattered, were visible on arrival. Flight Lieutenant J Šnajdr reported only small, isolated fires in Berlin on his arrival. They dropped their mixed load, creating two more fires near the aiming point. The crew of Sergeant V Korda bombed from between 13,000ft and 17,000ft, encountering fierce flak. On return to East Wretham, they reported, *'Saw one big red fire on arrival. We saw our bombs burst, fresh fires started visible from 60 miles'.* The last squadrons of 3 Group turned for home at 01:10hrs for the long flight home. The raid achieved only modest success. Most of the bombs fell in the sparsely populated southern districts. Following almost immediately behind the Berlin duo were the five crews given Bremen. They would have varied success.

### March 12th 1941: GY4772 – Bremen-Neuenland Focke-Wulf Flugzeubau AG

| Pilot | 2nd Pilot | Serial | Code | Bomb Load | Result |
|---|---|---|---|---|---|
| S/Ldr J Ocelka | Sgt A Jedounek | T2971 | KX-H | 1x1000lb+2x500lb+1x250 | Completed |
| Sgt J Doktor | Sgt H Dostál | R1015 | KX-L | 4x500lb+1x250lb+120x4lb | Early Return |
| F/O J Breitcetl | Sgt J Bernát | R1371 | KX-F | 6x500lb | Early Return |
| Sgt A Šedivý | Sgt J Čapka | T2561 | KX-A | 6x500lb | Completed |
| Sgt F Fencl | Sgt F Radina | R1378 | KX-K | 4x500lb+1x250lb+120x4lb | Alternative Target |

*Pilot, Sergeant Josef Čapka (2nd from right) with mechanics who were responsible for servicing of Wellington Mk.IC R1598/KX-C for "Cecilka". Sergeant Čapka flew 15 sorties of this aircraft. (Marie Lambourne)*

The squadron notched up two 'boomerangs'. The first back was the crew of Sergeant J Doktor, who experienced engine trouble soon after take-off. The bombs were jettisoned, and the crew landed at R.A.F Marham at 23:45hrs. Flying Officer J Breitcetl and crew were forced to abort when Wellington R1371 KX-F could not maintain height. The bombs were jettisoned, and the crew landed safely back at base at 01:30hrs. As the crew of Squadron Leader J Ocelka arrived over Bremen's dock area, two bombloads exploded, throwing up a sheet of flames. It was the only activity seen. The crew dropped their all HE load from just

10,000ft, which was seen to burst near an existing fire near the factory complex. These bombs created a column of greyish smoke visible for 40 miles. The bombload of Sergeant A Šedivý was observed to explode between the railway station and the centre of the target. Faulty map reading on the part of the navigator aboard Sergeant F Fencl's Wellington found the crew over Hamburg. They bombed the western part of the dock area, producing two large red fires, which they reported were visible from 20-30 miles.

### March 12th 1941: CC.29 – Boulogne

| Pilot | 2nd Pilot | Serial | Code | Bomb Load | Result |
|---|---|---|---|---|---|
| Sgt O Helma | Sgt K Schoř | R1451 | KX-P | 4x500lb+120x4lb | Completed |

The last crew to take-off were the first to land back at R.A.F Wretham at 23:08hrs. Sergeant O Helma, along with other 'Freshmen' crews and a solitary Short Stirling of 7 Squadron, carried out a reasonably accurate raid on Boulogne. Sergeant Helma had the satisfaction of watching his bombs explode between the quay and Basins 4 and 5. Overall, 3 Group H.Q seemed satisfied with the Group's performance. A total of 232,000lb of bombs had been dropped. 311 Squadron's contribution was 15,410 lbs, its best to date. While most of the attention was focused on the raids, the Training Flight was still busy with its trainee crews. On this night, the flight would suffer the loss of Wellington P9226, which suffered undercarriage failure on landing at 22:00hrs. The pilot, Sergeant Jaroslav Hájek and his instructor Sergeant Jaroslav Bala, were both cleared of blame. The commanding officers believed the undercarriage had been weakened due to the heavy landings it had suffered on the Training Flight.

| Date | 12/03/1941 |
|---|---|
| Mark | Mk.IA |
| Serial | P9226 |
| Code | KX-Z |
| Taken On Charge | 02/11/1940 via No.22 MU. |
| Manufacturer | Vickers Armstrong |
| Contract | 549268/36 |
| Pilot (s) | Sergeant Jaroslav Bala / Sergeant Jaroslav Hájek |
| Flight | Training – Dual Instruction |
| Time | 22:00hrs |
| Cause | Weakened Undercarriage |

The immobile Wellington had to be removed as it was causing an obstruction, it was no surprise given its condition, two badly damaged airscrews, undercarriage wrecked, port nacelle and fuel tanks holed. The Wellington once salvaged never returned to East Wretham. The squadron was not involved in the successful operation against the Blohm & Voss shipyards in Hamburg on the 13th. The Czechs did not have long to wait to operate again. On the morning of March 14th Form B.439 arrived at East Wretham. 311 Squadron would supply eight crews that night, the target was Gelsenkirchen (Nordstern) and its oil plants.

### March 14th 1941: GQ1509 – Gelsenkirchen (Nordstern) Gelsenberg-Benzin AG Plant

| Pilot | 2nd Pilot | Serial | Code | Bomb Load | Result |
|---|---|---|---|---|---|
| S/Ldr J Ocelka | Sgt A Jedounek | T2971 | KX-H | 6x500lb+1x250lb | Completed |
| F/Lt J Šnajdr | Sgt B Blatný | R1410 | KX-M | 6x500lb+1x250lb | Completed |
| Sgt V Korda | Sgt F Kráčmer | T2972 | KX-G | 6x500lb+1x250lb | Completed |
| Sgt A Šedivý | Sgt J Čapka | T2561 | KX-A | 6x500lb+1x250lb | Completed |
| F/O J Breitcetl | Sgt J Bernát | R1371 | KX-F | 6x500lb+1x250lb | Completed |
| Sgt F Fencl | Sgt F Radina | R1378 | KX-K | 6x500lb+1x250lb | Completed |
| Sgt J Doktor | Sgt H Dostál | R1015 | KX-L | 6x500lb+1x250lb | Completed |
| Sgt O Helma | Sgt K Schoř | R1451 | KX-P | 6x500lb+1x250lb | Completed |

*A stunning photograph of Wellington Mk.IC R1598 KX-C being service. This aircraft carried out the first raid at the 311 Squadron on 27th March 1941 and the final one on 28th January 1941. It recorded the total of 51 raids, the largest number a Wellington of 311 Squadron flew with the Bomber Command. (Václav Kolesa)*

A total of forty-six 3 Group Wellingtons would take part in the raid. Unlike the previous operation, there were, to the groundcrews credit, no abortive sorties. Conditions varied over the target depending on when the crews arrived. The squadron was part of the early wave. A few scattered fires were observed in the target area on arrival. One in particular witnessed by Squadron Leader J Ocelka was a fierce red conflagration and very visible. A similarly sized fire was reported by the crew of Sergeant F Fencl, 1 mile west of the target area. Sergeant J Doktor noted that 1 mile south of the oil plant, four small fires had begun to take hold. Despite the clear weather, the early crews had trouble identifying the target. This quickly changed when bombload after bombload was seen enveloping the oil plant producing vivid flashes and importantly large fires. Every crew of 311 Squadron claimed to have bombed the target or buildings nearby. It was an outstanding effort by the squadron crews who, despite fierce flak, pressed home their attacks. Half of the Group's Wellingtons were hampered by cloud and haze. Those not affected inflicted severe damage, but not as expected on the Gelsenkirchen Nordstern oil plant but on the Hibernia Mining Company-owned Hydrierwerk Scholven AG GE-Buer Coal liquefaction plant. This was hit by a reported 16 bombloads causing considerable damage, and production stopped completely. Also hit was the worker's housing estate. All the crews landed safely back at East Wretham. It had been a good night for the squadron, who dropped a record 26,000lb of bombs. On the 17$^{th}$, a raft of new aircrew arrived on 311 Squadron. All were navigators with varying degrees of experience. Pilot Officer Karel Náprstek arrived from C.I.G London, while Pilot Officers Václav Jelínek, Jaromír Brož, Václav Haňka, Adolf Koukal, Karel Sláma and Viktor Krcha arrived from the Czech Depot Wilmslow. Two navigators, Pilot Officer's Alois Tolar and František Dittrich arrived from 214 Squadron, based at R.A.F Stradishall. On arrival, they were transferred to the Operational Training Flight. A later-than-usual take-off would record the departure of six crews against the shipyards at Bremen on the 18$^{th}$. 3 Group would provide 33 of the 57 bombers despatched. One of the number was a Short Stirling Mk.I of 7 Squadron based at R.A.F Oakington. At the controls of N3652 was Squadron Leader Lynch-Blosse DFC, who would take the credit for being the first to bomb a German target with the Stirling. Group Captain Josef Berounský arrived on detachment from 3 Group H.Q on the 17$^{th}$, and his time stay would be anything but dull.

March 17$^{th}$/18$^{th}$ 1941: GR3586 – Bremen – Deutsche Schiff und Maschinenbau.AG

| Pilot | 2$^{nd}$ Pilot | Serial | Code | Bomb Load | Result |
|---|---|---|---|---|---|
| Sgt L Anderle | F/O J Šejbl | R1378 | KX-K | 6x500lb+1x250lb | Completed |
| Sgt A Šedivý | Sgt J Hájek | T2561 | KX-A | 6x500lb+1x250lb | Completed |
| S/Ldr Pickard DSO, DFC | Sgt J Filler | R1371 | KX-F | 6x500lb+1x250lb | Completed |
| Sgt F Fencl | Sgt F Radina | T2972 | KX-G | 6x500lb+1x250lb | Completed |
| Sgt J Doktor | Sgt H Dostál | R1015 | KX-L | 6x500lb+1x250lb | Completed |
| Sgt O Helma | Sgt K Schoř | R1451 | KX-P | 6x500lb+1x250lb | Completed |

Two pilots would be operating for the first time. 32-year-old Sergeant Josef Šejbl, who had just celebrated his birthday and would eventually command the squadron, and Sergeant Jaroslav Hájek. East Anglia was covered with a dense layer of 10/10th cloud, which did not bode well as the crews climbed for altitude. Thankfully, and to the delight of the crews, this cleared almost entirely over the North Sea, affording the crews excellent visibility which would continue over occupied Holland and northern Germany. The customary flak appeared over the Dutch coast and continued almost until Bremen. The first of 3 Group's Wellingtons arrived over the target at 03:15hrs. The flak, although accurate, appeared in less quantity than other visits. All six claimed to have bombed the target from between 14,000ft and 16,000ft. The excellent visibility allowed the crews to observe their handy work. Squadron Leader Pickard DSO DFC reported, *'Dropped in two sticks. Two bombs observed in docks south of target. Others observed in target area. Bursts seem unusually big, the last burst followed by small explosion'*. Another crew who claimed success was Sergeant F Fencl. They informed the Intelligence Officer back at East Wretham, *'All bombs dropped in target area. Started fires in northern part of target. Two burst in docks, 1 in quay south of target started red fire'*. The last of 3 Group's aircraft departed at 04:03hrs. Bremen appeared, at least to the crew's to have been hit hard. All the indications appeared to show the raid to have been a success. The only drama of the night occurred over East Wretham. As he approached the airfield, Sergeant Leo Anderle suffered a starboard engine fire aboard Wellington R1378 KX-K. Showing great skill and nerve, he brought the Wellington in to land. Touching down across the flare path, the undercarriage collapsed, bringing the Wellington to a sliding halt in the southeast corner of the airfield. The fire was quickly extinguished, and the shaken

*Wellington Mk.IC R1378/KX-K after belly landing at East Wretham on 18th March 1941. The aircraft was struck off charge but crew led by Sergeant Leo Anderle escaped just with slight bruises. (Pavel Vančata)*

crew collected and quickly taken to the SSQ. All but Sergeant Anderle and the Wireless Operator, Sergeant Jan Plzák, were slightly injured. Vickers Wellington R1378, which had been on the squadron less than a month suffered extensive damage and would not return to squadron service.

| Date | 18/03/1941 |
|---|---|
| Mark | Mk.Ic |
| Serial | R1378 |
| Code | KX-K |
| Taken On Charge | 22/02/1941 via No.38 MU. |
| Manufacturer | Vickers Armstrong |
| Contract | B.992424/39 |
| Pilot (s) | Sergeant Leo Anderle / Flying Officer Josef Šejbl |
| Flight | Operations |
| Time | 05:15hrs |
| Cause | Starboard engine fire |

311 (Czech) Squadron nearly suffered a catastrophe that would have resulted in a significant setback in the squadron's future. On the evening of the 18th, the squadron commanding officer, Wing Commander Mareš Toman, Squadron Leader Percy Pickard DSO, DFC, his wife, Dorothy, Flying Officer Arnošt Fantl, the squadron Operations Officer, and Pilot Officer Carlin, along with the visiting Group Captain Berounský decided that a visit to The Bull at Barton Mills for a few drinks would be a good way of unwinding. After what would have been an enjoyable evening, the group set about driving back for some well-deserved sleep. At the wheel was Wing Commander Mareš Toman. The following is from the book titled *'Wings of the Night'*[22] by Alexander Hamilton.

*The Bull Hotel at Barton Mills the scene of many a riotous party!*

---

[22]*Wings of Night. Alexander Hamilton ISBN 0 947554 34 3*

*Driving the large estate car was Wing Commander Toman with Dorothy in the front for whom special dispensation had been obtained as civilians were not allowed the use of RAF transport. Pick again, managed to bend the rules. Returning along the main road to Thetford, Dorothy in front saw some lights ahead. The car did not appear to be slowing down. She glanced across to the driver Wing Commander Toman and noticed that he was sound asleep! In a flash, Dorothy reached over towards the wheel and swung the steering wheel to the right. The estate car was much too close and with a screech of grinding metal they hit the obstacle ahead. It was a large pantechnicon furniture lorry. Toman was instantly awake and applied the brakes automatically, but the car hit the ditch on the righthand side, throwing Sidney Carlin out. Everywhere was broken glass, bent metal, and blood.*

Wing Commander Mareš Toman was the most seriously injured with reported broken ribs, the others all suffered cuts and bruises. Pilot Officer Sidney 'Timber toes' Carlin, an old friend of Pickard's, had his wooden force leg smashed[23]. If it had not been for the awareness of Dorothy Pickard, the outcome would have been much worse. Cologne was once again chalked on the operations board on the 19th, the squadron providing five crews. 'A' Flight's Squadron Leader Josef Schejbal took temporary command of the squadron in the absence of Wing Commander Mareš Toman. His first job was to brief the crews for Cologne that night.

### March 19th 1941 : Town 2 – Cologne

| Pilot | 2nd Pilot | Serial | Code | Bomb Load | Result |
|---|---|---|---|---|---|
| Sgt F Fencl | Sgt F Radina | T2972 | KX-G | 4x500lb+1x250lb+120x4lb | Completed |
| Sgt A Šedivý | Sgt J Hájek | T2561 | KX-A | 4x500lb+1x250lb+120x4lb | Completed |
| S/Ldr J Ocelka | Sgt J Čapka | T2971 | KX-H | 4x500lb+1x250lb+120x4lb | Completed |
| F/Lt J Šnajdr | Sgt B Blatný | R1410 | KX-M | 4x500lb+1x250lb+120x4lb | Completed |
| F/O J Breitcetl | Sgt J Bernát | R1371 | KX-F | 4x500lb+1x250lb+120x4lb | Completed |

Cologne would be the sole responsibility of 3 Group on this night. A total of 36 Wellingtons would be detailed and briefed. The Waterbeach-based 99 Squadron would provide six, 40 Squadron eight, Honington's 9 Squadron eight, while Mildenhall's 149 Squadron added nine crews. Flight Lieutenant J Šnajdr was the first away at 19:17hrs on his 13th operation. Conditions over Cologne were good, and once again, the squadron crews displayed their determination when each claimed to have found and bombed the target area. Both Sergeant F Fencl and Flying Officer J Breitcetl bombed the important Köln-Duetz railway station situated close to the eastern bank of the Rhine. Sergeant F Fencl identified the station by dropping a flare, which greatly assisted the bomb aimers accuracy. On leaving the target, the fires they had started began to spread. Sergeant A Šedivý and crew started four fires, while Squadron Leader J Ocelka's bombload started a large reddish blaze, which quickly took hold in the target area. This fire was still visible from 20 miles away. By 22:03hrs, the raid was over. Group H.Q seemed pleased with the results, they reported. *'Conditions over Cologne were good, and it would appear that much damage was done by our aircraft. Nearly all sorties reported fires, some of which were visible from 70 miles'.* All the crews landed back at East Wretham apart from Flight Lieutenant J Šnajdr and crew, who landed at R.A.F Honington. They had been tasked with photographing the raid. Unfortunately, the camera failed.

Early morning fog on the 20th postponed training until the afternoon. This did not stop two crews, along with passengers from being ordered to ferry back two new Wellingtons from Ternhill, Shropshire. It was soon after their departure that

---

[23] *Conflicting details on the severity of injuries. In his book MOONLIGHT FLYER, author John Gellner reports that Group Captain Berounský would be out for a week with his injuries while Wing Commander Toman, six weeks and Flying Officer Fantl two months.*

*A happy looking Sergeant Josef Cibulka, one of the most experienced Wireless Operators of 311 Squadron at the beginning, posing in front of the tent at RAF East Wretham. (Jaroslav Popelka)*

Form B.446 arrived requesting that 311 Squadron prepare eight crews for operations that night. Seven would attack Lorient, while a 'Freshmen' crew would visit Oostende, Belgium. This put the squadron in an awkward predicament, four of its pilots were already on their way to Ternhill. The problems continued late afternoon. It was at 16:15hrs the first of three accidents occurred. The first involved Wellington T2553 KX-B captained by Flying Officer J Šejbl who made a rather heavy landing after ferrying a Wellington from R.A.F Honington for operations that night. The undercarriage collapsed and the Wellington slid to a halt. This was the fourth accident the unfortunate Josef Šejbl had been involved in since his arrival.

| Date | 21/03/1941 |
|---|---|
| Mark | Mk.Ic |
| Serial | T2553 |
| Code | KX-B |
| Taken On Charge | 15/10/1940 via No.24 MU. |
| Manufacturer | Vickers |
| Contract | 38600/39 |
| Pilot (s) | Flying Officer Josef Šejbl |
| Flight | Ferry flight from Honington |
| Time | 16:15hrs |
| Cause | Pilot error |

The Wellington would be salvaged and repaired by No.43 (Maintenance) Group and returned to squadron service on June 6th. The next accident followed shortly after when the inexperienced Sergeant Styblík made a rather heavy landing damaging the tail wheel assembly to Wellington L7841 KX-S. He had at the time only 30 hours on the Wellington.

| Date | 21/03/1941 |
|---|---|
| Mark | Mk.IA |
| Serial | L7841 |
| Code | KX-T |
| Taken On Charge | 28/09/1940 via No.23 MU |
| Manufacturer | Vickers Armstrong |
| Contract | 992424/39 |
| Pilot (s) | Sergeant Miroslav Styblík |
| Flight | Unknown |
| Time | 16:25hrs |
| Cause | Inexperience. |

This incident was followed almost immediately when a wheel of Wellington P9224 went down a hole, damaging the propeller. The unfortunate pilot, 30-year-old Flying Officer František Pohlodek, was being guided by a groundcrew member when taxiing to a dispersal at East Wretham at 16:45hrs.

| Date | 21/03/1941 |
|---|---|
| Mark | Mk.IA |
| Serial | P9224 |
| Code | KX-T |
| Taken On Charge | 18/03/1941 via 115 Squadron. |
| Manufacturer | Vickers Armstrong |
| Contract | 549268/36 |
| Pilot (s) | Flying Officer František Pohlodek |
| Flight | Unknown |
| Time | 16:45hrs |
| Cause | Pilot not diligent. |

The Wellington's undercarriage and propeller would be repaired by No.43 (Maintenance) Group, and it would be once repaired transferred to the Czech Training Flight on July 7th 1941. Somehow, despite everything the squadron still managed to prepare and brief six crews for Lorient.

### March 21st 1941 : 'Polecat' Submarine Base – Lorient.

| Pilot | 2nd Pilot | Serial | Code | Bomb Load | Result |
|---|---|---|---|---|---|
| F/Lt J Šnajdr | Sgt B Blatný | R1410 | KX-M | 10x250lb SAP | Completed |
| Sgt A Šedivý | Sgt J Hájek | T2561 | KX-A | 10x250lb SAP | Completed |
| Sgt L Anderle | Sgt J Filler | T2990 | KX-T | 10x250lb SAP | Completed |
| Sgt F Fencl | Sgt F Radina | T2972 | KX-G | 10x250lb SAP | Completed |
| S/Ldr J Ocelka | Sgt J Čapka | T2971 | KX-H | 10x250lb SAP | Completed |
| Sgt O Helma | Sgt K Schoř | R1451 | KX-P | Withdrawn | Cancelled |

The problems continued when minutes before take-off Sergeant O Helma and crew were withdrawn with engine problems. Finally after a day of tremendous frustration, the first Wellington departed at 18:19hrs. Once again 3 Group H.Q committed only a small proportion of its available force, just 19 Wellingtons and a single Stirling were detailed on Lorient while six 'Freshmen' crews were given the docks at Oostende. Arriving over the target at 20:59hrs, the crews carried out a determined attack from between 12,000ft and 15,000ft in the face of fierce flak. Each crew dropped their all semi-armour-piercing bombs in one stick from the north-east bank of the entrance channel towards the Point de Cauden. These landed along the left bank covering shore, three dry-docks, the Smithery and power station. A number of fires were observed as the squadrons departed at 21:36hrs.

On the morning of the 23rd, R.A.F Honington and East Wretham received Form B.447 from Exning. The squadrons were to provide 13 Wellingtons, five of which would be supplied by 311 Squadron. The target was Berlin.

### March 23rd 1941 : 'Town 13B' – Berlin

| Pilot | 2nd Pilot | Serial | Code | Bomb Load | Result |
|---|---|---|---|---|---|
| Sgt L Anderle | Sgt J Bernát | T2990 | KX-T | 1x500lb+1x250lb+360x4lb | Completed |
| F/Lt J Šnajdr | Sgt B Blatný | R1410 | KX-M | 1x500lb+1x250lb+360x4lb | Completed |
| Sgt A Šedivý | Sgt J Hájek | T2561 | KX-A | 1x500lb+1x250lb+360x4lb | Completed |
| Sgt F Fencl | Sgt F Radina | R1371 | KX-F | 1x500lb+1x250lb+360x4lb | Completed |
| S/Ldr J Ocelka | Sgt J Čapka | T2971 | KX-H | 1x1000lb+2x500lb+1x250lb | Completed |

The weather throughout the 23rd was lousy. Rain followed by snow showers fell until mid-morning, followed by almost continuous drizzle. Despite the weather, the diligent ground crews laboured on their aircraft. It would be another low-key operation. 3 Group would provide just 33 Wellingtons which, on this occasion, would follow the Whitleys of 4 Group over Berlin. The crews were away by 19:42 hours on a bitterly cold night. Clouds covered most of the route, but fortunately, there were a few gaps large enough to identify Berlin. The squadron began arriving over the target at 22:50hrs and were almost immediately engaged by a tremendous barrage of flak. Flying at 16,000ft the crew of Sergeant L Anderle were fortunate. A burst of flak exploded close enough to damage the Wellingtons tail and rudder. All five crews located and bombed the target. Numerous bursts were noted and fires started. Intense flak and searchlight activity made accurate observation of the results difficult. On return from Berlin, the squadron devoted its attentions to training for the next three days. A 'Freshmen' trip was organised then cancelled on the 24th. Away from the airfield, a Station Dance was held at R.A.F Honington on the 25th. The following day operations were back on, with the main effort directed against Cologne. The Czechs would provide five crews, plus two 'Freshmen' crews to Dunkirk.

### March 27th 1941 : CC.25 – Dunkirk

| Pilot | 2nd Pilot | Serial | Code | Bomb Load | Result |
|---|---|---|---|---|---|
| Sgt J Čapka | F/O J Šejbl | R1598 | KX-C | 5x500lb+120x4lb | Complete |
| Sgt F Kráčmer | F/O F Sixta | R1599 | KX-J | 5x500lb+120x4lb | Complete |

Dunkirk was deemed the ideal target for both 'Freshmen' crews. Sergeant Josef Čapka would be captaining a crew for the first time after completing 12 operations as 2nd pilot. Also on his first outing was 27-year-old Flying Officer František Sixta, who would join the novice crew of Sergeant František Kráčmer on his first operation as skipper. Both crews were over Dunkirk between 20:55hrs and 21:01hrs. Sergeant J Čapka was over the docks at 15,000ft. On his arrival, a fire was observed south of the target, which the crew wisely chose to ignore. Instead they continued on until satisfied and watched with some pleasure as their bombload burst on the quay between docks 1 and 4, slightly north of the tidal basin. Also bombing dock 4 was Sergeant F Kráčmer, whose bombs were seen to explode on or around the docks, creating four fires. Within 2 hours and 44 minutes, both crews were back at their dispersals. It had been an excellent effort.

### March 27th 1941 : Town 2(A) – Cologne

| Pilot | 2nd Pilot | Serial | Code | Bomb Load | Result |
|---|---|---|---|---|---|
| F/Lt J Šnajdr | Sgt B Blatný | R1410 | KX-M | 6x500lb+1x250lb | Complete |
| F/O J Breitcetl | Sgt J Bernát | T2561 | KX-A | 6x500lb+1x250lb | Complete |
| Sgt O Helma | Sgt K Schoř | R1451 | KX-P | 6x500lb+1x250lb | Complete |
| S/Ldr J Ocelka | Sgt M Styblík | T2971 | KX-H | 6x500lb+1x250lb | Complete |
| Sgt J Doktor | Sgt H Dostál | R1015 | KX-L | 6x500lb+1x250lb | Alternative Target |

Squadron Leader J Ocelka would be joined on his 22nd operation by Sergeant Miroslav Styblík. Josef Ocelka was born on March 12th 1909. A dark, thickset, brooding individual, he graduated from the Military Academy in July 1931. Ocelka spent a number of years pre Czech occupation in various roles, first at infantry, later as an air observer, before completing his twin-engine pilot's course in January 1939. Like most of 311 Squadron, he escaped to France via Poland and spent a short time in the French Legion. With the French surrender, he sailed from Bordeaux to Falmouth in June 1940. An extremely popular individual whom everyone on the squadron respected. His courage and determination were never in doubt. A steadying and calming influence, especially during troubled times, he was one of the squadron's most outstanding pilots. All the Cologne crews were away by 19:51hrs in filthy weather, which continued almost all the way to the target. Sergeant J Doktor and crew were forced to bomb the docks at Flushing. Engine trouble over the North Sea meant they bombed the dock area at 20:51hrs. They were back at East Wretham by 22:17hrs, where a safe landing was made. The remaining crews were fortunate to find Cologne reasonably clear of clouds. They arrived over the target at 21:32hrs. Flight Lieutenant J Šnajdr bombed from 15,000ft, his bombs landing ½ mile west of the target area. Flying Officer J Breitcetl dropped his entire bombload in one stick. The first of which landed south of the vitally important Hohenzollern Bridge while the last exploded just south of the aiming point. The bombload of Sergeant O Helma was dropped from 16,000ft. These were seen to burst on the west bank of the Rhine between the Hohenzollern Bridge and the Rhine ferry, located close to the Hindenburg Bridge. All the crews returned safely. Squadron Leader J Ocelka landed at R.A.F Bircham Newton, having been airborne for almost 5 hours. On the 28th, an E.N.S.A unit arrived at Honington and put on a show titled *'The Cavendish Unit'*, which by all accounts was thoroughly enjoyed. That night, a Wellington from the R.A.F Feltwell-based 57 Squadron crashed on the airfield at 00:20hrs in poor visibility on return from Cologne. The pilot, Sergeant John Emmerson and crew survived. They had just returned from their first operation. Wellington Mk.Ic R1441 was written off. There was an unfortunate accident involving Sergeant Anderle on Friday, March 28th, resulting in the death of Private Albert Cox of the Suffolk Regiment, who was hit while cycling by a car driven by Anderle on the road between Bury and Thetford. It was a tragic incident for all concerned. On March 28th, both the *Scharnhorst* and *Gneisenau* were discovered in Brest, where they had taken refuge on the 22nd. Their discovery gave the Royal Navy, or specifically Coastal Command a chance to finish off what they had started out in the Atlantic. Almost immediately, R.A.F Bomber Command found itself regrettably involved and would continue to be so for the next ten months. While 37 Wellingtons tackled Brest, 3 Group H.Q despatched 14 'Freshmen' crews to the docks at Calais, two of this number being provided by 311 Squadron.

### March 30th 1941 : CC.37 – Calais

| Pilot | 2nd Pilot | Serial | Code | Bomb Load | Result |
|---|---|---|---|---|---|
| Sgt J Čapka | F/O J Šejbl | R1598 | KX-C | 5x500lb+2x250lb+120x4lb | Complete |
| Sgt F Kráčmer | F/O F Sixta | R1599 | KX-J | 5x500lb+2x250lb+120x4lb | Complete |

The Group's Wellingtons were over Calais between 21:00hrs and 22:30hrs. Fires were already visible when 311 Squadron arrived. Sergeant J Čapka deposited his mixed load from 15,000ft, creating two fires 200 yards apart between docks 5 and 6. The crew of Sergeant F Kráčmer reported their bombs were seen to burst, starting a large white flamed fire between docks 5 and 6, but north of another blaze. It was a return visit to Bremen on the 31st, seven crews were detailed and briefed. They would be joined by eight Wellingtons from both 9 and XV Squadrons, plus five from 214 Squadron. Six 3 Group Wellingtons would also attack Emden while five 'Freshmen' crews would visit Rotterdam.

### March 31st 1941 : GR3586 – Bremen – Deutsche Schiff und Maschinenbau.AG

| Pilot | 2nd Pilot | Serial | Code | Bomb Load | Result |
|---|---|---|---|---|---|
| Sgt A Šedivý | Sgt J Hájek | R1410 | KX-M | 6x500lb+1x250lb | Complete |
| Sgt J Doktor | Sgt H Dostál | R1516 | KX-U | 6x500lb+1x250lb | Complete |
| Sgt L Anderle | Sgt J Filler | T2990 | KX-T | 6x500lb+1x250lb | Complete |
| Sgt O Helma | Sgt K Schoř | R1451 | KX-P | 6x500lb+1x250lb | Complete |
| S/Ldr J Ocelka | Sgt M Styblík | R1598 | KX-C | 6x500lb+1x250lb | Complete |
| F/O J Breitcetl | Sgt J Bernát | R1532 | KX-R | 6x500lb+1x250lb | Complete |
| Sgt F Kráčmer | F/O F Sixta | T2972 | KX-G | 6x500lb+1x250lb | Complete |

The seven crews departed every 60 seconds. They quickly disappeared into menacing clouds that would stretch almost the entire route to Bremen. The first of the Group's bombers were over the target at 21:40hrs and were met by a barrage of heavy flak. Haze and searchlights added to the flak, made accurate observation of the results difficult, even for the Czechs, who pressed home their attacks. Two crews who were lucky to glimpse the results of their bombing were Sergeants A Šedivý and J Doktor at the controls of a factory fresh Wellington. Both reported their HE loads explode between the pier and Holz- und Fabrikenhafen timber harbour and Frei-Hafen I. The explosions were followed by red flames and fierce fires. The first crews were back at East Wretham just before midnight. It was not until 01:05hrs that the last crew captained by Sergeant F Kráčmer landed safely. Thus ended March 1941. It had been an exceptional month for the squadron, which had operated on 11 nights and flown 61 operations without loss.

*A fine photograph of a smiling Thomas Kirby-Green. He was born in Dowa, Nyasaland, where his father, Sir William Kirby-Green, was the British District Governor. Admired and respected by all, this exceptional officer tirelessly trained numerous novice Czech crews. On departing 311 Squadron, he was posted to the 40 Squadron for a second tour in 1941. Shot down, he could have spent the war as a PoW. However, his tenacity and sheer doggedness meant he would sent to Stalag Luft III. Sadly, he was murdered by the Gestapo in March 1944, one of the fifty re-captured after the Great Escape from Stalag Luft III.*

# April 1941 : Keeping Up the Pressure

Training for the first three days of April kept the squadron occupied. Operations against Brest were planned, then called off. On the 3rd, Group sent 50 bombers to Brest, but Honington, and East Wretham sat out the attack. 311 Squadron's turn would come the following night when Group H.Q requested the squadron provide eight crews for another crack at *Scharnhorst* and *Gneisenau*.

The now familiar captains' names were chalked on the Operations Board. Each Wellington would be loaded with Semi-Armour-Piercing 500-pounders and 250-pounders, plus a 250lb General Purpose for good measure. Over at R.A.F Honington, Air Vice Marshal Robert Saundby MC, DFC, AFC visited the station. That evening, an E.N.S.A concert featuring Jack Brandy and his band was held in the Airmen's Mess. The bruised and battered Wing Commander Mareš Toman returned to the squadron on April 3rd after recovering from injuries sustained in the road accident in March. His return was welcomed but created a delicate issue. Squadron Leader Josef Schejbal had proved himself to be a capable commanding officer since assuming temporary command. The Czech liaison officer at 3 Group H.Q, Acting Group Captain Berounský informed the C.I.G: *"I suggest Col K. Toman be transferred to London as soon as possible. Col. Toman has just returned from leave, no further leave can be granted and his position, unless he takes command of the squadron again, is uncertain. Major Schejbal has been in command of the squadron since March 19 and it is in the interest of the smooth running of the service at the squadron so that there is no change of command again in a short time."*

## April 4th 1941 : CC.49 – Brest

| Pilot | 2nd Pilot | Serial | Code | Bomb Load | Result |
|---|---|---|---|---|---|
| Sgt F Fencl | Sgt F Radina | R1466 | KX-D | 8x250lbSAP+2x500lbSAP | Complete |
| Sgt J Čapka | F/O J Šejbl | R1598 | KX-C | 8x250lbSAP+2x500lbSAP | Complete |
| Sgt F Kráčmer | F/O F Sixta | R1599 | KX-J | 8x250lbSAP+2x500lbSAP | Complete |
| Sgt L Anderle | Sgt J Filler | T2990 | KX-T | 8x250lbSAP+2x500lbSAP + 3 x Flash-bombs | Complete |
| Sgt A Šedivý | Sgt J Hájek | T2561 | KX-A | Withdrawn | Cancelled |
| Sgt J Doktor | Sgt H Dostál | R1516 | KX-U | 2x500lbSAP+6x250lbSAP | Complete |
| F/O J Breitcetl | Sgt J Bernát | R1532 | KX-R | 8x250lbSAP+2x500lbSAP | Complete |
| S/Ldr J Ocelka | Sgt M Styblík | T2971 | KX-H | 1x500lbSAP+8x250lbSAP | Complete |

The crew of Sergeant A Šedivý were withdrawn with a fractured hydraulic pipe just before take-off. A switch to the reserve Wellington was unsuccessful when the wireless set was found unserviceable.[24] Brest would be spared one bombload. The weather over the target area was reported as good but freezing as the first of the squadron's Wellingtons arrived at 21:12hrs. Bombing from between 10,000ft and 14,500ft, the majority of the squadron's bombs landed in and around Dock 1 and 2, producing several small fires. Both 1 and 2 docks appeared to have been the focal point of much of the bombing. Sergeant L Anderle watched fascinated as four of his bombs were seen to burst across the entrance of the dry dock producing a vivid flash. The squadron's time over Brest was brief. The last crew to bomb at 21:43hrs was Squadron Leader J Ocelka. All the crews returned safely to base, three however complained that the heating aboard their Wellington was inadequate. On the morning of the 5th, *Gneisenau's* dry dock was drained, an unexploded 500-pounder having been found. This discovery was sufficient to prompt the ship's captain, Kapitän-zur-See Otto Fein, to move *Gneisenau* quickly into Brest's main harbour. On the 5th it was the turn of the Czechs at R.A.F East Wretham to

---

[24] *There is some confusion about this crew. Various reports suggest the aircraft took off and landed almost immediately, while others state it did not take off. I have used Air25/62 as my reference.*

welcome Air Vice Marshal Robert Saundby MC, DFC, AFC who talked with the crews and at some length with temporary commanding officer, Squadron Leader Josef Schejbal. On conclusion of his visit he returned direct to Bomber Command H.Q. His visit coincided with yet another unwelcome visit by German bombers. Headquarters 3 Group reports that East Wretham was bombed on the 5th at around 2 a.m. and was made temporarily unserviceable due to five bombs falling near the flare-path. A pillbox on the western side of the aerodrome was also damaged. Strangely, neither R.A.F Honington Station Records Book, nor the 311 ORB reports the attack.

It was no surprise that with the arrival of Form B.460 on the morning of the 6th, Brest would once again be the target for that night. 3 Group would provide 46 Wellingtons, plus eight aircraft on Calais. The Group made up the majority of the attack force. 311 Squadron would provide its customary seven crews. During the day, Group Captain Richard Harrison DFC, AFC, arrived at R.A.F Honington. At the time, he was S.A.S.O 1 Group H.Q. The reason for his visit is unrecorded. He would, in February 1943, take command of 3 Group and skilfully guide the Group during some of its most turbulent and challenging times.

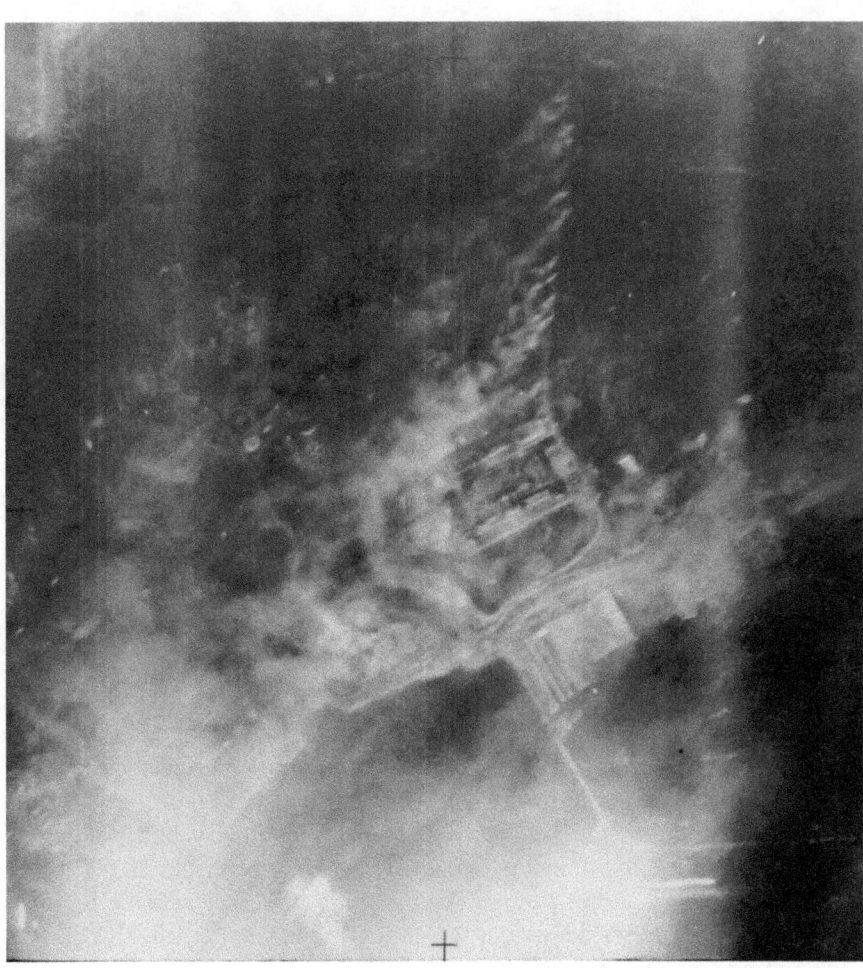

*Brest Harbour is on the receiving end. This target photo was taken by a No.3 Group Wellington on April 6th 1941.*

April 6th 1941: CC.49 – Brest

| Pilot | 2nd Pilot | Serial | Code | Bomb Load | Result |
|---|---|---|---|---|---|
| Sgt F Fencl | Sgt F Radina | R1466 | KX-D | 3x500lb SAP+ 7x250lb SAP | Completed |
| Sgt A Šedivý | Sgt J Hájek | T2990 | KX-T | 3x500lb SAP+ 7x250lb SAP | Completed |
| Sgt J Čapka | F/O J Šejbl | R1598 | KX-C | 3x500lb SAP+ 7x250lb SAP | Completed |
| Sgt F Kráčmer | F/O F Sixta | R1599 | KX-J | 3x500lb SAP+ 7x250lb SAP | Completed |
| Sgt J Doktor | Sgt H Dostál | R1516 | KX-U | 3x500lb SAP+ 7x250lb SAP | Completed |
| F/O J Breitcetl | Sgt J Bernát | R1371 | KX-F | 3x500lb SAP+ 7x250lb SAP | Completed |
| S/Ldr J Ocelka | Sgt M Styblík | T2971 | KX-H | 3x500lb SAP+ 7x250lb SAP | Completed |

Dense cloud got the better of the bombers over Brest on this occasion. With an impenetrable layer of 10/10th cloud covering the whole coast, the best the squadron crews could do was to drop on flak concentrations or reflections in the cloud, hoping their bombs would land in the target area. The first of the crews to attack was Sergeant A Šedivý at 22:22hrs from 11,500ft. His bombs were believed to have fell some 2 miles west of the aiming point centre. Sergeant F Kráčmer reported on departure at 22:27hrs one solitary red fire. It was a disappointing raid for all involved.

A photographic reconnaissance flight by 3 P.R.U on the 7th prompted a switch to northern Germany and the port of Kiel. Group H.Q requested a maximum effort from its twelve squadrons. In response, 115 bombers were made available. It was an excellent effort. Bomber Command gathered a force of 229 aircraft, representing the most significant effort yet sent against a single target. Half the force was provided by 3 Group and clearly shows the Group's prominence at the time. The Czechs of 311 Squadron were also up for the challenge, providing eight Wellingtons and crews.

### April 7th 1941 : Town 8(b) – Kiel

| Pilot | 2nd Pilot | Serial | Code | Bomb Load | Result |
|---|---|---|---|---|---|
| F/Lt J Šnajdr | Sgt B Blatný | R1410 | KX-M | 3x500lb+360x4lb | Completed |
| Sgt F Fencl | Sgt F Radina | R1466 | KX-D | 3x500lb+360x4lb | Completed |
| Sgt J Čapka | F/O J Šejbl | R1598 | KX-C | 3x500lb+360x4lb | Completed |
| Sgt F Kráčmer | F/O F Sixta | R1599 | KX-J | 3x500lb+360x4lb | Completed |
| Sgt J Doktor | Sgt H Dostál | R1516 | KX-U | 3x500lb+360x4lb | Completed |
| Sgt A Šedivý | Sgt V Netík | R1371 | KX-F | 3x500lb+360x4lb | Completed |
| F/O J Breitcetl | Sgt J Bernát | R1532 | KX-R | 5x500lb | Completed |
| S/Ldr J Ocelka | P/O V Korda | T2971 | KX-H | 5x500lb | Completed |

Sergeant Václav Netík, a 30-year-old from Dvůr Králové would be undertaking his first operational flight. He would join the experienced crew of Sergeant A Šedivý. Squadron Leader J Ocelka was joined by Pilot Officer Václav Korda in the right-hand seat. The first crew departed at 20:30hrs, the filthy weather that had protected Brest was thankfully not encountered, and as the squadron flew north, they enjoyed clear skies and bright moonlight. On departing over the English coast, 3 Group's force would set course for Rømø Island on Denmark's western coast. From there, they would turn east to a position north of Flensburg before making their run into the target from the north. Sergeants F Fencl and J Čapka were the first crews over Kiel at 23:00hrs. Each reported a large fire on arrival. Flak was described as intense, stirred up by the early bombers. Sergeant Fencl bombed from 18,000ft, he reported, *'Enormous fire observed on arrival near the aiming point. All bursts seen starting new fires which merged with existing'*. It would be almost 40 minutes before the next crews were over the target. Flight Lieutenant J Šnajdr was the first to attack. His bombload straddled the aiming point, the crew observing a large fire just astern of a battleship moored in the docks. Flying Officer J Breitcetl stated that the fires at Kiel were observed well before he arrived over the target. They reported numerous fires, however also noted, some were in the town outside the port area. The last squadron crew to bomb at 00:07hrs was Sergeant A Šedivý. They went in at the perilously low altitude of just 13,000ft. Back at East Wretham they recorded, *'Two rows of fires in target area with explosions taking place within. One large fire started and two smaller ones. Fires visible from 40 miles from town'*. All the squadron landed safely back at base. The only issue was experienced by Sergeant J Čapka, who could not lower his undercarriage due to a hydraulic issue. The undercarriage was eventually lowered with the use of the hand pump. Three crews reported bomb and incendiary hang-ups, which brought the total bombs dropped by 311 Squadron down to a still impressive 22,044lb. The raid, which occurred over a five-hour period, inflicted severe damage, especially around the eastern dock areas. Widespread damage was also reported in the Deutsche Werke and Germania Werft, plus residential and civilian housing was also severely affected. A total of 104 3 Group crews reported attacking the target, dropping in the process 319,674lb of bombs. The squadron sat out the attack on Kiel the following night and instead it was back to training.

At 22:55hrs on the evening of the 8th, the squadron came close to losing a crew while on a night training flight. Pilot Officer Jan Hrnčíř was giving dual training to Sergeant Jaroslav Nyč aboard Wellington P9230 KX-X when they were attacked by a Junkers Ju88 flown by Hptm Karl Theodor Hulshoff of Stab I./NJG2. Cannon shells tore into the Wellington damaging the Wellington's rudder, tail, rear turret and elevators. By a miracle, one of the shell's velocity was dramatically reduced before it struck Pilot Officer Jan Hrnčíř in the occipital region of the head, causing a depressed skull fracture. Sergeant Jaroslav Nyč, showing commendable awareness, managed to take control and bring the badly damaged Wellington down in one piece. The Wellington, which was reported to have its navigation lights on at the time, crash landed at West Toft, near Thetford. Jan Hrnčíř was rushed to Newmarket Hospital. His recovery was long, and it

*Wellington Mk.IA P9230/KX-X showing the tail damage caused by a German intruder night fighter on 8th April 1941. Pilot-Pupil Sergeant Jaroslav Nyč made a perfect belly landing by himself and escaped unhurt. The Pilot-Instructor Pilot Officer Jan Hrnčíř was taken in hospital with a gunshot wound to the back of the head.*

was not until October 16th 1941 he was discharged but would continuing receiving medical treatment until January 1942. Wellington P9230 was recovered and repaired at Brooklands, it did not return to squadron service.

| Date | 08/04/1941 |
|---|---|
| Mark | Mk.IA |
| Serial | P9230 |
| Code | KX-X |
| Taken On Charge | 09/09/1940 via 115 Squadron. |
| Manufacturer | Vickers Armstrong |
| Contract | 549268/36 |
| Pilot (s) | Pilot Officer Jan Hrnčíř / Sergeant Jaroslav Nyč |
| Flight | Training |
| Time | 22:55hrs |
| Cause | Intruder action |

On the 9th, a 3 Group Station Commanders conference was held at Exning to discuss the possibilities of the Group resuming daylight operations. It was not a prospect that anyone in the Group relished.

It was a return to Berlin on the 9th, with Vegesack and Emden also to be attacked. The old pairing of R.A.F Feltwell and R.A.F Honington, plus Waterbeach would provide the bulk of the Wellington force, while R.A.F Newmarket's 7 Squadron would supply three Short Stirlings. 311 Squadron would provide seven crews, five on Berlin, and two on the shipyards at Vegesack.

April 9th 1941 : Town 13A – Berlin

| Pilot | 2nd Pilot | Serial | Code | Bomb Load | Result |
|---|---|---|---|---|---|
| Sgt F Fencl | Sgt F Radina | R1466 | KX-D | 1x500lb+360x4lb | Alternative target |
| Sgt J Doktor | Sgt H Dostál | R1516 | KX-U | 1x500lb+360x4lb | Completed |
| Sgt O Helma | Sgt K Schoř | R1451 | KX-P | 1x500lb+360x4lb | Sent S.O.S |
| Sgt A Šedivý | Sgt J Hájek | R1371 | KX-F | 1x500lb+360x4lb | Completed |
| S/Ldr J Ocelka | Sgt B Blatný | T2971 | KX-H | 1x500lb+360x4lb | Completed |

The Berlin-bound crews made their way to the target via northern Holland. At the Dutch - German border the force encountered an abundance of searchlights and flak which appeared to claim one unfortunate bomber. Sergeant F Fencl was obliged to bomb Brunswick due to port engine trouble. Their bombload was dropped from 17,000ft and was seen to start two fires, which were visible from 20 miles away. Another crew experiencing trouble was Sergeant O Helma. They were engaged and damaged by flak near Hanover. The bomb-hatch door was badly buckled, and the crew, believing the aircraft was on fire, sent an S.O.S at 23:30hrs. However, it was premature, the bombs were jettisoned 5 miles north of Brunswick, and the crew safely returned. The weather conditions outbound and over the target were excellent, but it was bitterly cold, with a reported -22 degrees at 16,000ft. Berlin's flak was plentiful and accurate, while the city's numerous searchlights, working in small groups, were actively searching for individual bombers.

The first to bomb was Sergeant A Šedivý at 23:55hrs, followed 13 minutes later by Sergeant J Doktor and crew from 16,000ft. They reported, *'Nine fires seen in area. Explosion from 500-pounder was followed by two red explosions'*. The last crew over the target was Squadron Leader J Ocelka, who bombed at 00:20hrs. Both Sergeant J Doktor and Squadron Leader J Ocelka could have turned back early on the flight due to technical issues. Sergeant J Doktor lost his gyrocompass soon after take-off, plus the heating. Squadron Leader J Ocelka reported on his return the cockpit instrumentation was erratic, and they too lost their heating. The fact that they continued to Berlin reflects on their determination. There was one casualty. Sergeant Oldřich Helma suffered frostbite to both hands in the sub-zero conditions and would need specialist treatment. This would keep him off operations until mid-May. Around 60% of the attacking force claimed to have bombed Berlin, but of those, nearly a third were unable to confirm their accuracy due to the glare of the searchlights.

April 9<sup>th</sup> 1941 : GR3603 – Vegesack

| Pilot | 2<sup>nd</sup> Pilot | Serial | Code | Bomb Load | Result |
|---|---|---|---|---|---|
| Sgt J Čapka | F/O J Šejbl | T2972 | KX-G | 6x500lb+1x250lb | Completed |
| Sgt F Kráčmer | F/O F Sixta | R1599 | KX-J | 6x500lb+1x250lb | Completed |

The target for the two crews was the Bremer Vulkan Schiffbau Yards at Vegesack known for their production of U-boats. Located on the northern bank of the River Weser to the north-west of Bremen, it was No.3 on the list of important targets drawn up to attack. Sergeant J Čapka arrived over Vegesack at 22:55hrs. Conditions were ideal, excellent visibility and no clouds, but freezing between 13,000ft and 16,000ft. All apart from one 500-pounder which iced up was dropped from 12,000ft. Also carried were three flash bombs. The crew had been chosen to photograph the target, which they did successfully bringing back three excellent photographs. Sergeant F Kráčmer bombed at 23:08hrs. They reported on return, *'Small fires seen burning north of town centre, Southeast of railway station. Bombed in one stick, bombs straddled target. First burst on the south bank of river, two in river the rest on the target. One large red fire started'*. The last of the crews to land back at East Wretham was Squadron Leader J Ocelka at 03:25hrs. He had been airborne for seven bitterly cold hours.

Both 9 and 311 Squadron sat out the operations on April 10<sup>th</sup>, 12<sup>th</sup> and 14<sup>th</sup> against Brest. On the 12<sup>th</sup> the squadron was visited by the Czechoslovakian Military Attaché, Lieutenant Colonel Kalla. The following day both Wing Commander Mareš Toman and Lieutenant Colonel Kalla visited 311 Squadron Training Flight at R.A.F Honington.[25] It was not until the 15<sup>th</sup> that Form B.469 arrived informing the squadron that it would be returning to Kiel that night. Seven Wellingtons were armed, fuelled, and made ready.

April 15<sup>th</sup> 1941 : GR3589 – Friedrich Krupp Germania Werft AG – Kiel

| Pilot | 2<sup>nd</sup> Pilot | Serial | Code | Bomb Load | Result |
|---|---|---|---|---|---|
| F/Lt J Šnajdr | Sgt B Blatný | R1410 | KX-M | 2x500lb+1x250lb+360x4lb | Completed |
| Sgt J Doktor | Sgt H Dostál | R1516 | KX-U | 4x500lb+2x250lb | Completed |
| Sgt L Anderle | Sgt J Hájek | R1466 | KX-D | 4x500lb+2x250lb | Completed |
| Sgt F Kráčmer | F/O F Sixta | R1599 | KX-J | 4x500lb+2x250lb | Completed |
| S/Ldr J Ocelka | Sgt M Styblík | T2971 | KX-H | 4x500lb+1x250lb | Completed |
| P/O V Korda | Sgt K Schoř | T2972 | KX-G | 4x500lb+3x250lb | Completed |
| Sgt J Čapka | F/O J Šejbl | T2990 | KX-T | 4x500lb+3x250lb | Alternative target |

A total of 45 bombers would be committed by 3 Group for operations that night, of which four 'Freshmen' crews would bomb the docks at Boulogne. The squadron began taking off around 22:00hrs for the long flight to Kiel. The weather conditions were far from ideal, with thick cloud almost the entire route. When gaps did appear in the clouds, ground visibility was fair but hazy. Four crews were over Kiel between 00:50hrs and 00:56hrs. One of the Wellingtons was flown by Sergeant L Anderle. He had suffered port engine issues just before reaching the target. Undeterred, he pressed on, but was unable to positively identify the aiming point. They did however observe the Kieler Förde before bombing. They dropped their bombload from 14,000ft. A small gap in the clouds and what appeared to be a red fire was enough for the crew of Sergeant J Doktor to drop their all HE load at 00:56hrs. Pilot Officer V Korda encountered intense flak and searchlights when making his bomb run at 01:06hrs. He and his crew were among the few who reported their bombs bursting in the town's centre and the estuary's west bank. The last two crews encountered problems identifying the target area due to the 9/10th cloud and hostile reception. The crew of Sergeant J Čapka dropped a single 250-pounder on Kiel

---

[25] *These could be the same visits, dates differ between 311 ORB and Honington Station Records Book.*

*Pilot Officer Benedikt Blatný DFM survived the tour of operations with 311 Squadron (38 sorties) between 27th December 1940 and 23rd July 1941 and further service with 138 (Special Duties) Squadron from 7th October 1941 to 18th January 1942. Sadly, he was killed in a mid-air collision on 8th July 1943 while flying as a Pilot-Instructor with 32 SFTS at Moose Jaw, Saskatchewan, Canada. (Pavel Vančata)*

but were unwilling to drop the remainder due to filthy conditions. Showing tremendous pluck, they headed towards the Wilhelmshaven shipyards, where from 14,000ft, they deposited their remaining bombs. Flight Lieutenant J Šnajdr stayed over the target area longer than he should, hoping for a break in the cloud. He finally dropped his mixed bombload at 01:54hrs, reporting on return, *'Unable to identify the target due to 10/10th cloud. Identified area by E.T.A, intense flak and searchlight activity. Bombed in one stick in the centre of flak area. No results seen'*.

Cross-country training flights were the main pursuit on the 16th. Sadly, it would result in the loss of Wellington P9212 KX-J. The Wellington suffered engine failure when both engines inexplicably cut out at 150ft during a practice flight from R.A.F Honington. The pilots attempted to land back at Honington but hit a tree 1-mile north-west of the airfield and crashed, bursting into flames at 10:00hrs. Both pilots, Pilot Officer Stanislav Zeinert, the second pilot, Pilot Officer Miroslav Švic and the Wireless Operator, Sergeant Viktor Tégel, manage to scramble clear with minor abrasions. The veteran Wellington, which had previously flown with the NZ Flight, was burnt out.

| Date | 16/04/1941 |
| --- | --- |
| Mark | Mk.IA |
| Serial | P9212 |
| Code | KX-J |
| Taken On Charge | 11/10/1940 via NZ Flight Stradishall. |
| Manufacturer | Vickers Armstrong |
| Contract | 549268/36 |
| Pilot (s) | Pilot Officer Stanislav Zeinert / Pilot Officer Miroslav Švic |
| Flight | Training |
| Time | 10:00hrs |
| Cause | Engine failure. |

Bomber Command assembled a force of 118 aircraft on the evening of the 17th to send against Berlin. 3 Group contributed thirty-six Wellingtons and a solitary Stirling, along with eight Wellingtons briefed to attack Cologne. 311 Squadron cobbled together seven Wellingtons, five of which were briefed for Berlin.

### April 17th 1941 : Town 13(b) – Berlin

| Pilot | 2nd Pilot | Serial | Code | Bomb Load | Result |
| --- | --- | --- | --- | --- | --- |
| P/O V Korda | Sgt K Schoř | R1021 | KX-W | 4x500lb+1x250lb | Completed |
| S/Ldr J Ocelka | Sgt M Styblík | R1410 | KX-M | 1x500lb+1x250lb+360x4lb | Completed |
| Sgt L Anderle | Sgt A Rozum | T2990 | KX-T | 1x500lb+1x250lb+360x4lb | Completed |
| Sgt A Šedivý | Sgt J Hájek | R1371 | KX-F | 1x500lb+1x250lb+360x4lb | Completed |
| F/O J Breitcetl | Sgt J Bernát | R1532 | KX-R | 4x500lb+1x250lb | Early Return |

Flying his first operation since February 6th was 28-year-old Sergeant Alois Rozum. He would be joining the crew of Sergeant L Anderle. Flying Officer J Breitcetl was forced to abort the operation within sight of the English coast with a surging port engine which despite both pilots efforts, refused to cooperate. They landed back at East Wretham within 47 minutes of take-off. The night was noted for being particularly dark as the crews flew slightly north of Great Yarmouth and out over the North Sea. Passing south of Texel, the squadron headed toward Celle and then onto Berlin. Almost along the entire route, flak and searchlights were encountered. German defences were becoming increasingly well-organised. The first crew over Berlin was Sergeant L Anderle at 23:58hrs. Opposition over the target was abnormally subdued, and there appeared to be no signs of any bombing. As the crew dropped their mixed bombload in one stick from 15,600ft, another bombload exploded a few seconds before theirs in the target area. The crew observed their 500-pounder burst which was followed by a large yellow fire. It was almost 30 minutes before Sergeant A Šedivý and Squadron Leader J Ocelka started their bomb runs. Two fires, one of which was large, were started along the river in the centre of Berlin by Squadron Leader J Ocelka. At the same time, Sergeant A Šedivý was unable to precisely identify where his bombs landed due to industrial haze. His bombs created a small green fire, which slowly turned yellowish. The last crew over the target was Pilot Officer V Korda at 00:44hrs. The crew bombed successfully and submitted the following report on return. *'Three bursts observed just Northeast of River adjacent to A/P 'B'. Small fires noted just west of A/P 'B' and two large fires in the Wilmersdorf area, yellow and red in colour. Several small fires near the Tiergarten seen on leaving'.*

## April 17th 1941 : Town 2 (b) – Cologne

| Pilot | 2nd Pilot | Serial | Code | Bomb Load | Result |
|---|---|---|---|---|---|
| Sgt J Doktor | Sgt H Dostál | R1015 | KX-L | 3x500lb+1x250lb+360x4lb | Completed |
| Sgt F Kráčmer | F/O F Sixta | R1599 | KX-J | 7x500lb+1x250lb | MISSING |

It was en route to Cologne that the searchlight-coned Wellington flown by Sergeant F Kráčmer had the misfortune of being intercepted by Hptm Werner Streib of Stab I./NJG1 at 23:39hrs. The Wellington crashed 1 mile west of Baaksem, near Roermond, Holland taking with it the entire crew.

Weather conditions over Cologne were no better than Berlin, with dense cloud, limited visibility and industrial haze. Sergeant J Doktor and crew were over Cologne at 23:01hrs. The only sign of bombing was two small fires visible on arrival in the centre of the target, both reddish in colour. The bombs were dropped in one stick from 15,000ft. The three 500-pounders were seen to burst along the left bank of the river in the city centre. These produced two very small fires observed on leaving. On returning to R.A.F East Wretham, Sergeant J Doktor reported the following, *'At 22:40hrs saw an aircraft shot down by an enemy fighter and crashing in flames. One of the crew was seen to bail-out. Their position at the time was 5105N – 0620E'.*

## Vickers Wellington Mk.IC R1599 KX-J

| Manufacturer | Vickers Armstrong | |
|---|---|---|
| Contract | 992424/39 | |
| Taken on Charge | 21/03/1941 via No. 24 MU. | |
| Cat E Missing | 17/04/1941 | |
| Struck Of Charge | N/K | |
| Total Flying Hours | N/K | |
| Take-Off Time | 20:55hrs | |
| Bomb Load | 7x500lb+1x250lb | |
| **CREW** | | |
| Captain | Sergeant František **Kráčmer** 787244 RAFVR. Age 24. | Grave 12.A.7-8 |
| Second Pilot | Flying Officer František **Sixta** 82574 RAFVR. Age 27. | Grave 12.A.7-8 |
| Navigator | Pilot Officer Vladimír **Kubíček** 82613 RAFVR. Age 27. | Grave 12.A.9 |
| Wireless Operator | Pilot Officer Václav **Košulič** 82609 RAFVR. Age 25. | Grave 16.H.8 |
| Front Gunner | Sergeant Václav **Štětka** 787497 RAFVR. Age 25. | Grave 12.A.7-8 |
| Rear Gunner | Sergeant Rudolf **Lifczicz** 787533 RAFVR. Age 30. | Grave 12.A.7-8 |
| Posting History | Sgt F Kráčmer NFD. F/O F Sixta posted via No.11 O.T.U, 15/10/1940. | |
| Operations Flown | Sgt F Kráčmer 12 / F/O F Sixta 8 | |
| Buried | JONKERBOS WAR CEMETERY | |

Only one of the crew was originally identified. Pilot Officer Václav Košulič was initially buried in the Venlo Military Cemetery in Grave 82, Plot Military, Row 10. A 1947 No.2 M.R.E Unit report stated that Pilot Officer Vladimír Kubíček was in Grave 89, and his four companions in Grave 88-90, two being buried together in Grave 88. Rudolf Lifczicz, perhaps more than any, had so much to lose if taken PoW, he came from a Jewish family.

There was a real sense of shock throughout the squadron on the morning of the 18th. The loss of a crew was always hard, it was even harder when the squadron had not posted a crew missing since February 6th after a wonderful loss-free run. There was only one way to overcome the loss of friends, a trip to the Mess or local pubs, The Bull, or The Dog and Partridge. At 06:35hrs the following morning 311 Squadron provided three crews for a sea search. Reports that Wellington 'L' of 9 Squadron had ditched off Lowestoft returning from Berlin had been received. Sadly, despite an extensive search the only thing located was an empty dinghy. Sergeant R Stark and his all NCO crew were never found. The crews were back at East Wretham by 10:12hrs. That night, a planned operation to Mannheim was cancelled soon after briefing due to adverse weather. Bomber Command H.Q believed the weather had improved sufficiently to order an operation against Cologne and Rotterdam to be flown on the 20th. 311 Squadron would contribute just six crews to 3 Group's nightly total of 51 bombers. Light rain and thunderstorms swept R.A.F East Wretham throughout the morning and afternoon and looked like it would continue into the evening. The conditions were not encouraging. Regardless of the weather, six Wellingtons were made ready.

Over at Honington, the Training Flight came close to losing both a crew and a Wellington at 13:00hrs. While flying at 200ft, the Wellington of Sergeant Vilém Soukup experienced severe icing, cutting both engines. Prompt action by the young pilot resulted in both engines picking up, if but reluctantly. Wisely, the crew decided to return to Honington. Coming into land, Wellington L7841 KX-S hit the topmost branches of a tree. Showing steady nerves and some skill, the Wellington was brought in for landing. On inspection, both the tailplanes and the starboard elevator were damaged. The perspex bomb aimer panel was smashed, and the underside of the Wellington had extensive fabric and geodetic damage. But, more importantly, the crew was safe.

| Date | 20/04/1941 |
|---|---|
| Mark | Wellington Mk.Ic |
| Serial | L7841 |
| Code | KX-S |
| Taken On Charge | 28/09/1940 via No. 22 MU. |
| Manufacturer | Vickers Armstrong |
| Contract | 992424/39 |
| Pilot (s) | Sergeant Vilém Soukup / Sergeant Josef Horáček |
| Flight | Training |
| Time | 13:00hrs |
| Cause | Icing |

April 20th 1941 : Town 2(a) – Cologne

| Pilot | 2nd Pilot | Serial | Code | Bomb Load | Result |
|---|---|---|---|---|---|
| P/O V Korda | Sgt K Schoř | T2972 | KX-G | 7x500lb+1x250lb | Completed |
| Sgt L Anderle | Sgt M Styblík | T2990 | KX-T | 3x500lb+1x250lb+360x4lb | Completed |
| F/O J Breitcetl | Sgt J Bernát | R1451 | KX-P | 3x500lb+1x250lb+360x4lb | Completed |
| Sgt J Doktor | Sgt H Dostál | R1516 | KX-U | 7x500lb+1x250lb | Completed |
| S/Ldr J Ocelka | Sgt A Rozum | R1410 | KX-M | 7x500lb+1x250lb | Completed |
| Sgt A Šedivý | Sgt J Hájek | R1371 | KX-F | 3x500lb+1x250lb+360x4lb | Alternative Target |

The first crew departed at 20:30hrs into a distinctly hostile-looking night sky. Almost along the entire route to the target, the crews encountered either thunderstorms, solid cumulus cloud or severe icing. Somehow, all but one crew managed to find Cologne. Sergeant A Šedivý bombed Aachen as an alternative. They had encountered severe icing en route and reported a temperature of -26 degrees. Clouds prevented observation of their results, but a few incendiaries were seen to burst in the south-east of the town at 22:40hrs. Sergeant L Anderle and crew were the first over Cologne at 22:33hrs. They bombed from 16,000ft above an almost solid overcast. Squadron Leader J Ocelka and Sergeant J Doktor were over the target within a minute of each other. Squadron Leader Ocelka reported his bombs created one large fire and one smaller, red-coloured fire whilst Sergeant Doktor dropped his entire bombload in one stick. He confidently reported his 500-pounders bursting in the centre of Cologne. Nine minutes later, it was the turn of Flying Officer J Breitcetl and crew. They reported on return to East Wretham, *'Fires seen on arrival just south of Cathedral and east of the Cologne-Deutz Marshalling Yards'*. They bombed the glow seen below the cloud at 22:50hrs. The last to bomb and displaying gritty determination over Cologne were the crew of Pilot Officer Václav Korda. They spent thirty minutes over the target area in an attempt to identify the aiming point. To help them, they dropped their 'Flash-bomb', which did little but illuminate the clouds. Finally, having pushed their luck, they bombed the glow of fires in an area on the west bank of the river. On leaving Cologne, this fire was seen to be rapidly spreading. All the crews landed safely at base, thankful to be on the ground. For the next two days the squadron busied itself with training and cross-country flights. On the 22$^{nd}$, a cinema show in the Airmen's Upper Dinning Hall at R.A.F Honington played the film *'Wings of the Morning'* starring Henry Fonda and the French born actress, Annabella. The squadron was not required on the night of April 22$^{nd}$ for a small-scale attack on Brest flown by the squadrons of R.A.F Marham and Stradishall. All four squadrons failed to locate the enemy warships in the face of intense searchlights and flak.

On the 23$^{rd}$, it was the turn of Honington and East Wretham, plus Wellingtons supplied by Mildenhall's 149 Squadron. Also involved was R.A.F Newmarket and the Short Stirlings of 7 Squadron. This would be the first time the squadron Wellingtons would be loaded with the 113inch long forged steel 2000lb Mk.I Armour Piercing bomb, in addition to the customary 250-pounder S.A.P.

## April 23$^{rd}$ 1941 : CC.49 – Brest

| Pilot | 2$^{nd}$ Pilot | Serial | Code | Bomb Load | Result |
|---|---|---|---|---|---|
| P/O V Korda | Sgt V Bufka | T2972 | KX-G | 1x2000lb AP + 2x250lb SAP | Completed |
| F/O J Breitcetl | Sgt J Bernát | T2990 | KX-T | 1x2000lb AP + 2x250lb SAP | Completed |
| Sgt A Šedivý | Sgt J Hájek | R1371 | KX-F | 1x2000lb AP + 2x250lb SAP | Completed |
| Sgt F Fencl | Sgt F Radina | R1466 | KX-D | 1x2000lb AP + 2x250lb SAP | Completed |
| Sgt J Doktor | Sgt M Styblík | R1516 | KX-U | 1x2000lb AP + 2x250lb SAP | Completed |

The loading of the 2000-pounder was beyond the capacity of the armourers at East Wretham. Specialist lifting equipment was needed, and this was only available at R.A.F Honington. The five Wellingtons flew the short flight to Honington to be loaded. The squadron was visited by Air Marshal Karel Janoušek during the day. His visits were always a topic of conversation.

Pilot Officer V Korda was the first away from 311 Squadron at 19:30hrs. Within five minutes, the squadron was airborne. Over the sea, the crews encountered a thin layer of cloud and some scattered sleet. Approaching the French coast, considerable sea mist, which appeared like white waves, was observed. It was particularly dense around the entrance to Brest Harbour. Arriving over the target between 12,000ft and 15,000ft, the flak and searchlight defences were already active and in full swing. Two of the crews were unable to identify where in the docks their bombs actually landed due to a combination of flak, searchlights, sea mist and flares. Three however did. The crew of Pilot Officer V Korda circled the target for twenty minutes before they identified the aiming point after establishing their position using the west bay of the Rade-de Brest. They observed their bombs and numerous others bursting across the docks at 22:07hrs. The dock area was on the receiving end of Sergeant J Doktor's three bombs. Unsurprisingly, no fires were reported. The last crew to bomb was Sergeant F Fencl at 22:35hrs. They, like Pilot Officer Korda, were carrying a 'Flash-Bomb' which was dropped with their bombload and allowed them to observe with some satisfaction their AP and SAP bombs bursting across Dock No.1. Of the 21 Wellingtons despatched, sixteen reported having successfully bombed the target. Of the two Stirlings, one failed to take-off and the other returned early. All of 311 Squadron's crews landed back at East Wretham having dropped 15,000lbs of bombs.

The squadron sat out the attacks on Kiel on the 24th and 25th. Instead, it busied itself with the usual rounds of training and maintenance. On the 25th, Honington took delivery of three crews of 218 Squadron[26] as their home base at RAF Marham was under attack. One of the diverted Wellington was flown by the squadron commander, Wing Commander Herbert Kirkpatrick. Weather conditions seriously restricted the Group's contribution on the 26th against Hamburg. Only three squadrons would be involved, 9, 311 and 149, who provided just 23 Wellingtons between them. Five other Wellingtons would bomb Emden. These totals would be reduced when one Wellington returned early, and two failed to take-off.

*Two Czech Armourers are seen here manhandling a 250-pound General Purpose bomb. The Czech Ground crews had a difficult life, even by the standards of Bomber Command. Living accommodations were, at best, adequate in the summer months, but during the winter, they were horrifically difficult and demanding even for the tough Czechs. It is doubtful that their R.A.F equivalent would have worked under such primitive conditions for so long.*

---

[26] *218 ORB records two crews.*

## April 26th 1941 : Town 4 (B) – Hamburg

| Pilot | 2nd Pilot | Serial | Code | Bomb Load | Result |
|---|---|---|---|---|---|
| P/O V Korda | Sgt V Bufka | T2972 | KX-G | 2x500lb+1x250lb+360x4lb | Completed |
| F/Lt J Šnajdr | Sgt B Blatný | R1410 | KX-M | 2x500lb+1x250lb+360x4lb | Completed |
| F/O J Breitcetl | Sgt J Bernát | T2990 | KX-T | 6x500lb | Completed |
| Sgt J Doktor | Sgt M Styblík | R1015 | KX-L | 6x500lb | Completed |
| Sgt J Čapka | F/O J Šejbl | R1598 | KX-C | 6x500lb | Completed |
| Sgt F Fencl | Sgt F Radina | T2971 | KX-H | 6x500lb | Completed |

The squadron's Wellingtons were all safely airborne by 21:00hrs. Weather conditions were considerably better than forecast. The only issue was stronger than expected winds. It was these winds that very nearly caused the loss of one crew. The first crew bombed at 23:55hrs. Flying Officer J Breitcetl observed his stick of bombs bursting southeast of the artificial lakes at Binnenalster. Sergeant Fencl's 'Flash-bomb' exposed the area just south of Borgelde, a densely populated area of Hamburg. At 00:01hrs, Sergeant J Čapka reported their bombs bursting in line between the Alster and the Hamburg-Dammtor railway station. The last squadron crew departed at 00:16hrs and reported fires visible from over 25 miles on the return flight. It had been an excellent attack despite the tremendous flak barrage. All the crews managed to land back at R.A.F East Wretham except Sergeant Doktor, who drifted northwards due to the stronger-than-forecast winds. Inexperience on the part of the navigator and wireless set issues did not help as the crew found themselves over unfamiliar countryside.

The hopelessly lost crew were fortunate to observe a searchlight which they circled. Realising that the crew were in trouble or lost, the searchlight crew directed the beams towards R.A.F Church Fenton. Arriving at Church Fenton, the Czechs made three approaches to land, but each time, they were shown a red. Running out of fuel, Sergeant Doktor opted to return to the searchlight and search for somewhere to land nearby. Finally, with fuel now perilously low, the pilots lowered the undercarriage and brought the Wellington in for a forced landing in a field 2 ½ miles northwest of Wetherby. On landing on the soggy field, the aircraft tipped onto its nose before thumping back onto its undercarriage and rear wheel. The time was 03:45hrs. The Wellington was relatively unscathed, the underside of the front turret, bomb sight and cockpit windscreen were damaged. The only injury was suffered by the second pilot who fractured his tibia resulting in a trip to York Military Hospital. It was a remarkable piece of airmanship by Sergeant Doktor and his second pilot, Sergeant M Styblík, in landing the Wellington at night in a strange field.

| Date | 27/04/1941 |
|---|---|
| Mark | Wellington Mk.Ic |
| Serial | R1015 |
| Code | KX-L |
| Taken On Charge | 30/11/1940 via No. 48 MU. |
| Manufacturer | Vickers Armstrong |
| Contract | 992424/39 |
| Pilot (s) | Sergeant Jaroslav Doktor / Sergeant Miroslav Styblík |
| Flight | Operations |
| Time | 03:45hrs |
| Cause | Fuel shortage |

The crew were collected the next day by Pilot Officer V Korda. Wellington R1015 would be partially dismantled by a salvage team and sent for repair. It would eventually return to the squadron. The reluctance of R.A.F Church Fenton to switch on the flare path may have been due to the fatal crash of Defiant N1568 of 54 O.T.U at 02:08hrs. Sergeant Frederick Crozier was airborne on a night flying exercise when attacked by a Junkers Ju88 of NJG/2 flown by Lt Rudolf Pfeiffer. In an attempt to escape the night fighter, the Defiant flew low but clipped the top of some trees and crashed, bursting into flames on impact and killing the crew. With intruders known to be in the region, there was likely an unwillingness to illuminate the airfield. While the crews were on their way to Hamburg, Sergeant Alois Mžourek was up on a night training flight aboard Wellington R3234. On conclusion of the flight, the Wellington returned to East Wretham, where at 21:50hrs, a gust of wind took the young, inexperienced pilot by surprise while coming into land, resulting in a heavy landing on one wheel. The crew were uninjured, but the Wellington would require the attention of No.43 Group. It would eventually return to service with the Czech Training Flight.

| Date | 26/04/1941 |
|---|---|
| Mark | Mk.Ic |
| Serial | R3234 |
| Code | KX-N |
| Taken On Charge | 09/08/1940 via No. 23 MU. |
| Manufacturer | Vickers Armstrong |
| Contract | B3943/39 |
| Pilot (s) | Sergeant Alois Mžourek |
| Flight | Training Flight |
| Time | 21:50hrs |
| Cause | Gust of wind / pilot inexperience |

The last operation of the month was flown on the 28th. The squadron would provide just four crews against the *Scharnhorst* and *Gneisenau* in Brest. They would be joined by eight Wellingtons of 9 Squadron, eleven from 214 Squadron and three Stirlings from 7 Squadron.

April 28th 1941 : 'Toads' – Brest

| Pilot | 2nd Pilot | Serial | Code | Bomb Load | Result |
|---|---|---|---|---|---|
| P/O V Korda | F/O K Vildomec | T2972 | KX-G | 1x2000lb+2x500lb-1x250lb | Completed |
| Sgt F Fencl | Sgt F Radina | R1466 | KX-D | 1x2000lb+2x500lb-1x250lb | Completed |
| Sgt B Blatný | Sgt K Šťastný | R1371 | KX-F | 1x2000lb+2x500lb-1x250lb | Completed |
| Sgt J Bernát | Sgt K Schoř | R1451 | KX-P | 1x2000lb+2x500lb-1x250lb | Completed |

The operation would see some new names chalked on the Op's Board. Sergeants B Blatný and J Bernát would captain their own crews for the first time. Both were experienced second pilots, Josef Bernát having flown 20 operations while Benedikt Blatný had completed 17. While these two veterans were well into their operational tours, two were starting. Flying Officer Karel Vildomec, an experienced pilot who saw action in France with Groupe de Bombardement d'Assault II/54 and was awarded the *Croix de Guerre* and Sergeant Karel Šťastný, another veteran of France. Each Wellington was again flown to R.A.F Honington to be loaded with the 2000-pounder S.A.P. bombs. Those early over Brest encountered ground haze, which got worse as the raid progressed. Black smoke from either burning buildings or smoke pots also greatly hindered observation. Bombing from between 14,500ft and 16,000ft, each crew dropped in one stick. Only one was certain where their bombs landed. Sergeant B Blatný bombed at 22:33hrs, and he reported his bombs straddling Dock No.1. Fellow Sergeant František Fencl stated his bombs landed south of the mouth of the Elorn at 22:40hrs. On return, Group H.Q. were somewhat optimistic about the results but conceded that results were difficult to ascertain.

April had been a challenging month for the squadron. Despite a reduction in sorties flown, the crews had, on every occasion, pressed home their attacks with fearless dedication. The loss of a crew was always tragic, but given the targets attacked and the weather conditions encountered, the squadron had been fortunate in comparison to some squadrons. There were no issues with reliability, which was always good for morale. A total of 59 sorties had been flown against 12 targets during the month. There was just one early return reported due to a mechanical issue, a testament to the skill and dedication of the ground crews.

# May 1941 : Switch to Germany

Lousy weather kept flying to a minimum on May 1st and May 2nd. A few cross-country flights were flown, as was some night flying training. It was on one of these training flights that the unfortunate Sergeant Vilém Soukup was involved in yet another accident. Having completed his training flight, the pilot brought Wellington P9294 in for a landing at East Wretham at 22:40hrs. Unfortunately, he failed to lower the undercarriage fully, and the Wellington hit the ground with an almighty thud. It was a costly and basic error on the part of the pilot, who had clocked up a total of 65 hours on the Wellington, but of which only seven were at night.

| Date | 02/05/1941 |
|---|---|
| Mark | Mark IA |
| Serial | P9224 |
| Code | KX-T |
| Taken On Charge | 18/03/1941 via 115 Squadron. |
| Manufacturer | Vickers Armstrong |
| Contract | 549263/36 |
| Pilot (s) | Sergeant Vilém Soukup |
| Flight | Night Training Flight |
| Time | 22:24hrs |
| Cause | Pilot's failure to lower undercarriage |

The Wellington would undergo repair with No.43 Group and would return to service with the Czech Training Flight in May. That same night a 'Freshmen' operation by two crews was planned but was cancelled before briefing. The Group detailed and briefed 59 crews on the evening of May 3rd. The bulk of the bombers would be directed against the cruisers in Brest, while 16 Wellingtons, including six from 311 Squadron would bomb Cologne. Two 'Freshmen' crews would attack the oil tanks at Rotterdam.

## May 3rd 1941 : 'Trout 'A' – Cologne

| Pilot | 2nd Pilot | Serial | Code | Bomb Load | Result |
|---|---|---|---|---|---|
| Sgt V Bufka | Sgt A Rozum | T2990 | KX-T | 3x500lb+1x250lb+360x4lb | Complete |
| Sgt B Blatný | Sgt K Šťastný | R1371 | KX-F | 1x1000lb+6x500lb | Alternative Target |
| Sgt F Fencl | Sgt F Radina | R1466 | KX-D | 1x1000lb+6x500lb | Complete |
| Sgt J Doktor | Sgt H Dostál | R1516 | KX-U | 1x1000lb+6x500lb | Early Return |
| Sgt J Bernát | Sgt K Schoř | R1451 | KX-P | 1x1000lb+6x500lb | Complete |
| Sgt A Šedivý | Sgt J Hájek | R1371 | KX-F | 1x1000lb+6x500lb | Complete |

Sergeant J Doktor and crew suffered a catalogue of mechanical issues aboard Wellington R1516 KX-U while en route. The starboard engine began to run rough, the W/T packed up, as well as the heating. They opted to bomb the docks at Dunkirk from 11,000ft at 23:07hrs. Their bombs landed in the centre of the docks, creating a yellowish-red fire. The squadron was over Cologne between 23:45hrs and 00:10hrs. Conditions were not favourable with 9/10th cloud obscuring the target. One crew opted for a secondary target when they were unable to identify the aiming point. Sergeant J Bernát chose to attack Bonn, creating two fires close to the city centre at 00:01hrs. A small gap in the cloud gave the crew of Sergeant B Blatný the briefest glimpse of Cologne. They dropped their bombs in two sticks at 00:08hrs. The flak defences over Bonn were used to make a timed run on Cologne by the crew of Sergeant F Fencl, who dropped their entire load in one stick. Sergeant V Bufka were the only ones who seemed confident about the raid's success. They reported on return, *'three bursts in target area adjoining aiming point 'A' then incendiaries. On leaving target area seven very large fires seen accompanied by three large red explosions 3 minutes later'*.

*Sergeant Karel Schoř ((Pavel Vančata)*

### May 3rd 1941 : Z.3 (d) – Rotterdam Dock

| Pilot | 2nd Pilot | Serial | Code | Bomb Load | Result |
|---|---|---|---|---|---|
| F/O J Breitcetl | F/O F Pohlodek | T2561 | KX-A | 16x250lb | Complete |
| F/Lt J Šnajdr | Sgt J Nyč | R1021 | KX-W | 16x250lb | Complete |

Two new pilots would be occupying the right-hand seats on this night. Flying Officer František Pohlodek and Sergeant Jaroslav Nyč would benefit from the experience of two of the most gifted pilots on the squadron. Thick haze over Rotterdam made locating the oil tanks extremely difficult. Flying Officer J Breitcetl dropped his entire load from 14,000ft. All the bombs were seen to burst just south of the actual target. Fifteen minutes later it was the turn of Flight Lieutenant J Šnajdr. His bombload was seen to explode by the rear gunner at 23:20hrs, however, the precise location was uncertain. Although the squadron had dropped an impressive 29,880lb of bombs on this night, they had struggled with the weather conditions over both targets.

Bomber Command turned its attention to Mannheim on the 5th, all four heavy bomber groups would be involved and gather a force of 141 aircraft. 311 Squadron would provide six crews, plus an additional 'Freshmen' crew on the docks at Cherbourg.

### May 5th 1941 : Mannheim

| Pilot | 2nd Pilot | Serial | Code | Bomb Load | Result |
|---|---|---|---|---|---|
| Sgt F Fencl | F/O K Vildomec | R1466 | KX-D | 2x500lb+1x250lb+360x4lb | Complete |
| Sgt V Bufka | Sgt A Rozum | T2561 | KX-A | 2x500lb+1x250lb+360x4lb | Complete |
| Sgt J Bernát | Sgt K Schoř | R1451 | KX-P | 1x1000lb+5x500lb | Complete |
| Sgt B Blatný | Sgt K Šťastný | T2971 | KX-H | 6x500lb+1x250lb | Complete |
| Sgt A Šedivý | Sgt J Hájek | R1371 | KX-F | 6x500lb+1x250lb | Complete |
| Sgt J Čapka | F/O J Šejbl | R1516 | KX-U | 6x500lb+1x250lb | Alternative target |

A defective wireless set aboard Wellington R1516 KX-U forced Sergeant J Čapka to bomb the docks at Dunkirk at 23:15hrs. The 311 Squadron ORB[27] reports that the crew had an inconclusive encounter with a German night fighter, it records, *'Three bursts from our aircraft, went into a shallow dive and disappeared into cloud'*. On return to R.A.F East Wretham, it was discovered that the aerial leads had been broken.

Mannheim was covered by an almost impenetrable blanket of cloud on the squadron's arrival. Fires and explosions were observed below the cloud base. Sergeant Fencl reported, *'A large fire reflected in the clouds and yellow explosions seen on arrival'*. Sergeant V Bufka dropped his mixed load in one stick. He reported that the bursts were seen in the reflection of the clouds. He also commented, *'many flares from other aircraft and flak rendered observation difficult'*. The only crew to glimpse Mannheim was Sergeant B Blatný. A gap in the cloud over the River Decka allowed them to confirm their position before bombing just south of the aiming point. A total of 121 returning crews claimed to have bombed in the general area of Mannheim, the vast majority through cloud. A local report claimed that only around twenty-five bombs had hit the city, causing only modest damage, mostly to residential property.

### May 5th 1941 : CC.16 – Cherbourg

| Pilot | 2nd Pilot | Serial | Code | Bomb Load | Result |
|---|---|---|---|---|---|
| F/Lt J Šnajdr | Sgt J Nyč | R1021 | KX-W | 5x500lb+1x250lb+120x4lb | Alternative Target |

---

[27] *311 ORB Form 540 Air27/1686*

The crew headed south for the long, lonely flight to Cherbourg. On arrival, they found the target covered in clouds and thick fog, making any possibility of accurate bombing impossible. Unwilling to return with a full bombload, they bombed the dock area of Calais from 14,000ft at 00:26hrs. Three bomb bursts were observed through the cloud, but little else. Night flying training on the 5th was interrupted at 01:21hrs when a Junkers Ju88C bombed the airfield. Four small bombs were dropped, two on a dispersal area and two landed off the airfield. One Wellington received minor damage to its fabric and three army huts were slightly damaged. It was a timely reminder that these intruders still posed a real threat.

Flying training was carried out throughout the daylight hours on the 6th. However, appalling weather towards evening prevented any night training. It meant just one thing, a night in the various Messes or a trip to the local pubs. The Honington and East Wretham pairing sat out the attack on Hamburg on the 6th and, surprisingly, the attack on Brest on the 7th. The squadron had been scheduled to operate against Brest but was ordered to stand down early in the morning. Bomber Command called for a maximum effort from its squadrons on the 8th, and its squadrons did not disappoint. It would be a record-breaking night with an unprecedented 364 sorties flown. 3 Group put up a maximum effort of ninety-eight Wellingtons for the Blohm & Voss works at Hamburg, plus three Wellingtons and two Stirlings for Berlin and two Wellingtons for Emden. The Stradishall-based 214 Squadron provided the most significant contribution with a respectable 13 crews. At East Wretham, the recent lay-off had meant that 311 Squadron could detail eight crews, two of which would visit Emden.

### May 8th 1941 : GR3587 – Blohm & Voss Shipyards – Hamburg

| Pilot | 2nd Pilot | Serial | Code | Bomb Load | Result |
|---|---|---|---|---|---|
| Sgt V Bufka | Sgt A Rozum | T2990 | KX-T | 2x500lb+1x250lb+360x4lb | Complete |
| Sgt B Blatný | Sgt K Šťastný | T2971 | KX-H | 2x500lb+1x250lb+360x4lb | Complete |
| Sgt F Fencl | F/O K Vildomec | R1466 | KX-D | 2x500lb+1x250lb+360x4lb | Complete |
| Sgt A Šedivý | Sgt J Hájek | R1371 | KX-F | 3x500lb SAP + 3x500lb | Alternative target |
| Sgt J Čapka | F/O J Šejbl | R1516 | KX-U | 3x500lb SAP + 3x500lb | Complete |
| Sgt J Bernát | Sgt K Schoř | R1451 | KX-P | 3x500lb SAP + 3x500lb | Alternative target |

Once again, it was an all-sergeant-led operation with the crew of Sergeant Vilém Bufka the first to leave at 21:15hrs. Disappointingly, two crews were forced to look for alternative targets due to engine issues. Sergeant J Bernát's troubles started soon after take-off. Twenty-two miles northeast of North Walsham, Norfolk, the starboard engine aboard Wellington R1451 KX-P started surging, making any thoughts of reaching Hamburg foolhardy. The crew could have turned for home but instead headed south and attacked the docks and shipping at Dunkirk at 23:59hrs. The next crew to bomb an alternative target was Sergeant A Šedivý. The port engine oil temperature rose to an alarming level, giving the crew no option but to turn for home. They bombed a flak position on the island of Langeoog, one of the East Frisian Islands, at 00:28hrs. There were a few anxious minutes for the crew. While pumping oil into the port engine, the overload oil pump handle suddenly snapped, all the while the overworked starboard engines temperature started showing worrying signs of rising. Thankfully, they along with Sergeant J Bernát, made a safe return.

Arriving over Hamburg between 14,000ft and 15,000ft, the squadron was confronted with a target already ablaze. Flak was reported as heavy and searchlights were active, and flares were in abundance. Violent detonations were clearly visible in the darkness. These were the new 4000-pounders, which one crew described as *'a pleasing explosion and blue flashes'*. In good visibility, the squadron bombed, and the report submitted by Sergeant V Bufka back at R.A.F East Wretham sets the scene over Hamburg. *'One big fire seen south of the Binnen-Alster, smaller fires in docks. All bursts seen both sides of the river just east of the target. One big fire started on south bank of river and smaller fires from incendiaries started'*. Most of the participating crews claimed good bombing results. Hamburg reported a total of 83 fires, of which 38 were classified as large. One in particular was noted, the Deutsche Erdolwerke oil depot was seen ablaze with flames rising up hundreds of feet into the air. For the residents of Hamburg, it was the first taste of what was to come. 185 people were killed, 518 injured and 1,966 bombed out. Both the crews given Emden were successful. Fires had already taken hold when they arrived over the target.

## May 8th 1941 : 'Herring' - Emden

| Pilot | 2nd Pilot | Serial | Code | Bomb Load | Result |
|---|---|---|---|---|---|
| F/O J Breitcetl | F/O F Pohlodek | T2561 | KX-A | 5x500lb+2x250lb+120x4lb | Complete |
| F/Lt J Šnajdr | Sgt J Nyč | T2972 | KX-G | 5x500lb+2x250lb+120x4lb | Complete |

Large fires were reported in the area around Alter Binnenhafen, where both crews reported dropping their mixed bombload. Keen to exploit the success over Hamburg and the clear weather, Bomber Command turned its attention towards the twin cities of Mannheim and Ludwigshafen, which were separated by the east and west banks of the Rhine. Mannheim had been attacked on the 8th by both 1 and 4 Group who inflicted some worthwhile damage. Now it was 3 Group's turn.

R.A.F Honington was bombed during the early hours of May 9th disturbing the sleep of many. A suspected Junkers Ju88C dropped nine 50kg and 20 incendiaries, the majority of which landed outside the airfield. During the day, 311 Squadron prepared six crews for Mannheim, most of whom had flown against Hamburg the previous night.

## May 9th 1941 : 'Chub' – Mannheim

| Pilot | 2nd Pilot | Serial | Code | Bomb Load | Result |
|---|---|---|---|---|---|
| Sgt V Bufka | Sgt A Rozum | T2990 | KX-T | 3x500lb+360x4lb | Complete |
| Sgt B Blatný | Sgt K Šťastný | T2971 | KX-H | 3x500lb+360x4lb | Complete |
| Sgt J Bernát | Sgt K Schoř | T2972 | KX-G | 3x500lb+360x4lb | Complete |
| Sgt A Šedivý | Sgt J Hájek | T2561 | KX-A | 3x500lb+360x4lb + Flash-bomb | Complete |
| Sgt F Fencl | F/O K Vildomec | R1516 | KX-U | 7x500lb + Flash-bomb | Complete |
| Sgt J Čapka | F/O J Šejbl | R1021 | KX-W | 7x500lb | Complete |

3 Group would supply 50 Wellingtons drawn from 9, 40, 57, 75(NZ), 214 and 311 Squadrons. Of these, three failed to take-off, and four returned early. It was not one of 3 Groups better nights. Thankfully, the Czechs had a clean sheet. All six crews were over Mannheim between 00:45hrs and 01:18hrs. The first to arrive was Sergeant V Bufka at 00:45hrs. They found a subdued target. There was little activity from the city's defences and no tangible signs of bombing. The crew bombed in one stick from 14,000ft and, in good visibility, reported their bombs bursting near the aiming point, producing two large yellow and red fires. When Sergeant B Blatný arrived six minutes later, fires were visible dotted around the aiming point. By then, Mannheim's defences were in full swing, but compared to Hamburg's, the searchlight and flak appeared noticeably weaker. Even after six minutes, the momentum of the bombing had increased dramatically, as was the destruction being inflicted below. On his arrival, Sergeant J Bernát noted two medium fires in the town centre, one large fire southwest of the town centre and a sway of incendiary fires in the docks. The crew dropped their mixed load from 14,000ft onto the aiming point, adding to the rapidly growing fires. By the time the last crew bombed at 01:18hrs, Mannheim was ablaze. Sergeant A Šedivý dropped his entire load in one stick, which was seen to burst north of Neckarufer in the centre of the town. He reported on return to East Wretham, *'Results difficult to observe owing to extensive light from the fires'*. The last crew landed safely at 04:3hrs, the squadron had dropped 18,760lb of bombs and incendiaries. Two crews did well to complete the operation. Sergeant J Bernát had to cope with a faulty port engine, a leaky oil tank and an unserviceable wireless set, while Sergeant Fencl completed the operation with an unusable front turret due to a short circuit in the electrics. Local sources provided a catalogue of damage that included fifty-three military, industrial, commercial and residential buildings on both sides of the river damaged or destroyed. They reported that 3,500 people had been bombed out of their homes. On the 10th, a message was received from the C-in-C Bomber Command, which read: *'I CONGRATULATE THE COMMAND ON LAST NIGHTS SUCCESSFUL OPERATION ALSO UPON THE OUTSTANDING NUMBERS OPERATED BY THE GROUP'*. [28]

---

[28] *The message was in all probability in response to the operation of May 8th.*

*Three pilots "who wrote the squadron history". L-R: Sergeant Karel Pospíchal who finished a tour on Wellingtons flying with both Bomber Command and the Coastal Command. He started his second tour on 22nd June 1944 flying Liberators. He was injured in a crash on 29th October 1944 and never returned to operational flying. Sergeant Alois Šiška became a PoW after a successful ditching while aboard Wellington Mk.IC T2553 KX-B in the Channel while returning from Wilhelmshaven on 28th December 1941. Sergeant Stanislav Linka drowned in Wellington Mk.IX Z8966 KX-E after a successful ditching in Irish Sea while returning from Kiel on 16th November 1941. (Pavel Vančata)*

The squadron sat out the attack on Hamburg on the 10th but was heavily involved in the follow-up attack on the 11th, providing seven crews plus a 'Freshmen' crew on Dieppe. Flying with Pilot Officer V Korda was Wing Commander Mareš Toman on what would be his last operation with his beloved 311 (Czech) Squadron. Three pilots would be making their operational debut. Their arrival was welcomed as the frequency of operations was putting a tremendous strain on the small pool of operational crews. The new pilots were Sergeants Adolf Musálek, Alois Mžourek and Václav Ryba.

<u>May 11th 1941 : GR3587 – Blohm & Voss Shipyards – Hamburg</u>

| Pilot | 2nd Pilot | Serial | Code | Bomb Load | Result |
|---|---|---|---|---|---|
| Sgt B Blatný | Sgt A Musálek | T2971 | KX-H | 2x500lb+1x250lb+360x4lb | Complete |
| Sgt F Fencl | F/O K Vildomec | R1466 | KX-D | 2x500lb+1x250lb+360x4lb | Complete |
| P/O V Korda | W/Cdr Toman | T2972 | KX-G | 2x500lb+1x250lb+360x4lb | Complete |
| Sgt A Šedivý | Sgt A Mžourek | R1371 | KX-F | 6x500lb | Complete |
| F/Lt J Šnajdr | Sgt J Nyč | R1021 | KX-W | 6x500lb | Complete |
| Sgt J Čapka | F/O J Šejbl | R1598 | KX-C | 6x500lb | Alternative Target |
| Sgt J Bernát | Sgt V Ryba | R1532 | KX-R | 6x500lb | Early Return |

The squadron took off in staggered intervals on what would be a busy night. The first away at 22:16hrs was the Wellington flown by Pilot Officer V Korda. The crew of Sergeant B Blatný were the last to depart at 22:49hrs. 3 Group contributed seventy-five Wellingtons and a solitary Stirling for the return visit to Hamburg. The Group once again provided most of the attacking force briefed to attack the Blohm & Voss shipyards. Eight crews would also visit Merignac airfield, Rotterdam oil installation, and Dieppe.

Sergeant J Bernát and crew were attacked twice within three minutes by what was reported to be a Junkers Ju88 soon after crossing the English coast off Orford Ness. The first attack at 23:15hrs was from behind and slightly above, the Ju88 firing a single burst, which passed harmlessly under the Wellington. The rear gunner, Sergeant Jindřich Beneš, immediately returned fire with two short bursts at 400 yards. This was followed by a second attack at 23:18hrs from the same height but from the port bow, on this occasion firing a short burst which passed over the port engine. Sergeant J Bernát took violent evasive action by putting the nose of the Wellington down at the same time jettisoning the bombs followed all the while by the Ju88. The rear gunner engaged with a third burst of machine gun fire. It was enough to deter the German pilot. The crew returned to RAF East Wretham.[29]

Another crew that would not reach Hamburg was Sergeant J Čapka. They suffered complete W/T failure within an hour of take-off but continued on before depositing the entire HE load in one stick on Bremerhaven producing one large fire at 01:10hrs. The first bombers were over Hamburg at 01:25hrs and greeted with ideal bombing conditions. Fires had already taken hold when 311 Squadron arrived. Several small fires were observed in the town centre, while a large red fire producing dense smoke was reported just north of the target area. Flight Lieutenant J Šnajdr was over Hamburg at 16,000ft. He reported scattered bombing 2 miles south-south-west of the Blohm & Voss shipyards. Opposition over Hamburg was fierce, flak of every calibre was bursting between 14,000ft and 18,000ft, and searchlights criss-crossed the night sky. The crew of Sergeant B Blatný were one of the squadron crews tasked with taking photographs. Approaching Hamburg at 15,000ft, the pilot commenced a dive down to 14,000ft for bombing. It was seconds after the mix load was dropped when the Wellington was caught in a searchlight beam. This solitary beam was quickly joined by a multitude of others until both pilots were utterly dazzled. Explosions rocked the Wellington, tossing it about like a model. Both heavy and medium flak, then seemed to concentrate their fire on the Wellington exploding all around the bomber. Instinctively Sergeant Blatný took aggressive evasive action, throwing the Wellington around like a fighter in an effort to dodge the flak and lose the searchlights. It was during one of these manoeuvres that a flak splinter entered the cockpit and struck Sergeant Blatný on the back of his head, rendering him temporarily unconscious. Quickly recovering, he continued to try and shake off his antagonists. The cockpit instrumentation was undamaged apart from the left boost dial, and the engines appeared to be running smoothly. Everything seemed in order until the Wireless Operator, Pilot Officer Josef Doubrava, informed the pilot that oil was leaking from the emergency oil tank used to operate the undercarriage. Finally, clear of the murderous flak and searchlights, Sergeant Blatný handed over control to his 2nd pilot to receive first aid treatment from the observer. Still groggy and concerned he may lose consciousness, he gave the inexperienced 2nd pilot and navigator constant instructions and details on landing without an undercarriage or flaps. Thankfully, the crew managed to cross northern Germany out over the relative safety of the North Sea without further trouble. Once in R.A.F East Wretham's circuit, the crew tried everything possible to lower the undercarriage, but it stubbornly refused too fully lower. Sergeant Blatný ordered the second pilot to land with the undercarriage partially lowered. The fuel supply and magnetos were switched off moments before the Wellington force landed. The Wellington made a near-perfect landing, slithering to a halt at 05:00hrs. Sergeant Benedikt Blatný was taken to SSQ for treatment. Sergeant Adolf Musálek's first operation was over. This young pilot was rightly praised for his actions on this night. It was later established that the Navigator and Wireless Operator were struck by splinters but miraculously uninjured.

---

[29] *The 311 ORB states the crew bombed Dieppe. However, the 3 Group HQ Records Book, 3 Group Form E and RAF Honington Summary of events state the crew jettisoned their bomb load.*

| Date | 12/05/1941 |
|---|---|
| Mark | Mk.IC |
| Serial | T2971 |
| Code | KX-H |
| Taken On Charge | 15/01/1941 |
| Manufacturer | Vicker Armstrong |
| Contract | 38600/39 |
| Pilot (s) | Sergeant Benedikt Blatný / Sergeant Adolf Musálek |
| Flight | Operations |
| Time | 05:00hrs |
| Cause | Flak damage |

Pilot Officer John Gellner, in his book, 'Moonlight Flyer' records the following, *'The plane had ninety-three holes. The luck of the crew on this night was incredible. One splinter went through the wing, just between the two fuel tanks, while another went clean through the cowling of the aircrew'.*

After extensive repair, Wellington T2971 would return to 311 Squadron in July.

### May 11th 1941 : CC31 - Dieppe

| Pilot | 2nd Pilot | Serial | Code | Bomb Load | Result |
|---|---|---|---|---|---|
| F/O J Breitcetl | F/O F Pohlodek | T2561 | KX-A | 17x250lb SAP | Complete |

The crew of Flying Officer J Breitcetl took off at 22:35hrs. They arrived over Dieppe docks at 00:20hrs to find it covered in sea haze. Dropping the semi-armour-piercing bombs in three sticks from 13,000ft the crew witnessed 15 bursts in the dock area complex. Opposition was slight, it was trouble free operation as they turned for home at 00:40hrs.

Training exercises were flown on the 12th in between rain showers. Operations were planned for the 13th and 14th, but both were cancelled before take-off. The hugely respected and admired Squadron Leader Percy 'Pick' Pickard DSO DFC left 311 Squadron at this time. He was transferred to 9 Squadron for a vacant flight commander role on the 14th. His contribution to the squadron's success especially in the early days was immeasurable. His time thankfully coincided with the squadron having a number of exceptional individuals like Josef Šnajdr and Josef Ocelka who did everything possible to help him. He would be replaced at the end of the month by another Englishman, Squadron Leader Kenneth Batchelor. On the 15th, the squadron was informed it was to bomb Kiel, although this signal was later amended and changed to Hanover. The squadron would prepare eight Wellingtons. There was high ranking addition to the crew list. Group Captain Josef Berounský would occupy the front turret aboard the Wellington captained by Pilot Officer Korda.

### May 15th 1941 : Eel – Hanover

| Pilot | 2nd Pilot | Serial | Code | Bomb Load | Result |
|---|---|---|---|---|---|
| P/O V Korda | Sgt J Horáček | R1451 | KX-P | 3x500lb+360x4lb | Complete |
| F/Lt J Šnajdr | Sgt J Nyč | R1410 | KX-M | 3x500lb+360x4lb | Complete |
| Sgt A Šedivý | Sgt A Mžourek | R1371 | KX-F | 3x500lb+360x4lb | Alternative |
| Sgt J Čapka | F/O J Šejbl | R1598 | KX-C | 3x500lb+360x4lb | Complete |
| Sgt K Schoř | Sgt K Šťastný | R1516 | KX-U | 3x500lb+360x4lb | Complete |
| Sgt J Bernát | Sgt V Soukup | R1532 | KX-R | 7x500lb | Complete |
| F/Lt J Breitcetl | F/O F Pohlodek | T2561 | KX-A | 7x500lb | Alternative |
| Sgt F Fencl | Sgt V Ryba | R1466 | KX-D | 7x500lb | Complete |

Sergeant Karel Schoř would be responsible for captaining a crew for the first time on this operation. He had previously completed 15 operations in the second pilot role. The unlucky Sergeant Josef Horáček and Vilém Soukup would conduct

*Sergeant Václav Pánek talks to Squadron Leader P. C. Pickard DSO, DFC in presence of Acting Squadron Leader Ocelka DFC at Wretham Hall in 1941. None of this trio will see the end of the war. Sgt Pánek would complete 33 sorties at 311 Squadron between 12th August 1941 and 12th February 1942. He would join 138 (Special Duties) Squadron on 22nd June 1942 just to went missing as Second Pilot of Handley Page Halifax A Mk.II W1002/NF-Y on 10th December 1942. (Pavel Vančata)*

their first operation. Both were involved in the crash on April 20th, and Vilém Soukup had almost written off Wellington P9224 on May 2nd when he forgot to lower the undercarriage. Bomber Command gathered a force of 101 aircraft on this night, 3 Group would provide a modest 36 Wellingtons and a solitary Stirling of XV Squadron against Hanover,

while four Stirlings of 7 Squadron, and two from XV Squadron would attempt to bomb Berlin. The force made its way across a freezingly cold northern Holland and Germany. Visibility en-route was hampered by 5/10th cloud. Two crews were unable to reach Hanover, both experiencing engine trouble. Flight Lieutenant J Breitcetl and crew bombed Munster at 00:20hrs when the oil temperature rose alarmingly in both engines aboard Wellington T2561 KX-A. The conditions were so severe the hand oil pump froze solid in the extreme -32 degrees. Munster was again on the receiving end when Sergeant A Šedivý dropped his entire load at 00:45hrs. The port engine aboard Wellington R1371 KX-F started showing worrying signs of trouble two hours after take-off. Over Germany it started to run erratically, and again, an increase in oil temperature meant the operation had to be abandoned. Both crews reported their bombs bursting in the centre of Munster.

Despite Bomber Command's instructions to aim for two specific targets, the main post office and telephone exchange, most crews knew this was just a cover for an attack aimed at the city centre. On reaching Hanover the crews were greeted by haze. One medium-sized fire with vivid red flames was observed in the south-east of the city. Bombing heights within the squadron varied. Sergeant J Čapka dropped their mixed bombload from 14,500ft 1½ miles from the city centre. The three 500-pounders were seen to burst, and the twinkling lights of the 4-pounders walked across what appeared to be a built-up area. Sergeant F Fencl recorded bombing from the unheard-of altitude of 19,000ft in Wellington R1466 KX-D. They bombed in one stick but were coned by searchlights and subjected to intense flak. Unable to shake them off, Sergeant Fencl put the nose of the Wellington down and dived, finally pulling out below 500ft. He flew the rest of the journey back at this altitude. The remaining crews bombed in the general target area and reported fires, some of which were decoys, but a glow remained on the horizon for some seventy miles into the return flight. While skirting the town of Nienburg at 02:00hrs on the return flight, the port engine of Wellington R1410 KX-M, flown by Flight Lieutenant J Šnajdr, suddenly cut and caught fire at 14,000ft. The fire was successfully extinguished, and the crew flew the remaining journey on the starboard engine. The first to land from the crews that reached Hanover was Sergeant J Bernát at 03:13hrs. There would be almost a three-hour wait until the last crew were reported safe. The crew of Sergeant Fencl somehow managed to survive flying across occupied Europe below 500ft and reach the English coast desperately low on fuel due to holes in the petrol tanks courtesy of the Hanover flak. With the fuel almost exhausted, Sergeant František Fencl made a near-perfect forced landing purportedly at Thurton, a village in Norfolk lying 8½ miles southeast of Norwich, at 05:15hrs. There is however an alternative landing site, which the author believes is the correct location. A witness to the events reported, *'A Wellington force-land wheels down near Thorpe Rectory, flying in a SE direction. It passed under HT cables but he thought the pilot "opened up again" to clear the Thorpe-Norton Sub course road and the aircraft ran over a ditch where the tail wheel broke off.'* The witness was convinced the crew were Czechs.

Flight Lieutenant J Šnajdr had nursed his one remaining engine almost 300 miles, then with his fuel almost gone, the crew made a belly landing in a field near Chequers Woods, Little Bromley, Essex, at 06:00hrs. The R.O.C log Centre reports the following entry: *'0553hrs Bromley RDF reports plane crashed at Shepherd's Farm, Little Bromley, crew O.K. Later report states plane crashed on Staceys Farm 1 miles south of Manningtree - not seriously damaged. A/C No. K.X.M R1410 from 311 bomber squadron Wretham Nr Thetford. U/C and props damaged.'*

| Date | 16/05/1941 |
|---|---|
| Mark | Mk.IC |
| Serial | R1466 |
| Code | KX-D |
| Taken On Charge | 14/02/1941 via No. 48 MU. |
| Manufacturer | Vickers Armstrong |
| Contract | 992424/39 |
| Pilot (s) | Sergeant František Fencl / Sergeant Václav Ryba |
| Flight | Operations |
| Time | 05:15hrs |
| Cause | Fuel shortage / Battle damage |

Both Wellingtons would require the specialist care of No.43 Group. Wellington R1466 would return to service briefly with 75(NZ) Squadron while R1410 would pound the runways with No.12 O.T.U.

| Date | 16/05/1941 |
|---|---|
| Mark | Mk.IC |
| Serial | R1410 |
| Code | KX-M |
| Taken On Charge | 26/02/1941 via No. 37 MU. |
| Manufacturer | Vickers Armstrong |
| Contract | 992424/39 |
| Pilot (s) | Flight Lieutenant Josef Šnajdr / Sergeant Jaroslav Nyč |
| Flight | Operations |
| Time | 06:00hrs |
| Cause | Fuel shortage / Battle damage |

From a squadron perspective the operation had been a challenging one. Two abandoned sorties, two crash landings, one aircraft returning with most of its incendiaries due to icing and another Wellington with intercom failure and an overheating engine. On the plus side, all the crews were at least safe.

On the 16th, two 'Freshmen' crews were given the opportunity to bomb Boulogne Docks. Sergeant Jaroslav Hájek would be captaining a crew for the first time.

### May 16th 1941 : CC29(a) – Boulogne

| Pilot | 2nd Pilot | Serial | Code | Bomb Load | Result |
|---|---|---|---|---|---|
| Sgt J Hájek | Sgt V Netík | R1046 | KX-E | 6x500lb+2x250lb +120x4lb | Complete |
| Sgt O Helma | Sgt A Musálek | R1532 | KX-R | 6x500lb+2x250lb +120x4lb | Early Return |

While 30 plus Wellingtons of 3 Group were pounding Cologne, 11 crews kept the pressure on Boulogne. One crew abandoned soon after take-off. The gyro horizons indicator aboard Wellington R1532 began to play up almost as soon as the aircraft became airborne. The crew landed back at East Wretham within the hour. Sergeant J Hájek circled Boulogne for ten minutes in an attempt to identify the docks, sea haze preventing the crew from identifying any features. Eventually, they bombed the flak and searchlight defences believed to be positioned on the north bank of the harbour at 23:30hrs. Both the Training Flight and Operations Flight spent the 17th training or carrying out routine air-tests. The following day 3 Group offered just 28 Wellingtons to attack Kiel. Both 9 and 149 Squadrons would provide 11 Wellingtons, with the Czechs offering six, although only five took off. The squadron reported an avoidable accident on the afternoon of the 18th. Sergeant Oldřich Jambor was taxiing Wellington R3206 when he collided with a lifting jack at 15:30hrs, damaging the Wellington's tailwheel. At the time, he was being guided by two mechanics, but the unfortunate pilot was deemed responsible. The luckless Czech only had 25 flying hours on the Wellington at the time of the incident. Jambor enlisted in the Czech Air Force and completed his training in Piešťany and Prague between 1935–1936. After the occupation in June 1939, he left for Poland, from where he took a ship to France, where he joined the Foreign Legion. He was then taken by ship to Oran, from there to Sidi bel Abbes, where he was assigned to the 1st Regiment of the Legion. After the outbreak of the war, he returned to France, where he was assigned to an air group and underwent bomber pilot training in Toulouse. After the defeat of France, he left Port Vendres and returned to Oran on the ship *'Meonia'* on June 9th 1940. From there, they continued by train to Casablanca and arrived in Gibraltar on the ship *'Gib-el-Dersa'*. On July 2nd 1940, he set sail for England aboard the *'Neuralia'* landing in Liverpool on July 12th. Thirteen days later, he was accepted into the R.A.F at Cosford.

| Date | 18/05/1941 |
|---|---|
| Mark | Mk.Ic |
| Serial | R3206 |
| Code | KX-B |
| Taken On Charge | 01/05/1941 via No. 20 MU. |
| Manufacturer | Vickers Armstrong |
| Contract | B3913/39 |
| Pilot (s) | Sergeant Oldřich Jambor |
| Flight | Training |
| Time | 15:30hrs |
| Cause | Hit obstacle. |

It would be another disappointing night for 3 Group. Of the 28 Wellingtons detailed, five failed to take-off, and five returned early. The Mildenhall-based 149 had three returns, with 311 (Czech) Squadron reporting two crews abandoning.

## May 18th 1941 : GR3589 (Python) – Fried-Krupp Germania Werft AG – Kiel

| Pilot | 2nd Pilot | Serial | Code | Bomb Load | Result |
|---|---|---|---|---|---|
| Sgt J Čapka | F/O J Šejbl | R1598 | KX-C | 1x100lb+3x500lb | Complete |
| Sgt O Helma | Sgt A Musálek | R1532 | KX-R | 2x500lb+360x4lb | Early Return |
| Sgt J Hájek | Sgt V Netík | R1015 | KX-L | 2x500lb+360x4lb | Complete |
| F/O F Pohlodek | F/O K Vildomec | T2561 | KX-A | 5x500lb+1x250lb | Complete |
| Sgt K Schoř | Sgt K Šťastný | R1516 | KX-U | 5x500lb+1x250lb | Early Return |

Sergeants O Helma and K Schoř encountered icing over the North Sea and aborted the operation. Sergeant K Schoř jettisoned his bombload at 23:25hrs, 70 miles northeast of North Walsham, while Sergeant Helma dropped his bombs at 23:45hrs in the same area. They both landed back at East Wretham within two minutes of each other. It was a disappointing effort.

The remaining crews found Kiel covered by haze, making any observation difficult. What was not difficult to see was the lack of any fires. Flying Officer F Pohlodek, flying his first operation as captain, dropped his bombload in one stick southeast to northeast across the target area. Four bursts were seen. Sergeant J Hájek dropped his mixed load from 16,000ft over what was believed to be the aiming point. The crew of Sergeant J Čapka summed up the operation, *'Observation of the results too difficult to ascertain'*. Wireless trouble aboard R1015 KX-L on the return flight meant that Sergeant J Hájek landed at R.A.F Alconbury. There was some concern about the fate of Flying Officer F Pohlodek and crew who eventually landed at 05:43hrs. They had encountered their fair share of problems during the operation. The climb and glide instruments refused to work, and worrying issues with the starboard engine were compounded when the starboard engine glycol pump froze-up. However, importantly they got home, if but rather late. It was a disappointing night for the command, Group and squadron. The squadron finally parted company with Wing Commander Karel Mareš Toman, on the 20th. He was transferred to the C.I.G in London. In August he was promoted to Acting Group Captain. Post-war his life was far from happy, ill and on the verge of poverty, humiliated and constantly persecuted, he died prematurely in a hospital in Prague at the age of 61 in 1960.

The operation planned for the 20th was cancelled. This again happened on the 21st, severe weather keeping the command practically grounded. On the 22nd, the squadron reported another accident involving one of its crews. Sergeant Miroslav Jindra was airborne in Wellington L7841 on a training flight from East Wretham. All was well until at 15:00hrs when he attempted a landing. Struggling with the gusty conditions he made a rather heavy landing tail-first resulting in a collapsed tailwheel. The subsequent investigation found that strong and gusty winds may have been a contributing factor.

| Date | 22/05/1941 |
|---|---|
| Mark | Mk.Ic |
| Serial | L7841 |
| Code | KX-S |
| Taken On Charge | 28/09/1940 via No. 22 MU. |
| Manufacturer | Vickers Armstrong |
| Contract | 992424/39 |
| Pilot (s) | Sergeant Miroslav Jindra |
| Flight | Training |
| Time | 15:00hrs |
| Cause | Heavy Landing |

The Wellington had only just returned to squadron service. Once repaired by No.43 Group the aircraft would be passed onto the Czech Flight. Arriving on the 22nd, was a welcomed batch of pilots. Sergeants, František Dostál, Oldřich Hlobil, Vladimír Pára, Jan Janek and Alois Tománek.

The squadron sat out the operation on the 23rd to Cologne and on the 24th the weather was so bad that Bomber Command H.Q cancelled the planned operation. Finally on the 25th, the weather improved and immediately the squadron set in motion more training. An Air-firing practice flight was organised late morning. Tragically, it would end with the deaths of four crewmen. Taking off from R.A.F Langham at 11:30hrs, Pilot Officer Stanislav Zeinert struck the uttermost branches of some trees resulting in the aircraft crashing in a field near the airfield with such force it broke in two and instantly burst into flames. Witnesses confirmed that the Wellington appeared to be in trouble almost immediately after take-off. Tragically, apart from the usual crew, two passengers were also on board.

### Vickers Wellington Mk.IA N3010 KX-L

| | | |
|---|---|---|
| Manufacturer | Vickers Armstrong | |
| Contract | 549268/36 | |
| Taken on Charge | 20/09/1940 via 214 Squadron. | |
| Cat E Missing | N/A | |
| Struck Of Charge | N/K | |
| Total Flying Hours | 303hrs.15mins | |
| Take-Off Time | 11:30 | |
| Bomb Load | N/A | |
| | | |
| | **CREW** | |
| Captain | Pilot Officer Stanislav **Zeinert** 82648 RAFVR. Age 25. | Grave 587/20. |
| Second Pilot | Pilot Officer Miloslav **Švic** 82578 RAFVR. Age 24. | Grave 588/23. |
| Navigator | Pilot Officer Josef **Čermák** 82589 RAFVR | Injured |
| Wireless Operator | Pilot Officer Miroslav **Vild** 82646 RAFVR | Injured |
| Front Gunner | Sergeant František **Dušek** 787825 RAFVR. Age 20. | Grave 586/21 |
| Rear Gunner | Sergeant Maxmilián **Stoček** 787890. Age 22. | Grave 585/22 |
| | | |
| | **PASSENGERS** | |
| | Pilot Officer Fleming Voltelin Van De Bijl 65685 (Dutch) | Injured |
| | Corporal Thomas Eric Litherland 847742 (British) | Injured |
| | | |
| Posting History | Not Known | |
| Operations Flown | Pilots Nil on 311 Squadron. | |
| Buried | EAST WRETHAM (ST. ETHELBERT) CHURCHYARD | |

The crew's two gunners appeared to have been killed at the scene or very soon after the crash. The survivors were initially taken to SSQ Bircham Newton and then to the Cromer & District Hospital. Pilot Officer Stanislav Zeinert died of his injuries on the afternoon of May 26th. Pilot Officer Miloslav Švic clung to life until June 4th when he succumbed to his injuries. On May 30th, three of the crew were buried together at East Wretham Churchyard, followed on June 9th by Pilot Officer Švic, who was buried beside his friends. The investigation into the crash blamed the young Pilot, stating, *'Pilot failed to correct the initial swing and attempted to take-off crosswind. Pulled aircraft off the ground steeply to clear hedge, stalled, hit trees'*. Corporal Thomas Eric Litherland was reported to be on strength of R.A.F Marham, while Pilot Officer Fleming Voltelin Van de Bijl was the new Squadron Education Officer.

On May 27th, the squadron was ordered to prepare eight Wellingtons for an attack on Cologne, plus three 'Freshmen' crews would be visiting Boulogne Harbour. Wellington KX-L R1015 was withdrawn before take-off with I.F.F issues, the other cancelled aircraft, or cause is not recorded. 311 Squadron would be the only 3 Group contribution on Cologne, they would be joining 46 Whitleys of 4 Group.

## May 27th 1941 : 'Trout' – Cologne

| Pilot | 2nd Pilot | Serial | Code | Bomb Load | Result |
|---|---|---|---|---|---|
| Sgt V Bufka | Sgt A Rozum | T2990 | KX-T | 1x1000lb+5x500lb+1x250lb | Early Return |
| P/O V Korda | Sgt J Horáček | T2972 | KX-G | 7x500lb+1x250lb | Complete |
| Sgt J Bernát | Sgt A Musálek | R1532 | KX-R | 7x500lb+1x250lb | Complete |
| F/O F Pohlodek | F/O K Vildomec | T2561 | KX-A | 7x500lb+1x250lb | Complete |
| Sgt F Fencl | Sgt J Filler | R1804 | KX-D | 1x1000lb+5x500lb+1x250lb | Complete |
| Sgt J Nyč | Sgt K Šťastný | R1451 | KX-P | 1x1000lb+5x500lb+1x250lb | Complete |
| Sgt J Čapka | F/O J Šejbl | R1598 | KX-C | 7x500lb+1x250lb | Complete |

The first crew away left around 23:00hrs into a wonderfully clear but freezing night. There was the usual early return. Within six minutes of take-off, Sergeant Bufka was back in the circuit with an overheating starboard engine having jettisoned his bombload. The first of the squadron crews arrived over Cologne at 01:35hrs and reported little activity and strangely no fires. Those following a few minutes behind found some scattered fires, but disappointingly on the outskirts of the city. Flak seemed subdued, but the numerous searchlights made visual identification difficult in the crystal clear conditions. Sergeant J Čapka dropped all his HE load from 14,000ft at 01:58hrs. On return, the crew reported, *'Bombs hung up[30] therefore jettisoned in the target area. Bursts seen near the Koln-Deutz Railway Station. Rear Gunner reported very large explosion followed by a big fire with huge black smoke'*.

## May 27th 1941 : CC29 – Boulogne

| Pilot | 2nd Pilot | Serial | Code | Bomb Load | Result |
|---|---|---|---|---|---|
| F/Lt J Šnajdr | Sgt V Netík | R1021 | KX-W | 7x500lb+1x250lb+120x4lb | Complete |
| F/Lt J Breitcetl | Sgt V Ryba | R1777 | KX-M | 7x500lb+1x250lb+120x4lb | Complete |

Visibility over Boulogne was fair. Both crews dropped in two sticks from 14,500ft. Flight Lieutenant J Šnajdr's first stick landed in the dock area, while the second stick added to the rubble in the town itself. Five of Flight Lieutenant Breitcetl's bombs were seen to burst across Dock 4 and 5. The rear gunner reported that fires were seen to *'spring up'*. All the crews were safely back by 04:50hrs, having delivered a new record in bombs dropped, 30,960lb. It was on the 27th that Squadron Leader Ken Batchelor was posted from 9 Squadron to take over the vacant role of Liaison Officer.

A day and night of good weather was taken full advantage of by both flights on the 28th, with a comprehensive programme of day and night training carried out. Operations were ordered on the 29th and 30th, but both were cancelled due to weather. The funerals of Pilot Officer Stanislav Zeinert, Sergeants František Dušek and Maxmilián Stoček took place on the 30th in the small churchyard at East Wretham. Squadron's padre Squadron Leader František Pouchlý conducted the service, and all the squadron flying crews attended. On the last day of the month, a 'Freshmen' operation was planned, but again, harsh weather resulted in a cancellation. May 1941 had found the squadron in the thick of the action over Germany. The month was not without its dramas. Three forced landings with battle damage, three flying accidents, and one fatal crash was a constant reminder that death or injury was not confined to operations over Germany. A lapse in concentration even when carrying out the most basic flying, could prove fatal. There were however reasons to celebrate, May brought a flurry of awards. One MBE was awarded to Pilot Officer Vladimír Nedvěd for his action on December 16th 1940. Sixteen squadron members were awarded the Czech Medal for Valour, including Sergeants A Musálek, K Šťastný and J Nyč, Flying Officers K Vildomec and F Pohlodek and Group Captain Berounský. A total of 18 squadron members were awarded the Czech War Cross. These included Sergeants A Musálek, K Schoř, Pilot Officer J Gellner, and Flying Officer J Šejbl. Three R.A.F airmen were also honoured with the Czech Medal of Valour, awarded as per London Gazette dated May 16th 1941. Wireless Operator Instructors Flight Sergeant Patrick Hennigan DFM, Sergeant Leo Judson and Sergeant Ernest Robb, Navigation Instructor.

---

[30] *Icing*

# June 1941 : German Production

Rain and low clouds on June 1st were enough to keep 311 Squadron on the ground during the day and sufficiently bad for 3 Group H.Q to keep its squadrons grounded that night. Form B.512 arrived on the morning of June 2nd. The weather had improved enough to detail 59 crews against Dusseldorf while another 12 would visit Berlin. The Czechs would offer eight crews, on what would turn out to be a disappointing night. Flying as co-pilot with Pilot Officer Korda was Wing Commander Josef Schejbal who had not flown operationally since October 21st 1940.

### June 2nd 1941 : 'Perch B' – Dusseldorf

| Pilot | 2nd Pilot | Serial | Code | Bomb Load | Result |
|---|---|---|---|---|---|
| P/O V Korda | W/Cdr J Schejbal | T2972 | KX-G | 3x500lb+1x250lb+360x4lb | Complete |
| Sgt V Bufka | Sgt A Rozum | T2990 | KX-T | 3x500lb+1x250lb+360x4lb | Alternative target |
| Sgt F Fencl | Sgt J Filler | R1804 | KX-D | 1x1000lb+5x500lb | Alternative target |
| Sgt J Nyč | Sgt K Šťastný | R1718 | KX-N | 7x500lb+1x250lb | Jettisoned |
| Sgt J Doktor | Sgt H Dostál | R1516 | KX-U | 7x500lb+1x250lb | Alternative target |
| Sgt A Šedivý | Sgt A Mžourek | R1371 | KX-F | 7x500lb+1x250lb | Complete |
| Sgt J Čapka | F/O J Šejbl | R1598 | KX-C | 7x500lb+1x250lb | Complete |
| Sgt K Schoř | Sgt A Musálek | R1451 | KX-P | 7x500lb+1x250lb | Complete |

Almost immediately after take-off, the crews entered the first layer of cloud at 1000ft. This persisted up to 5,000ft where it suddenly cleared. While climbing to operational height, the crews again encountered yet more clouds at 9,000ft, and this almost solid layer continued up until 15,000ft. Four crews reported mechanical issues or icing en route, and three would bomb an alternative target. Sergeant V Bufka bombed what was believed to be Aeltre airfield, Belgium at 00:30hrs due to an overheating engine. Wireless trouble aboard Wellington R1804 KX-D meant the crew of Sergeant F Fencl deposited their bombs on the docks at Oostende at 23:55hrs. Icing resulted in Sergeant J Nyč jettisoning his bombload in the sea 20 miles west of Brussels. Finally, Wellington R1516 KX-U suffered engine issues en route. Unwilling to just jettison, Sergeant J Doktor and crew sought out a cloud-covered Dunkirk. It was not a promising start. The remaining crews arrived over what was believed to be Dusseldorf around 01:10hrs, clouds having prevented accurate navigation. Crews either dropped on flak and searchlight concentrations or dropped on ETA. Some bursts were observed, and the rear gunner aboard Pilot Officer V Korda's Wellington reported a brief glimpse of burning houses but was just uncertain where. The journey home was equally challenging for the navigators. The last crew to land was Sergeant J Čapka at 04:35hrs. He and his crew had been aloft for 5 hours forty-nine minutes. A Polish manned Wellington of 304 (Silesian) Squadron landed at R.A.F East Wretham at 03:45hrs skippered by Squadron Leader Beill. They returned to their home base of R.A.F Syerston late afternoon.

Bad weather thwarted the operation planned on the 3rd for three 'Freshmen' crews. It also cancelled any night flying training. The following day, an improvement in the weather allowed the squadron to enjoy extensive training throughout the day and night. Bombing practice, gunnery, and cross-country flights were all successfully completed. A planned operation was however cancelled. On the 6th, the squadron remained grounded again due to yet more harsh weather. Finally on the 7th, 3 Group put together a small force of forty-three Wellingtons and seven Stirlings to attack the heavy cruiser *Prinz Eugen* at berth at Brest. 311 Squadron would offer nine Wellington crews.

### June 7th 1941 : 'Prinz Eugen' – CC.49 – Brest

| Pilot | 2nd Pilot | Serial | Code | Bomb Load | Result |
|---|---|---|---|---|---|
| Sgt A Šedivý | Sgt A Mžourek | R1371 | KX-F | 6x500lb+1x250lb | Complete |
| P/O V Korda | Sgt J Horáček | T2972 | KX-G | 6x500lb+1x250lb | Complete |
| Sgt V Bufka | W/Cdr J Schejbal | T2990 | KX-T | 6x500lb+1x250lb | Complete |
| Sgt F Fencl | Sgt J Filler | R1804 | KX-D | 6x500lb+1x250lb | Complete |

| Sgt K Schoř | Sgt A Musálek | R1451 | KX-P | 6x500lb+1x250lb | Early Return |
| Sgt J Čapka | F/O J Šejbl | R1598 | KX-C | 6x500lb+1x250lb | Complete |
| F/O F Pohlodek | F/O K Vildomec | T2561 | KX-A | 6x500lb+1x250lb | Complete |
| Sgt J Doktor | Sgt H Dostál | R1777 | KX-M | 6x500lb+1x250lb | Complete |
| Sgt J Nyč | Sgt K Šťastný | R1718 | KX-N | 6x500lb+1x250lb | Complete |

Once more, Wing Commander J Schejbal's name was to be found on the crew list for that night. On this occasion, he chose to fly with the crew of the relatively inexperienced Sergeant Vilém Bufka. The squadrons of 9 and 311 (Czech) were scheduled to be over Brest first, followed by 214 and 149 Squadron and the Stirlings of 7 Squadron. Weather conditions were marginal as the squadron climbed away from East Wretham airfield and set course for the French coast near St-Malo. At 23:00hrs, the AOC withdrew the 13 Wellingtons of 149 Squadron due to weather fears, leaving just the Wellingtons of 9, 214 and 311 Squadrons, plus the Stirlings to represent 3 Group. Sergeant K Schoř aborted the operation on reaching the English coast, a faulty A.S.I and altimeter aboard Wellington R1451 KX-P being the cause. The first crew to arrive a few minutes after midnight was Sergeant F Fencl. With little cloud, and just slight haze they dropped their all S.A.P load from 18,000ft and were able to confirm that they burst across Dock 8. This dock was also bombed by Pilot Officer V Korda at 00:35hrs, dropping his entire load in one stick. Four bursts were witnessed northwest of the dock. Sergeant V Bufka was at 13,500ft when he dropped his load in two sticks in the northern part of the target area. They reported that an effective smokescreen prevented observation of other bursts. Two crews claimed that part of their load hit Dock 6. Flying Officer F Pohlodek at 01:10hrs and Sergeant J Doktor at 01:35hrs. The latter reported on return, *'Bursts seen close to northwest side of docks starting red fire'*. Lousy weather over East Anglia meant that many of the crews were diverted. Four from 311 Squadron found themselves lodgers. Flying Officer Pohlodek landed at R.A.F Newmarket, Sergeant Doktor at R.A.F Stradishall damaging his Wellington. Sergeant Nyč landed at R.A.F Boscombe Down, while Sergeant Šedivý was welcomed at R.A.F Abington.

The undercarriage of Wellington R1777 KX-M collapsed on landing at R.A.F Stradishall at 04:30hrs, the result of damage caused by a heavy bump when retracting the wheels too early on take-off at R.A.F East Wretham. The Wellington would be repaired by the specialists of No.43 Group and would return to the squadron late July 1941.

| Date | 08/06/1941 |
|---|---|
| Mark | Mk.Ic |
| Serial | R1777 |
| Code | KX-M |
| Taken On Charge | 18/05/1941 via No. 45 MU. |
| Manufacturer | Vickers Armstrong (Chester) |
| Contract | 992424/39 |
| Pilot (s) | Sergeant Jaroslav Doktor / Sergeant Hugo Dostál |
| Flight | Operational |
| Time | 04:30hrs |
| Cause | Undercarriage damaged on take-off |

Soon after the crews had left for Brest, Sergeant Jambor was airborne in Wellington R1269 KX-G of the Training Flight on a night dual training sortie. While attempting to land at East Wretham at 23:45hrs, the inexperienced pilot forgot to lower the Wellingtons flaps which resulted in him overshooting his landing and running into a hedge.

| Date | 07/06/1941 |
|---|---|
| Mark | Mk.Ic |
| Serial | R1269 |
| Code | KX-G |
| Taken On Charge | 31/12/1941 via 115 Squadron after being repaired by No.43 Group |
| Manufacturer | Vickers Armstrong (Chester) |
| Contract | 992424/39 |
| Pilot (s) | Sergeant Oldřich Jambor / ? |
| Flight | Training (Czech Training Flight) |
| Time | 23:45hrs |
| Cause | Overshot landing, failure to lower flaps. |

June 8th, and bad weather ensured there was only limited training throughout the day. A planned 'Freshmen' operation was cancelled at 15:00hrs. The restrictions on flying continued into the following day. On the 10th, 3 Group attacked Brest with 48 Wellingtons drawn from the squadrons of R.A.F Feltwell and Marham. An effective smoke screen and haze prevented any damage to the *Prinz Eugen*. 311 Squadron had another stand-down at 14:45hrs, cancelling a planned 'Freshmen' operation.

*The rather splendid-looking Wretham Hall was home to the officers of No.311 (Czech) during their time with the RAF Bomber Command. Designed by Sir Reginald Blomfield to replace the original Hall, it was sadly demolished soon after the war's end.*

Finally, on the 11th, Form B.521 was received at R.A.F East Wretham. The Ruhr city of Düsseldorf would be the primary target for 3 Group providing 60 Wellingtons and seven Stirlings. A further 19 Wellingtons would attack the docks at

Boulogne. The influx of pilots, Wellingtons and a few days non-operational meant that 311 Squadron would be able to provide a respectable 12 crews on this night, a new record in aircraft despatched and bombs dropped. Nine would bomb Düsseldorf, while three 'Freshmen' crews would attack the docks at Boulogne.

### June 11th 1941 : 'Perch A' – Düsseldorf

| Pilot | 2nd Pilot | Serial | Code | Bomb Load | Result |
|---|---|---|---|---|---|
| P/O V Korda | Sgt J Horáček | W5682 | KX-Y | 3x500lb+1x250lb+360x4lb | Alternative target |
| Sgt A Šedivý | Sgt A Mžourek | R1371 | KX-F | 3x500lb+1x250lb+360x4lb | Alternative target |
| Sgt F Fencl | Sgt V Netík | R1804 | KX-D | 3x500lb+1x250lb+360x4lb | Complete |
| Sgt J Čapka | F/Lt J Šejbl | R1598 | KX-C | 1x1000lb+5x500lb | Complete |
| Sgt K Schoř | Sgt O Helma | W5711 | KX-H | 1x1000lb+5x500lb | Complete |
| Sgt J Nyč | Sgt K Šťastný | R1718 | KX-N | 1x1000lb+5x500lb | Complete |
| Sgt J Bernát | Sgt A Musálek | R1532 | KX-R | 7x500lb | Complete |
| F/O F Pohlodek | F/O K Vildomec | T2561 | KX-A | 7x500lb | Alternative target |
| Sgt J Doktor | Sgt H Dostál | R1777 | KX-M | 7x500lb | Early Return |

All 12 crews were airborne between 23:00hrs and 23:19hrs. Within minutes of take-off, Sergeant Doktor requested permission to land. The artificial horizon aboard Wellington R1777 KX-M was found to be unserviceable. Germany was shrouded in thick cloud, and any prospect of a clear target was quickly vanishing. Three crews were obliged to bomb alternative targets. Pilot Officer Korda bombed Duisburg at 01:20hrs from 13,000ft, creating a *'big fire in the middle of the town that was visible from over 40 miles away'*. Cologne was on the receiving end of the crew of Sergeant A Šedivý, who started one very large fire north-north-east of the town centre. Another crew who chose Cologne was Flying Officer F Pohlodek, he dropped all his HE load from 14,000ft at 01:45hrs. They observed three explosions in the northwest part of town starting a pinkish-coloured fire. The remaining crews reported bombing the primary with varying results. Inevitably, the raid lacked accuracy. Sergeant J Bernát claimed he bombed the centre of Düsseldorf, starting one fire. Sergeant Nyč reported, *'Six medium fires seen between the target and Neuss'*. Sergeant Fencl was the last to bomb at 02:00hrs, and ominously, he reported that nothing was seen on his arrival. 3 Group H.Q conceded that Düsseldorf was covered in thick cloud, but more in hope reported *'All stations reported large fires, especially a large oil fire on the east side of the river'*. There are no reports from Düsseldorf. There were however from Cologne, located some 20 miles to the south, they reported extensive damage on this night.

### June 11th 1941 : CC.29A – Boulogne Docks

| Pilot | 2nd Pilot | Serial | Code | Bomb Load | Result |
|---|---|---|---|---|---|
| Sgt J Hájek | Sgt V Soukup | R1015 | KX-L | 6x500lb+120x4lb | Complete |
| F/Lt J Breitcetl | Sgt V Ryba | T2972 | KX-G | 6x500lb+120x4lb | Complete |
| S/Ldr J Šnajdr | F/O J Stránský | R1021 | KX-W | 6x500lb+120x4lb | Complete |

One pilot would kick off his operational tour on this night, Flying Officer Josef Stránský. He was born on December 10th 1914, in the small village of Borová. Soon after the Nazi occupation, he, like so many brave Czechs, fled to Poland in July 1939, then travelled to France. While in France, he began training at Toulouse, flying the twin-engine Marcel Bloch M.B.200B, a disastrous deathtrap of an aircraft only the French could design. Before he could be assigned to a combat unit, the French surrendered. He arrived in Liverpool in July 1940. On October 28th 1940, he was posted to 24 Squadron R.A.F. The squadron operated a motley collection of ex-civil and military twin-engine aircraft in the communication and transport role. On March 14th, 1941, he was transferred to 311 Squadron due to the urgent need of replacement crews.

The three 'Freshmen' crews were all safely away and heading towards the south coast of England by 23:19hrs. The term 'Freshmen' was liberally used on 311 Squadron and not entirely appropriate when considering the experience of the three pilots operating. Sergeant Hájek had completed 17 operations, while the recently promoted Squadron Leader

*All smiles, Jo Capka sits in the pilot seat of KX-C – R1598 V for Victory. An impressive 50 operations are recorded on this old war-horse. (Photo via John Costin)*

Šnajdr had finished 25 and Flight Lieutenant Breitcetl had flown 26. The flights were more 'hands-on instructional' in nature and would eventually become known as 'second dickie' trips. Conditions over Brest were reasonably clear. The usual haze and smoke screens were encountered, but the dock area was visible. The crews arrived just after midnight, flying between 14,000ft and 14,500ft. Sergeant Hájek reported fires started at Dock 4, 5 and 9 while Squadron Leader J Šnajdr started one large reddish fire on the eastern edge of Dock 5. The crew of Flight Lieutenant J Breitcetl dropped their bombs on existing fires seen close to Dock 4. By 05:25hrs, all the squadron crews had safely landed. It had been a disappointing night over Düsseldorf, with thick clouds hampering accurate bombing, but there was reason to celebrate. All 12 crews returned, and the squadron dropped 37,510lb of bombs, a new record for a raid.

By the time, the returning crews had been debriefed and breakfasted, Form B.522 had arrived at East Wretham, informing the squadron it would be required for operations again that night. Headquarters 3 Group detailed 90 Wellingtons on the vital railway complex at Hamm, three on Emden, while the Short Stirlings of 7 Squadron would join forces with 4 Group on an attack against Huls. The Czechs at East Wretham would detail 12 crews. Throughout the day the ground crews worked feverously to make ready the Wellingtons hindered all the while by heavy showers.

### June 12th 1941 : GH593 – Railway Marshalling Yards – Hamm

| Pilot | 2nd Pilot | Serial | Code | Bomb Load | Result |
|---|---|---|---|---|---|
| Sgt J Čapka | F/Lt J Šejbl | R1598 | KX-C | 1x1000lb+5x500lb | Bombed Flak & S/Ls |
| Sgt K Schoř | Sgt O Helma | R1451 | KX-P | 7x500lb+1x250lb | Complete |
| Sgt B Blatný | Sgt V Soukup | W5711 | KX-H | 7x500lb+1x250lb | Complete |
| S/Ldr J Šnajdr | Sgt V Netík | R1021 | KX-W | 7x500lb+1x250lb | Complete |
| Sgt A Šedivý | Sgt A Mžourek | R1371 | KX-F | 7x500lb+1x250lb | Bombed Flak & S/Ls |
| Sgt F Fencl | Sgt V Ryba | R1015 | KX-L | 7x500lb+1x250lb | Bombed Flak & S/Ls |
| F/O F Pohlodek | F/O K Vildomec | T2561 | KX-A | 7x500lb+1x250lb | Bombed Flak & S/Ls |
| Sgt J Bernát | Sgt A Musálek | R1532 | KX-R | 7x500lb+1x250lb | Bombed Flak & S/Ls |
| Sgt J Doktor | Sgt H Dostál | T2990 | KX-T | 7x500lb+1x250lb | Bombed Flak & S/Ls |
| P/O V Korda | Sgt J Horáček | T2972 | KX-G | 7x500lb+1x250lb | Alternative Target |
| Sgt J Nyč | Sgt K Šťastný | R1718 | KX-N | 1x1000lb+5x500lb | Bombed Flak & S/Ls |
| S/Ldr J Ocelka | Sgt J Hájek | W5682 | KX-Y | 7x500lb+1x250lb | Bombed Flak & S/Ls |

Operating for the first time since April 20th was Squadron Leader J Ocelka. He would be joined by Sergeant J Hájek aboard the almost factory-fresh Wellington W5682 KX-Y. All the squadron crews were safely away from East Wretham into what promised to be a clear night. Across the Group, a disappointing eight crews were withdrawn, and a further seven failed to take-off due to various mechanical issues. This, added to the early return of five crews, made a worrying 20% reduction in the Group's contribution. On this night, the Czechs of 311 Squadron had a clean slate with no early returns. It was an excellent effort. The good weather encountered over England and the North Sea quickly deteriorated over the continent, and on reaching Hamm, the whole of the northern Ruhr was covered by thick haze. The majority of the squadron crews bombed on either E.T.A or on the abundance of flak and searchlight positions. Pilot Officer Korda and crew claimed to have bombed the Krupps Works at Essen at 01:41hrs and reported their bombs created *'Rapidly spreading fires'*. Despite the haze, two crews claimed to have bombed the primary and witnessed the results. Sergeant B Blatný reported depositing his bombs at 01:40hrs, *'Bombed target and saw on leaving one small fire in target area'*. The crew of Sergeant K Schoř bombed from 16,500ft at 01:48hrs, and they reported, *'Target bombed, purple, red irregular explosions caused spreading sparks over a wide area and lighting up the cloud base. Explosions did not persist'*. Some returning 3 Group crews reported large fires in the eastern and western ends of the marshalling yards. In contrast, the majority could only report bombing in the general target area in the face of heavy and light flak and searchlights. A local Hamm source reported just six bombs exploding in the town.

On the 14th, Flight Lieutenant J Breitcetl and a party from 311 Squadron attended the War Weapons Week parade held in the market town of Attleborough, Norfolk. The party were inspected by Field Marshal Ironside, a veteran of the Boer War, the Great War and the Russian Campaign. The Czechs must have made an impression on this old war-horse as he congratulated Flight Lieutenant J Breitcetl on the efficiency of his men. A 'Freshmen' operation was organised for the 15th, 311 Squadron providing just two crews.

### June 15th 1941 : CC.25 – Dunkirk

| Pilot | 2nd Pilot | Serial | Code | Bomb Load | Result |
|---|---|---|---|---|---|
| F/Lt J Breitcetl | Sgt V Ryba | R1021 | KX-W | 7x500lb+120x4lb | Complete |
| Sgt J Hájek | Sgt V Soukup | R1451 | KX-P | 7x500lb+120x4lb | Complete |

The main effort of the night was directed against the Cologne-Gereon Marshalling Yards by 52 Wellingtons of 3 Group. A further 10 Wellingtons would visit Dunkirk, while just seven would tackle Hanover. Flight Lieutenant J Breitcetl was the first of the pair to bomb at 00:18hrs. He dropped in one stick from 12,500ft. Three bursts were seen, creating three large and one smaller fire. Two minutes later, it was the turn of Sergeant J Hájek. He bombed searchlights and flak positions due to 10/10th cloud cover. Both crews were back at East Wretham within three hours. It was a return visit to Düsseldorf on the 16th. The raid would be an all-3 Group effort with 74 Wellingtons and eight Stirlings detailed. Eleven crews would be provided by 311 Squadron on what would be a disappointing night. Three of the briefed force detailed failed to take-off due to technical faults. Over at R.A.F Honington, Wellington L7871 'B' captained by Sergeant A.L Roberts caught fire on the flare path, resulting in only four of the 11 Wellingtons taking off. To add to the night's problems, seven bombers returned early, three of which were from 311 Squadron.

### June 16th 1941 : 'Perch A' – Düsseldorf

| Pilot | 2nd Pilot | Serial | Code | Bomb Load | Result |
|---|---|---|---|---|---|
| P/O V Korda | Sgt J Horáček | R1598 | KX-C | 3x500lb+1x250lb+360x4lb | Early Return |
| F/Lt J Breitcetl | Sgt V Ryba | T2972 | KX-G | 1x1000lb+5x500lb+1x250lb | E/A encounter |
| Sgt K Schoř | Sgt O Helma | R1451 | KX-P | 7x500lb+1x250lb | Alternative Target |
| S/Ldr J Ocelka | Sgt B Blatný | W5682 | KX-Y | 3x500lb+1x250lb+360x4lb | Duty Carried Out |
| Sgt V Bufka | Sgt A Rozum | T2990 | KX-T | 3x500lb+1x250lb+360x4lb | Duty Carried Out |
| Sgt J Bernát | Sgt A Musálek | R1532 | KX-R | 1x1000lb+5x500lb+1x250lb | Duty Carried Out |
| Sgt F Fencl | Sgt K Šťastný | R1804 | KX-D | 3x500lb+1x250lb+360x4lb | Duty Carried Out |
| Sgt A Šedivý | Sgt A Mžourek | R1371 | KX-F | 7x500lb+1x250lb | Duty Carried Out |
| Sgt J Nyč | Sgt V Netík | R1021 | KX-W | 7x500lb+1x250lb | Duty Carried Out |
| Sgt J Hájek | Sgt V Soukup | R1015 | KX-L | 7x500lb+1x250lb | Duty Carried Out |
| Sgt J Doktor | Sgt H Dostál | R1718 | KX-N | 1x1000lb+5x500lb+1x250lb | Duty Carried Out |

The first of the early returns on this night was Flight Lieutenant Breitcetl and crew, who reported an inconclusive encounter with a single-engined fighter, believed to be a Bf109, 10 miles north of Dunkirk, Northern France.[31] Evasive action and an alert rear gunner, Sergeant Pavel Svoboda saved the crew, the bombload was however jettisoned. The crew landed back at R.A.F East Wretham within two hours. At 01:50hrs, the crew of Pilot Officer V Korda touched down, having bombed the airfield at Zwevezele, located 15 miles south of Brugge at 00:29hrs. They had experienced two overheating engines aboard Wellington R1598 KX-C. Another crew who experienced engine issues was Sergeant K Schoř. Unable to climb above 9,000ft, they opted to bomb the docks at Oostende at 01:11hrs. Their bombs were seen to explode in the outer harbour area.

Visibility over Düsseldorf was poor, and there was a surprising lack of flak as the first of the squadron crews bombed at 01:47hrs. Sergeant V Bufka reported no fires were visible on his arrival. He dropped his mixed load from 14,000ft, the bursts from which were seen between the Rhine River and the aiming point. At 01:53hrs Squadron Leader J Ocelka dropped his mixed load from 14,300ft. His three 500-pounders were observed exploding, and the incendiaries starting a small fire in the centre of the target. All the squadron crews that reached Düsseldorf stated that they bombed the target, but the results of their bombing were almost impossible to determine due to the clouds. 311 Squadron were safely back at R.A.F East Wretham by 04:46hrs. 3 Group H.Q were cautious with their assessment of the raid. They acknowledged that visibility over the target was poor but chose to accept what the crews reported that large fires had been started in the target area. However, there was one cautionary report. One Station Commander considered the lack of flak suspicious and that the fires seen may have been dummies. Records show that only two heavy bombs fell in the southern Düsseldorf suburb of Wersten, well south of the aiming point, destroying a dozen houses. Numerous villages and towns miles from Düsseldorf reported bombing. There must have been some nagging doubt at Bomber Command H.Q about the effectiveness of the previous night's operation. On the morning of the 17th, Düsseldorf was once again the Group's main target for that night. Cologne, would also be re-visited by Hampdens and Whitleys of 4 and 5 Groups. 311 Squadron would contribute just five crews to the total force of 59. The previous two nights of operations having taken its toll of aircraft and crews.

---

[31] *Other sources record Gravelines. Dr Theo Boiten, author and historian reports that there was no single engine fighter activity on this night.*

## June 17th 1941 : 'Perch A' – Düsseldorf

| Pilot | 2nd Pilot | Serial | Code | Bomb Load | Result |
|---|---|---|---|---|---|
| P/O V Korda | W/Cdr J Schejbal | T2972 | KX-G | 1x1000lb+5x500lb | Duty Carried Out |
| Sgt J Nyč | Sgt K Šťastný | T2990 | KX-T | 7x500lb+1x250lb | Duty Carried Out |
| Sgt J Bernát | Sgt A Musálek | R1015 | KX-L | 1x1000lb+5x500lb | Alternative Target |
| Sgt K Schoř | Sgt O Helma | R1451 | KX-P | 7x500lb+1x250lb | Duty Carried Out |
| S/Ldr J Ocelka | Sgt B Blatný | W5682 | KX-Y | 3x500lb+1x250lb+360x4lb | Duty Carried Out |

It was another night of early returns throughout the Group. Nine crews turned back for various reasons. These, coupled with the three that failed to take-off, represented a reduction in the Group's contribution of nearly 20%. One of these early returns was from 311 Squadron. Starboard engine problems aboard Wellington R1015 KX-L meant that Sergeant J Bernát and crew wisely decided to bomb the docks at Dunkirk. They were back at R.A.F East Wretham within two hours of take-off.

The squadron commander, Wing Commander J Schejbal once again joined the crew of Pilot Officer Korda. Ground haze was encountered over the target and Düsseldorf's defenders put up a steady flow of flak and searchlights. Sergeant J Nyč summed up the raid, *'Scattered fires all over the town on arrival. Bursts seen between the river and the aiming point. One large red fire started'*. All four Czech crews claimed to have bombed. The staff at 3 Group H.Q were slightly more realistic in their appraisal of the raid, they recorded the following, *'There was haze over both Dusseldorf and Hanover, but most aircraft on Dusseldorf could pick up the river and bombed near the target. Very few fires started however, and all went out quickly.'* There are no reports of damage from Düsseldorf. The squadron sat out the raid on Brest on the 18th but occupied itself with the usual round of cross-country and training flights. The threat of fog was the main reason that Group detailed just 29 Wellingtons for an all 3 Group effort against Cologne on the 19th. The Czechs at R.A.F East Wretham would provide ten crews. At briefing they were warned that there would be a good chance of being diverted on return.

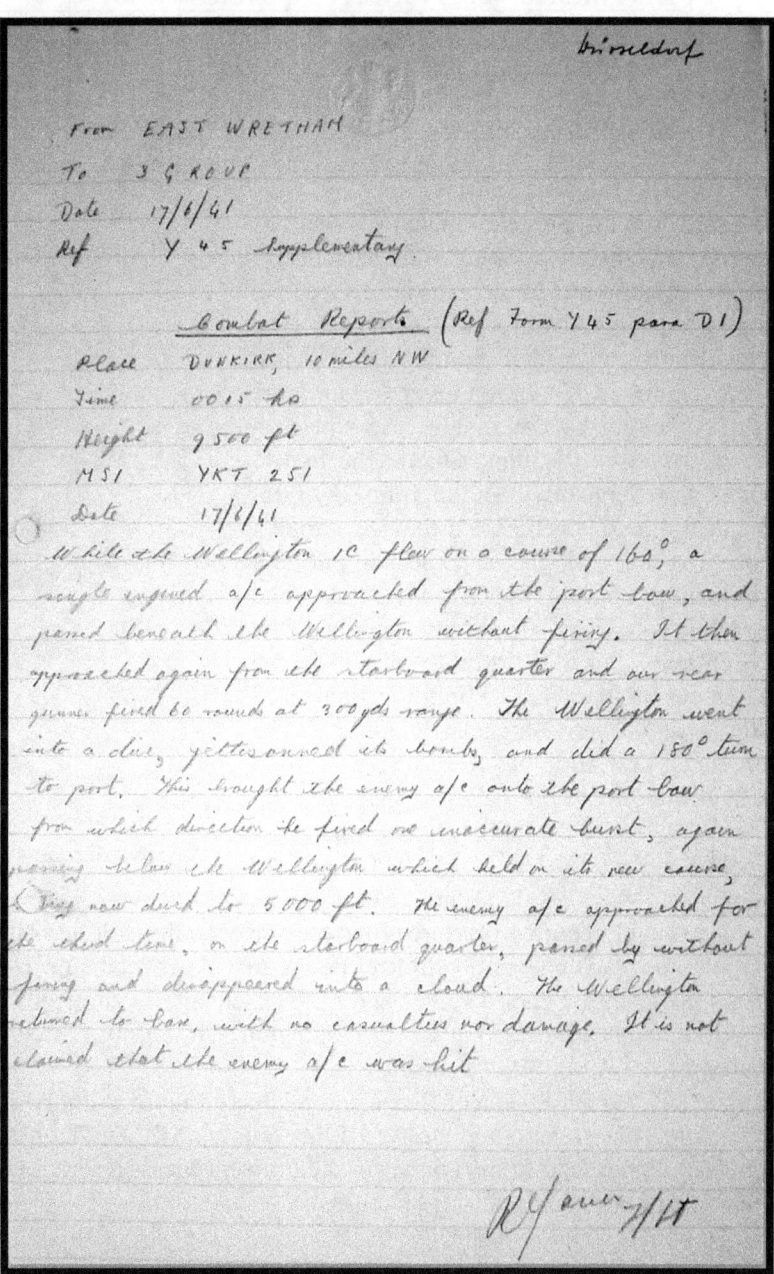

*A copy of the combat report submitted by Flight Lieutenant Breitcetl and crew recording the fighter encounter on June 16th 1941. (Pavel Vančata)*

## June 19th 1941 : 'Trout A' – Cologne

| Pilot | 2nd Pilot | Serial | Code | Bomb Load | Result |
|---|---|---|---|---|---|
| Sgt J Nyč | Sgt K Šťastný | R1718 | KX-N | 6x500lb | Duty Carried Out |
| Sgt V Bufka | Sgt A Rozum | T2990 | KX-T | 3x500lb+360x4lb | Duty Carried Out |
| Sgt F Fencl | Sgt V Netík | R1804 | KX-D | 3x500lb+360x4lb | Duty Carried Out |
| Sgt K Schoř | Sgt O Helma | R1451 | KX-P | 3x500lb+1x250lb+360x4lb | Duty Carried Out |
| F/Lt J Breitcetl | Sgt V Ryba | R1046 | KX-E | 3x500lb+360x4lb | Duty Carried Out |
| P/O V Korda | Sgt B Blatný | T2972 | KX-G | 1x1000lb+4x500lb | Duty Carried Out |
| Sgt J Bernát | Sgt A Musálek | R1532 | KX-R | 1x1000lb+4x500lb | Duty Carried Out |
| Sgt J Hájek | Sgt V Soukup | R1015 | KX-L | 1x1000lb+4x500lb | Duty Carried Out |
| Sgt A Šedivý | Sgt A Mžourek | R1371 | KX-F | 6x500lb+1x250lb | Duty Carried Out |
| Sgt J Doktor | Sgt H Dostál | R1718 | KX-N | 6x500lb+1x250lb | Early Return |

The squadron lost Sergeant J Doktor's services almost immediately after take-off. The cockpit instrumentation was found to be unserviceable, and inexplicably, the I.F.F. self-detonated! The crew landed with their bombload after just twenty minutes. Conditions were again hazy over the target as the first of the squadron crews arrived at 01:30hrs. One of the first to bomb was Sergeant V Bufka from 13,500ft. His mixed load was reported to have started a small red fire. Cologne, as expected, put up a frightening barrage of both flak and searchlights. Haze thankfully subdued the searchlights but did little to diminish the accuracy of the flak. Flight Lieutenant J Breitcetl was over Cologne at 01:34hrs, and he reported, *'first stick, no results seen, second stick yellow fires and explosions visible from 15 miles'*. Two of the last crews over the target were Sergeant J Hájek at 02:05hrs and Sergeant J Nyč at 02:10hrs. Surprisingly,

*A rare photograph showing a 311 Squadron Wellington at dispersal at RAF East Wretham. (John Costin)*

Jaroslav Hájek reported just three fires seen on arrival, while Jaroslav Nyč's report was brief, *'Bombed south-east part of town'*, he made no mention of fires on his arrival. As predicted, most crews were diverted due to fog on return. Three crews, Sergeant F Fencl, J Hájek, and Flight Lieutenant J Breitcetl, landed at R.A.F Oakington. Sergeants K Schoř and V Bufka managed to land at R.A.F Honington, while Pilot Officer V Korda and Sergeant A Šedivý found a fog-free Waterbeach.

On the 20th, Czech President Beneš and a large group of military advisors plus cameramen descended on R.A.F East Wretham. They were joined by AVM J.E.A Baldwin CB, DSO, OBE and Air Commadore A. Boyle CMG, OBE MC at a formal dinner held in his honour in the Officers Mess. It was a political visit, and little time was spent talking with members of the squadron.

*The second 311 Squadron Commanding Officer Acting Wing Commander Josef Schejbal accompanied Czechoslovak president Dr. Edvard Beneš during his visit to East Wretham on 20ᵗʰ June 20 1941. (Pavel Vančata)*

*Czechoslovak president Dr. Edvard Beneš addresses members of the 311 Squadron outside Wretham Hall on 20th June 1941. Seated in armchairs and chairs L-R: Minister of Foreign Affairs Jan Masaryk, (above him stands a guest – Sgt Stanislav Plzák, fighter pilot of 19 Squadron), gen. Antonín Hasal-Nižborský, Inspector of Czechoslovak Air Force in the RAF AVM Karel Janoušek KCB, ?, Minister of National Defence div. gen. Sergěj Ingr, Dr. Edvard Beneš, ?, plk. František Moravec, A/W/Cdr Josef Schejbal, F/O Lubomír Svátek. Sitting on the ground in front of AVM Janoušek KCB L-R: Pilots Sgt Leo Anderle, Sgt Alois Šedivý, Sgt Oldřich Helma, ?, Sgt Josef Bernát. (Pavel Vančata)*

*Deep in conversation. Acting Squadron Leader Ocelka with 3 Group Commanding Officer AVM J. E. A. Baldwin CB, DSO, OBE, outside Wretham Hall on 20 June 1941. RAF Station Honington Commanding Officer A/G/Cpt John Astley Gray DFC stands between them in the background. (Jaroslav Popelka)*

After a two day break, the squadron was ordered to prepare nine crews for a small scale raid directed against Bremen on the 22$^{nd}$. 3 Group would offer 43 Wellingtons which would be joined by Hampdens of 5 Group.

## June 22nd 1941 : Bremen

| Pilot | 2nd Pilot | Serial | Code | Bomb Load | Result |
|---|---|---|---|---|---|
| F/Lt J Breitcetl | Sgt V Ryba | R1021 | KX-W | 1x1000lb+4x500lb | Duty Carried Out |
| F/Sgt F Fencl | Sgt V Netík | R1804 | KX-D | 3x500lb+360x4lb | Duty Carried Out |
| F/Sgt A Šedivý | Sgt A Mžourek | R1371 | KX-F | 1x1000lb+4x500lb | Duty Carried Out |
| P/O V Korda | Sgt J Horáček | R1598 | KX-C | 3x500lb+360x4lb | Duty Carried Out |
| F/Sgt V Bufka | F/Sgt A Rozum | T2990 | KX-T | 3x500lb+360x4lb | MISSING |
| F/Sgt J Nyč | Sgt K Šťastný | R1718 | KX-N | 6x500lb | Alternative Targets |
| F/Sgt J Hájek | Sgt H Dostál | R1015 | KX-L | 1x1000lb+4x500lb | Duty Carried Out |
| F/Sgt K Schoř | Sgt O Helma | R1451 | KX-P | 6x500lb | Alternative Targets |
| Sgt B Blatný | Sgt A Musálek | W5682 | KX-Y | 3x500lb+360x4lb | Alternative Targets |

The customary Wellington withdrawals just before take-off due to mechanical issues were recorded, as were seven early returns from the 3 Group contribution. One of which was from 311 Squadron.

Conditions over the airfield were generally clear as the Wellingtons slowly gained altitude before crossing the English coast. The small force passed Den Helder, which offered its usual welcome of flak and searchlights and headed towards Bremen. Flight Sergeant K Schoř had nursed his starboard engine across the North Sea. Deciding that to continue on to Bremen was pushing their luck, the crew bombed Emden at 01:25hrs. Another crew who bombed Emden was Sergeant B Blatný. They also reported engine trouble but managed to drop their bombload from 10,500ft in the centre of the city, creating a fire. The squadron was over Bremen between 01:25hrs and 02:03hrs. Above the target, the sky was filled with thousands of heavy flak bursts. If this was bad enough, below the heavy flak were streams of vicious multi-coloured light flak, hosing the sky in almost every direction. Two Wellingtons were hit. Pilot Officer K Korda's Wellington R1598 KX-C was badly holed as they bombed at 01:50hrs, thankfully without injury to the crew. The luckiest crew was that of Flight Sergeant J Nyč. Unable to identify the aiming point due to haze, they bombed on E.T.A and flak flashes below. They were unlucky, a searchlight beam coned them and subjected the crew to a barrage of flak. It was only the skill of both pilots that saved the crew. Violent evasive action was taken, which thankfully had the desired result, but not before Wellington R1718 KX-N suffered severe damage. Those that identified the target reported a number of fires a mile southeast of aiming point 'B' which Flight Sergeant Fencl reported could be seen from 35 miles. It was on the return route that tragedy struck when Flight Sergeant V Bufka and crew were intercepted by Oblt Egmont Prinz zur Lippe-Weissenfeld of 4./NJG1 flying at Bf110 from Bergen airfield. All the crew apart from the pilot were killed when the blazing Wellington T2990 KX-T ploughed deep into the soft ground in the Kostverloren Polder at Nieuwe Niedorp (Noord-Holland), 10 miles northeast of Alkmaar at 02:14hrs. The crew would remain buried in the wreckage until 2021. Given the depth the Wellington buried itself into the soil, no recovery was attempted by the Germans. The local villagers would eventually adopt the crash site. A memorial mound of earth with shrubs surmounted by an 8ft cross marked the final resting place of the crew. The impact area was surrounded by green painted timber posts with heavy iron chains. It would be classified as a field grave post-war. The crew's names would be added to the memorial to the missing. The following is an abbreviated version of an interview Warrant Officer V Bufka gave in 1945.

*'The attacker fired several long rounds. The aircraft was hit and caught fire. From the struggle of the 2nd pilot it was obvious to me, that he could no longer control the plane which started descending fast. I gave the order to abandon the stricken aircraft via the Intercom but received no acknowledgement. I decided to check on the W/O and the navigator but could not reach them for the raging flames and heat. I returned to the cockpit where the 2nd pilot was struggling to get out of the plane which was falling in a spiral. I also noticed the open hatch into the front turret but could not see Hejna anywhere. I realised that I had to get out fast. The 2nd pilot was half-way out of the plane at that time. I put on a parachute but could not exit the aircraft because of the draught. I opened the chute inside and threw it out of the hole. The last thing I remember was a knock on my head. I woke up in hospital two days later. I had a broken leg and several cuts on my head and torso'.*

In the panic and chaos of the final minutes of Wellington T2990, Vilém Bufka reported that two of his crew were successful in parachuting out of the blazing Wellington. Sadly none would survive. Whilst serving with 311 Squadron, the rear gunner, Karel Valach met an English girl Doreen Francis Todd and they were married on December 10th 1940, in the Roman Catholic Church in Bury St Edmunds, where she lived. Doreen gave birth to a son in December 1941, and she named him Karel, after his father. Doreen had previously been romantically involved with Vilém Bufka, but while in hospital following a motorcycle accident he made the mistake of asking his rear gunner to find Doreen and explain he was safe and in hospital. His, gunner however obviously had other intentions!

While in his hospital bed, Flight Sergeant Bufka was visited by Oblt Egmont Prinz zur Lippe-Weissenfeld who talked with him about the encounter.

*Sergeant Vilém Bufka started tour of operations on 27th December 1940 as second pilot of Flight Lieutenant Josef Šnajdr's crew bombing Le Havre. He flew as a Skipper on 2nd and 16th January 1941 but while returning from the second raid, he suffered frostbites and had to be hospitalized. Returning to the squadron, he flew the first two sorties on 23rd and 26th April again as second pilot before taking over the crew of Sergeant Leo Anderle on 3rd May. He failed to return from his sixteenth raid and he was the only survivor from Wellington Mk.IC T2990/KX-T. (Pavel Vančata)*

## Vickers Wellington Mk.IC T2990 KX-T

| Manufacturer | Vickers Armstrong | |
|---|---|---|
| Contract | 38600/39 | |
| Taken on Charge | 18/03/1941 via No. 38 MU | |
| Cat E Missing | 18/06/1941 | |
| Struck Of Charge | N/K | |
| Total Flying Hours | N/K | |
| Take-Off Time | 23:16hrs | |
| Bomb Load | 3x500lb+360x4lb IB | |
| | **CREW** | |
| Captain | Flight Sergeant Vilém **Bufka** 787572 RAFVR. Age 25. | PoW |
| Second Pilot | Flight Sergeant Alois **Rozum** 787169 RAFVR. Age 28. | Coll. grave 24. AA. 1-5. |
| Navigator | Pilot Officer Vilém **Konštacký** 82608 RAFVR. Age 26. | Coll. grave 24. AA. 1-5. |
| Wireless Operator | Pilot Officer Leonard **Smrček** 82639 RAFVR. Age 25. | Coll. grave 24. AA. 1-5. |
| Front Gunner | Flight Sergeant Jan **Hejna** 787204 RAFVR. Age 26. | Coll. grave 24. AA. 1-5. |
| Rear Gunner | Flight Sergeant Karel **Valach** 787551 RAFVR. Age 23. | Coll. grave 24. AA. 1-5. |
| Posting History | F/Sgt V Bufka Unknown<br>F/Sgt A Rozum via Cosford Depot, 29/07/1940 | |
| Operations Flown | F/Sgt V Bufka : 14 / F/Sgt A Rozum : 12 | |
| Buried | BERGEN-OP-ZOOM WAR CEMETERY | |

### Miss Todd's Wedding.

Photo: H. I. Jarman.

Principals in a wedding at Bury St. Edmund's Roman Catholic Church reported last week: Air Gunner Sergt. Charles Valach, D.F.M., and Miss Doreen Frances Todd, elder daughter of Mr. and Mrs. Rowland G. Todd. Langton House. Bury St. Edmund's.

### Made Escape in German Plane

A Czech airman, who has had remarkable adventures, was the bridegroom in an interesting wedding at Bury St. Edmund's, when Air-Gunner Sergt. Karel Valach, D.F.M., of a Czech Squadron serving with the R.A.F., was married to Miss Doreen Frances Todd, elder daughter of Mr. and Mrs. Rowland G. Todd. Langton House, Bury St. Edmund's.

Sergt. Valach was trained as an air-gunner and ground engineer at a Czech flying school. After the German occupation of Czecho-Slovakia in June, 1939, he seized a German 'plane with which he escaped to Poland, where he joined the Polish Air Force. As an air-gunner, he took part in many operational flights during the German-Polish war, and on one occasion he was shot down by the Germans. When Poland was occupied by the Nazis, he escaped again—this time to Russia, whence he eventually made his way to France, having a hazardous journey through South-East Europe. At Lyons he rejoined the Polish Air Force.

Upon the defection of France he evacuated to North Africa, from which country he was able to reach Britain and join the R.A.F. He has taken part in several operational flights over Germany, and has been awarded the Czecho-Slovak Distinguished Flying Medal.

*Happier times. A newspaper photograph reporting the wedding of Sergeant Valach and his wife Doreen and the accompanying article. (Author)*

*The dramatic end of Wellington T2990 KX-T. The aircraft came down at 0213hrs at near the village of Nieuwe Niedorp north east of Alkmaar. Vilem Bufka was taken prisoner of war. Right : Crash site.*

*Recovered wreckage of Wellington T2990. Crumpled geodetics and the dinghy which clearly shows the aircraft serial number (right).*

In June 2021, the wreckage of Vickers Wellington T2990 was finally recovered from a meadow in the community of Hollands Kroon (the Netherlands). An expert team from the Royal Netherlands Armed Forces carried out the excavation and during the work human remains were found, and it has since been established that these belong to the five missing crewmembers. The national program for the recovery of aircraft wrecks with missing crewmembers was established in 2018 after an initiative of the Dutch Air War Study Group 1939-1945. Its aims are simple, to recover all aircraft wrecks which may still hold the remains of missing crewmembers and whose recovery is classified as 'favourable'. The Dutch government bears the costs of these recoveries, and they should be praised and thanked for this wonderful gesture.

The last of the squadron crews to land was Flight Lieutenant J Breitcetl at 05:15hrs, well over his expected return time. There was no initial concern about the crew of Flight Sergeant V Bufka. He could have landed away and suffered W/T failure, it happened all the time. However, as the morning stretched on and all the enquiries drew a blank, the sad realisation that the crew were missing swept the station. It had been almost nine weeks since the squadron lost a crew on operations, it was inevitable the run of luck would end, nevertheless, it still hit the whole squadron hard. It would not be until a Telegram from the International Red Cross eventually arrived was the crew's fate known. German reports stated that Bufka had survived to become a PoW. Sadly, the squadron also learnt that his crew had all been killed.

There was some good news, and it came at the best possible time. Four crews on the Training Flight were considered suitably trained and sufficiently skilled to take their place on the Operations Flight. There was no flying on the 23rd, operationally or training-wise. On the 24th, the AOC 3 Group AVM. J.E.A Baldwin CB, DSO, OBE presented the DFC to Squadron Leader Josef Ocelka and the MBE to Vladimír Nedvěd. It was an incredibly proud squadron that stood watching as the medals were pinned to the chests of their friends. On the 25th, 3 Group H.Q detailed 39 Wellington crews for a return trip to Bremen, 311 Squadron allocating seven crews.

## June 25th 1941 : 'Salmon B' – Bremen

| Pilot | 2nd Pilot | Serial | Code | Bomb Load | Result |
|---|---|---|---|---|---|
| F/Sgt J Nyč | Sgt K Šťastný | W5682 | KX-Y | 2x500lb+1x250lb+360x4lb | Alternative target |
| F/Sgt F Fencl | Sgt V Netík | R1804 | KX-D | 3x500lb+360x4lb | Jettisoned |
| F/Sgt J Čapka | F/Lt J Šejbl | R1598 | KX-C | 3x500lb+360x4lb | Jettisoned |
| F/Sgt K Schoř | Sgt O Helma | R1451 | KX-P | 6x500lb | Duty Carried Out |
| F/Lt J Breitcetl | Sgt V Ryba | R1021 | KX-W | 6x500lb+1x250lb | Jettisoned |
| F/Sgt J Bernát | Sgt A Musálek | R1532 | KX-R | 1x1000lb+4x500lb | Alternative target |
| F/Sgt J Hájek | Sgt V Soukup | R1015 | KX-L | 1x1000lb+4x500lb | Jettisoned |

Little did the crews realise as they climbed for altitude that the night would be a battle of survival against the elements. Over the North Sea, the squadron encountered a violent electrical storm and severe icing. Two Wellingtons, skippered by Flight Sergeant J Čapka and Flight Sergeant J Hájek, were both struck by lightning forcing each to jettison. Flight Sergeant Fencl's cockpit instruments were severely affected on reaching the Dutch coast due to the electrical storm, static causing the instruments to go haywire aboard the Wellington. The crew of Flight Lieutenant J Breitcetl lost an engine due to icing and jettisoned his bombload 5 miles northwest of Bramsche, Germany. The weather got the better of Flight Sergeant J Nyč who bombed Emden at 01:25hrs, no results were observed. Flight Sergeant J Bernát and crew were unable to reach Bremen. They bombed flak and searchlight concentrations along the coast between Oostende and Dunkirk at 02:10hrs. Only the crew of Flight Sergeant K Schoř claimed to have reached and bombed the target. They bombed at 01:38hrs from 16,200ft on E.T.A and a combination of flak bursts and searchlights. Somehow, all the squadron returned safely. Of the 39 crews detailed, a total of 18 returned early, and only seven crews reached and possibly bombed Bremen. It had been a harrowing night and questions were asked why the raid went ahead. There followed a general stand down on the squadron from operations for the remainder of the month. 3 Group operated against Bremen again on the 27th and 29th, but the Czechs were not detailed. Training was again the main activity, and this would result in the month's final accident. On the afternoon of June 28th, Sergeant Alois Šiška began to taxi Wellington N2775 but had insufficient brake pressure resulting in the Wellington hitting a hedge at East Wretham.

| Serial | N2775 |
|---|---|
| Code | KX-T |
| Taken On Charge | 06/05/1941 via 149 Squadron. |
| Manufacturer | Vickers Armstrong (Chester). |
| Contract | B.124362/40 |
| Pilot (s) | Sergeant Alois Šiška |
| Flight | Training |
| Time | 12:15hrs |
| Cause | Hit hedge |

Damage must have been sufficiently bad to require the attention of No.43 Group. The Wellington returned to see service with the Czech Training Flight on July 3rd. On the 30th, the Operations Record Book reports the appointment of Squadron Leader Josef Ocelka DFC as the new squadron commander vice Wing Commander J Schejbal. He was a popular choice with everybody on the squadron. Courageous and respected, he was admired and liked by all ranks. On 3rd July 1941, Wing Commander J Schejbal would be posted to the C.I.G in London and became an organisation staff officer (Wing Commander post). On 28th October 1941, he was promoted to the rank of Acting Group Captain and became the Deputy Inspector General.

# July 1941 : Keeping the Pressure On!

Up early on July 1st the ground crews trudged to their wet and windy dispersals. Until daybreak, thunderstorms had battered the East Anglian airfields. It was not a promising start to a new month and the British summer. The Czechs along with 9, 99 and 149 Squadron would be airborne that night. The Honington pairing of 9 and 311 were given *Prinz Eugen,* while 149 would bomb *Scharnhorst,* finally, 99 Squadron would attack the *Gneisenau.* Four 'Freshmen' crews from 311 would also be active, they were ordered to bomb the docks at Cherbourg.

### July 1st 1941 : 'Skunk' – *Prinz Eugen* – Brest

| Pilot | 2nd Pilot | Serial | Code | Bomb Load | Result |
|---|---|---|---|---|---|
| P/O V Korda | Sgt J Horáček | R1015 | KX-L | 1x1000lb+4x500lb+1x250lb | Duty Carried Out |
| F/Sgt F Fencl | Sgt V Netík | R1804 | KX-D | 1x1000lb+4x500lb+1x250lb | Duty Carried Out |
| F/Sgt A Šedivý | Sgt A Mžourek | R1371 | KX-F | 1x1000lb+4x500lb+1x250lb | Duty Carried Out |
| F/Sgt J Bernát | Sgt S Linka | R1532 | KX-R | 7x500lb | Duty Carried Out |
| Sgt A Musálek | Sgt V Ryba | W5682 | KX-Y | 7x500lb | Duty Carried Out |
| F/Lt J Šejbl | Sgt M Šebela | X3221 | KX-O | 7x500lb | Duty Carried Out |
| F/Sgt K Schoř | Sgt M Jindra | R1451 | KX-P | 7x500lb | Duty Carried Out |
| F/O F Pohlodek | Sgt O Jambor | T2561 | KX-A | 7x500lb | Duty Carried Out |

The night would be remarkable for a number of reasons. Seven pilots would be making their operational debut, four against Brest. Sergeants Stanislav Linka, Miroslav Jindra, Oldřich Jambor and Metoděj Šebela. Šebela's route to 311 Squadron was long and precarious. He learnt to fly privately at an aero club before he enlisted in the Air Force, where he completed training as a pilot of two-seater aircraft. He served in the 8th Observation Squadron of the 2nd Air Regiment during the mobilisation, flying the Czech built Letov Š-328. One of the few who stayed after the occupation, he would briefly serve the Protectorate Government Army. Disillusioned, he was among the last to escape from his occupied homeland in May 1940. His route to England and the R.A.F was long and took him through Slovakia and Hungary to Yugoslavia, where at the end of May, he was drafted into the Czechoslovak Foreign Army at the French Consulate in Belgrade. There followed a journey through Lebanon to Palestine, where on June 29th 1940, he was presented to the Czechoslovak Infantry Regiment in the Az Sumeiriya camp. With the urgent need for trained pilots in England, he left with a small group of other airmen on September 10th, 1940. They set out on the *Reina del Pacifico* from Suez via Aden, Port Elizabeth, Cape Town, Freetown, and Gibraltar. They eventually arrived in Liverpool on October 25th 1940.

Apart from the debut pilots, two others would be carrying out their first as captains after learning their trade sitting beside experienced skippers. The recently promoted Flight Lieutenant Josef Šejbl had completed 21 operations, the majority with Josef Čapka, while Sergeant A Musálek had flown 13. The first to depart were the eight crews attacking Brest. The weather over the airfield and to the target was fortunately better than forecast. The crews departed over the Dorset coast and headed towards Brittany. Arriving over the target between 12,000ft and 16,000ft, they could just pick out the docks below. At briefing, the crews had been told that the *Prinz Eugen* was moored in Dock 8. On this night, the usually reliable smoke screen was ineffective, giving the crews an ideal opportunity to inflict some damage, or if their luck was in destroy the heavy cruiser. Brest was strangely quiet as the first of the squadron's bombs started to fall around 01:00hrs. Flight Sergeant A Šedivý was one of the first. He reported no fires, just flak as he dropped his entire load in one stick which were seen to burst across Dock 5,6 and 8. Another early crew was that of Pilot Officer V Korda, who later claimed that their 1000-pounder could have struck the *Prinz Eugen*. One of the last to bomb at 02:16hrs was Flight Sergeant K Schoř. They reported on return, *'Bombed west of Dock 1 in long stick, all bursts seen'.* Back at de-briefing at East Wretham it was evident the vast majority of the bombs had landed on Dock 6 and Dock 8. It was later established that *Prinz Eugen* was struck by an armour-piercing bomb that destroyed the control centre under the bridge. The explosion killed 60 men and wounded more than 40 others. The loss of the control centre rendered the main guns useless and repairs lasted until the end of the year.

The following was written by John Gellner, navigator to Pilot Officer V Korda about this raid.

*'On Tuesday, July 1, I took part in an air raid on Brest, which will forever remain memorable for me. Our old 2972 is still under repair in Waterbeach, so we flew on 1015. Crew: P/O Korda – Sgt Horáček – me – P/O Parolek – Sgt Valeš – Sgt Kovařík. The night was clear, and the full-grown moon shone on our way. We had to be the first to reach our destination, because Brest was completely silent when we arrived. It was only after we were above the city that the searchlights came on and the batteries started firing. Our destination was the heavy cruiser Prinz Eugen anchored on the east side of the pier at dock No. 8. One of Brest's means of defence was a smoke screen, but the south wind was unfavourable to the Germans, blowing the smoke from the generators over the mainland and leaving the harbour clean. We were very careful this time, and after a short cruising over the target to determine the exact position of Dock No. 8, we carried out a raid from west to east along the coast. The cruiser itself could not be seen against the dark surface of the water, but we did receive a direct hit from our 1,000-pound (454 kg) bomb at Dock No. 8, and the end of our line of 500-pound (227 kg) bombs could have hit the ship, given the direction of flight and the spacing between the bombs. This is more than likely, as we have seen all the bombs that follow at 1,000 pounds explode on land. Of course, when we returned, we did not report that we had hit the ship, we only reported a direct hit on the dock and other details that made us believe we might have hit it. Two days later, the airfield commander (G/Cpt Gray) called and conveyed to us the congratulations of the headquarters on hitting Prinz Eugen. The ship was about to leave the harbour the next morning and was hit in the stern at the exact time of our bombardment, and the rudder and one screw were damaged. A later more detailed report spoke of two strikes that killed a total of fifty people aboard the ship. Brest turned out to be a real mousetrap for German warships, and although the fact that we hit the ship is only conjecture and not fully confirmed, I am glad that the trap also snapped on Prinz Eugen'.*[32]

<u>July 1st 1941 : CC.16 – Cherbourg</u>

| Pilot | 2nd Pilot | Serial | Code | Bomb Load | Result |
|---|---|---|---|---|---|
| F/O K Vildomec | Sgt V Procházka | R1718 | KX-N | 6x500lb+120x4lb | Duty Carried Out |
| F/Sgt J Čapka | F/O J Stránský | R1021 | KX-W | 6x500lb+120x4lb | Duty Carried Out |
| Sgt B Blatný | P/O J Nejezchleba | W5711 | KX-H | 6x500lb+120x4lb | Duty Carried Out |
| Sgt O Helma | Sgt A Plocek | R1516 | KX-U | 6x500lb+120x4lb | Duty Carried Out / Crashed |

The Cherbourg crews departed immediately behind their colleagues attacking Brest. They too headed for the rugged Dorset coast before the long sea crossing. Three new pilots would be sitting beside experienced captains for operational experience on this raid, Sergeant Václav Procházka, Sergeant Antonín Plocek and Pilot Officer Josef Nejezchleba. The flight to the Brittany coast was completed without incident in what turned out to be a beautiful clear night.

Flying Officer K Vildomec arrived over the docks to find the flak defences already in full swing on his arrival at 00:30hrs. The crew dropped their mixed load in one stick from 15,000ft and observed them explode across the outer harbour. They were followed by Sergeant B Blatný, whose bombs also exploded across the outer harbour. Arriving at 00:45hrs, the crew of Flight Sergeant J Čapka were rocked by flak and harassed by searchlights but were successful in dropping their bombload across Dock 8. On the return trip, the crew's rear gunner reported, *'Four large fires and explosions red in colour'*. These fires were reported to be still visible by Pilot Officer Josef Nejezchleba on reaching the English coast. The crew of Sergeant O Helma would have been looking forward to their operational breakfast and some sleep as they crossed the Wiltshire coast at 16,000ft. But, unbeknown to the crew, they were being stalked by a Bristol Beaufighter Mk.IF T4638 NG-F from No. 604 (County of Middlesex) Squadron from R.A.F Middle Wallop and flown by the Rhodesian squadron commanding officer, Wing Commander Henry Appleton DFC and his radar operator, Pilot Officer Derek Jackson. On patrol since 23:30hrs, the Beaufighter closed in on the unsuspecting Czechs. Flying at 12,000ft, well above the usual altitude for returning bomber aircraft, Wing Commander Appleton DFC positioned

---

[32] *Taken via Wikipedia page.*

himself behind the lumbering Wellington. It has been reported that the I.F.F. and wireless set aboard Wellington R1516 KX-U were both unserviceable, which would have contributed to the tragic events that were about to unfold. Satisfied that the bomber was hostile, Wing Commander Appleton DFC opened fire at an unidentified aircraft at 01.46hrs on July 2nd. The crew's fate was sealed as the 20mm cannons tore into the Wellington's geodetics, tearing large chunks away. The crew, taken entirely unawares, had little, if any, time to react. The Wellington crashed at Lower Mere Part Farm, 3 miles southeast of Mere Village, killing all the crew. The Wellington disintegrated in the air with the wreckage was spread over a radius of ¾ mile. The main fuselage burst into flames while the engines buried themselves in the ground. A telegram from nearby R.A.F Station Old Sarum sent at 04.30hrs informing the Air Ministry and No. 311 Squadron about the Wellington's crash, stating that it was *'probably shot down by another aircraft as there is evidence of machine gun fire in the air immediately before the crash'*.

The rear turret appeared to have broken off in the air and was initially unaccounted for. It was discovered by the farmer's son only the next day by chance while checking the cattle on the other bank of the river. Still strapped inside the rear turret was Sergeant Lančík. Wing Commander Appleton[33] DFC landed at 02:05hrs. It is not known when Wing Commander Appleton realised his tragic error. The following afternoon, a distraught Charles Appleton flew to R.A.F Honington and met with the Station Commander and members of 311 Squadron. The remains of Sergeant Lančík were cremated in the crematorium in Southampton. His father, former mayor of Moravian town Přerov, was serving with the Czechoslovak Army in Leamington Spa, wanted to take the urn home after the war. The R.A.F Bomber Command's monthly air accident summary lists the loss of Wellington R1516/KX-U among avoidable incidents, *'Aircraft shot down by British fighter when flying at 12,000ft over Britain without Resin Lights contrary to regulations'*.

<u>Vickers Wellington Mk.IC R1516 - KX-U</u> (The Broughton Wellington)

| Manufacturer | Vicker Armstrong | |
|---|---|---|
| Contract | 992424/39 | |
| Taken on Charge | 26/03/1941 via No. 24 MU. | |
| Cat E Missing | N/A | |
| Struck Of Charge | N/K | |
| Total Flying Hours | N/K | |
| Take-Off Time | 22:02hrs | |
| Bomb Load | 6x500lb+120x4lb IB | |
| | **CREW** | |
| Captain | Sergeant Oldřich **Helma** 787190 RAFVR. Age 25. | Sec. 6. Coll. grave 121 |
| Second Pilot | Sergeant Antonín **Plocek** 787355 RAFVR. Age 19. | Sec. 6. grave 122. |
| Navigator | Pilot Officer Richard **Hapala** 82601 RAFVR. Age 23. | Sec. 6. grave 123. |
| Wireless Operator | Sergeant Adolf **Dolejš** 787820 RAFVR. Age 28. | Sec. 6. Coll. grave 121 |
| Front Gunner | Sergeant Jaroslav **Petrucha** 787873 RAFVR. Age 21. | Sec. 6. Coll. grave 121 |
| Rear Gunner | Sergeant Jaroslav **Lančík** 787859 RAFVR. Age 20. | CREMATED - ASHES REPATRIATED TO CZECHOSLOVAKIA. |
| | | |
| Posting History | Sgt O Helma posted via 11 O.T.U, 11/10/1940. Sgt A Plocek posted via RAF Hendon, 14/03/1941. | |
| Operations Flown | Sgt O Helma 18 / Sgt A Plocek 0 | |
| Buried | SALISBURY (DEVIZES ROAD) CEMETERY TOWN CEMETERY IN HULÍN | |

---

[33] *Wing Commander Appleton DFC would continue to operate, he would loss a leg in a crash while operating over North Africa, but return to operations reaching the rank of Group Captain. He was killed flying a Hawker Typhoon ZY-G MN928 in August 1944, shot down by flak near Flers, France. Group Captain Appleton OBE, DSO, DFC MiD, Czech Military Cross.*

Five of the crew would be buried at 15:00hrs on Monday 7th, at the Devizes Road Cemetery, attended by members of the squadron. Sergeant Helma's operations on 311 Squadron began on March 1st 1941. On April 9th, he suffered severe frostbite to both hands. He returned to operations on May 16th, but he returned early on his first two raids. He then joined the crew of Sergeant Schoř in the second pilot role, where he completed a further seven operations. He returned to the captain's seat on July 1st.

*A poor quality photograph of Wellington R1516 KX-U. This particular Wellington was adorned with 'The Broughton Wellington'. (John Costin)*

Flying his first operation since his arrival on the squadron back in March 1941 was Sergeant Antonín Plocek. He was already an experienced aviator, having amassed 1,150 flying hours in the Czechoslovak Air Force. Like most, he escaped to France, and on the French surrender, he sailed from Bordeaux aboard the Polish cargo ship *'Robur 3'*, arriving at Falmouth on June 22nd 1940. The squadron came close to losing another crew when Flying Officer F Pohlodek ventured over Coventry[34] and hit a balloon cable at 04:40hrs, but luck was on their side, they survived and managed to land at Granthan. Why they were over a heavily defended town like Coventry is unclear, but navigational error was the probable cause.

On the 3rd, Form B.543 arrived. The squadron would provide six crews for a raid on Essen, plus three 'Freshmen' crews would also be airborne. The squadron notched up its first accident of the month on the late afternoon of the 3rd. Sergeant Bohuslav Hradil and crew were airborne on a practice flight aboard Wellington N2775 KX-T. On returning to R.A.F East Wretham, the pilot made a rather heavy landing, resulting in the Wellingtons undercarriage collapsing.

---

[34] *No.3 Group HQ Records Book reports incident near Rugby.*

| Date | 03/07/1941 |
|---|---|
| Mark | Mk.IC |
| Serial | N2775 |
| Code | KX-T |
| Taken On Charge | 06/05/1941 via 149 Squadron. |
| Manufacturer | Vickers Armstrong (Chester) |
| Contract | B.124362/40 |
| Pilot (s) | Sergeant Bohuslav Hradil |
| Flight | Training, war-load climb. |
| Time | 16:00hrs |
| Cause | Heavy Landing |

This was the second accident involving N2775 in less than a month, after it had only returned from repair after hitting a hedge on May 6th. The crew were shaken but survived.

### July 3rd 1941 : GQ1838 – Krupps AG Salzer & Amalle Mine – Essen

| Pilot | 2nd Pilot | Serial | Code | Bomb Load | Result |
|---|---|---|---|---|---|
| S/Ldr J Ocelka | Sgt V Netík | W5682 | KX-Y | 3x500lb+1x250lb+360x4lb | Duty Carried Out |
| F/Sgt J Bernát | Sgt S Linka | R1532 | KX-R | 7x500lb+1x250lb | Alternative Target |
| F/Lt J Šejbl | Sgt M Šebela | R1718 | KX-N | 1x1000lb+5x500lb | Duty Carried Out |
| F/Sgt K Schoř | Sgt M Jindra | R1451 | KX-P | 1x1000lb+5x500lb | Duty Carried Out |
| Sgt A Musálek | Sgt V Ryba | R1598 | KX-C | 3x500lb+1x250lb+360x4lb | Duty Carried Out |

A force of 68 3 Group Wellington crews were detailed to attack Krupps. Seven would fail to take-off for various reasons. Two of 311 Squadron Wellington had a slight collision while taxying for take-off which resulted in a reduction in the number of both forces: Flying Officer F Pohlodek X3221 KX-O (Essen) and Flight Sergeant J Čapka R1021 KX-W (Gilze-Rijen aerodrome). Squadron Leader J Ocelka was the first away from East Wretham at 23:32hrs. Various mechanical issues would plague him and another crew. Squadron Leader J Ocelka lost his intercom soon after take-off while engine issues aboard Wellington R1532 KX-R meant that Sergeant J Bernát could not climb above 8000ft. Wisely, the crew opted for a slightly safer target, and they dropped their bombload at 01:35hrs on the Noordzee Canal northwest of Amsterdam, starting three fires which quickly spread to create one large blaze with billowing black smoke. Squadron Leader Ocelka decided to continue on to Essen. Weather conditions en route and over the target were cloudless, but there was considerable ground haze over the Ruhr. The four crews were over Essen between 01:33hrs and 01:48hrs. Each claimed to have bombed the target and started small fires.

Only half the attacking force claimed to have bombed Essen. Those who did were generally confident that the operation was a success. Yet, local sources reported light housing damage, while Duisburg, Bochum, Dortmund, Wuppertal, and Hagen also reported bomb damage.

### July 3rd 1941 : Gilze Rijen aerodrome

| Pilot | 2nd Pilot | Serial | Code | Bomb Load | Result |
|---|---|---|---|---|---|
| F/O K Vildomec | Sgt V Procházka | R1371 | KX-F | 16x250lb | Duty Carried Out |
| Sgt B Blatný | F/O J Nejezchleba | W5711 | KX-H | 16x250lb | Duty Carried Out |

3 Group despatched seven 'Freshmen' crews to Gilze-Rijen aerodrome in southern Holland. The two crews supplied by 311 Squadron had flown to Cherbourg on the 1st. Sergeant B Blatný was the first to arrive at 00:50hrs and encountered dense mist in the area of the airfield. There is some confusion about what happened next. The Squadron ORB records the crew seeking an alternative target, the Eindhoven airfield, which they bombed with unseen results. 3 Group records

and the 'Form E' state the crew bombed the primary from 13,000ft. There was no confusion about Flying Officer K Vildomec and crew. They arrived at Gilze-Rijen aerodrome at 01:20hrs. Aided by a flashing beacon and boundary lights, the crew dropped their first stick of bombs, which were seen to fall just north of the airfield. The second stick landed across the airfield and close to some buildings. Ground mist prevented the crew from making an accurate observation of the results. The weather on the 4th, low clouds, drizzle, and fog, greatly restricted any flying throughout the day. With an improvement in conditions, a number of crews were over Langham Ranges on Air-Firing practice on the 5th. While they were honing their gunnery skills, the squadron ground crews were preparing ten Wellingtons for operations that night.

### July 5th 1941 : Rudd 'A' – Munster

| Pilot | 2nd Pilot | Serial | Code | Bomb Load | Result |
|---|---|---|---|---|---|
| P/O V Korda | Sgt J Horáček | W5682 | KX-Y | 3x1000lb+1x250lb+360x4lb | Duty Carried Out |
| F/Sgt F Fencl | Sgt V Netík | R1804 | KX-D | 3x1000lb+1x250lb+360x4lb | Duty Carried Out |
| F/Sgt K Schoř | Sgt M Jindra | R1451 | KX-P | 1x1000lb+5x500lb+1x250lb | Duty Carried Out |
| F/Sgt A Šedivý | Sgt A Mžourek | R1371 | KX-F | 1x1000lb+5x500lb+1x250lb | Duty Carried Out |
| F/Lt J Šejbl | Sgt M Šebela | R1598 | KX-C | 1x1000lb+5x500lb+1x250lb | Duty Carried Out |
| F/Sgt J Bernát | Sgt S Linka | R1532 | KX-R | 8x500lb | Duty Carried Out |
| F/O F Pohlodek | Sgt O Jambor | R1458 | KX-Z | 8x500lb | Duty Carried Out |
| Sgt A Musálek | Sgt V Ryba | R1718 | KX-N | 8x500lb | Duty Carried Out |
| Sgt B Blatný | P/O J Nejezchleba | W5711 | KX-H | 7x500lb+1x250lb | Duty Carried Out |
| F/O K Vildomec | Sgt V Procházka | R1015 | KX-L | 7x500lb+1x250lb | Duty Carried Out |

The Group detailed a total of 66 Wellingtons for the operation, of which three failed to take-off and four returned early with various mechanical issues. There were no such issues with 311 Squadron, who reported a clean sheet. The first of the crews arrived at 01:15hrs to find Munster clear of clouds and abnormally quiet. Flight Sergeants F Fencl and K Schoř were the first to bomb, and they reported that their mixed loads landed close to the aiming point, starting fires. Flight Sergeant K Schoř was one of the few crews reporting flak damage. A burst had punched a large hole in the starboard wing, holed a fuel tank, smashed the flaps and punctured the tyre, but somehow missed the engine. Both Sergeant B Blatný and Flying Officer K Vildomec bombed at 01:28hrs. Each reported their bombs bursting in the target area, creating a series of fires. The last crew to bomb was skippered by Flying Officer F Pohlodek, flying at 13,000ft. They dropped their 500-pounders in one stick at 01:50hrs. Back at R.A.F East Wretham they reported, *'Bombed target area, fires started east and south of town. Aircraft shaken at 13,000ft by apparently terrific explosion on the ground, two minutes after bombing'*.

By the time the last crew turned for home, Munster was ablaze, smoke was drifting up to 9,000ft, and fires were clearly seen from as far as 90 miles on the return journey. It was on the return journey that trouble in the form of prowling night fighters from NJG1 and NJG3 was encountered. The first crew in trouble was Sergeant Josef Bernát at 02:15hrs. The crew were flying at 12,000ft when what was reported to be a 'single engine' fighter opened fire, damaging the port wing and starting a fire in the port wing root. Giving control to the second pilot, Sergeant Stanislav Linka, Sergeant Bernát, and the crews Wireless Operator Flight Sergeant Rudolf Haering set about tackling the raging fire, which had begun to spread setting alight a sizeable portion of the fuselage fabric, cabin curtains and rest bunk. Returning to the cockpit, Sergeant Bernát took control, while Sergeant Linka, joined by Pilot Officer Jaroslav Kula, the Navigator and Flight Sergeant Haering, continued battling the blaze. With the aid of the last of the extinguishers, they succeeded in putting out the fire. At the rear, machine gun and cannon fire had ruptured the hydraulics to the rear turret, resulting in the rear gunner, Sergeant Jindřich Beneš, manipulating the turret by hand and fending off their antagonist. With the Wellington's

*Another handwritten Combat Report. This report describes the encounter by the crew of Sergeant Josef Bernát. (Pavel Vančata)*

hydraulics out of action, a belly landing was made at East Wretham at 04:32hrs. It had been an excellent show by the crew, especially the young gunner, who had stayed calm and alert throughout the encounter.[35]

| Date | 06/07/1941 |
|---|---|
| Mark | Mk.IC |
| Serial | R1532 |
| Code | KX-R |
| Taken On Charge | 21/03/1941 via No. 24 MU. |
| Manufacturer | Vickers Armstrong (Chester) |
| Contract | 992424/39 |
| Pilot (s) | Sergeant Josef Bernát / Sergeant Stanislav Linka |
| Flight | Operations |
| Time | 04:32hrs |
| Cause | Night Fighter |

Damage to Wellington R1532 was extensive. It was in November 1941 that the aircraft, now fully repaired by No.43 Group, returned to front-line service on the squadron.

---

[35] Reports differ on where the crew actually belly landed, RAF Honington is also mentioned.

Another squadron crew was also in action. Flight Sergeant K Schoř was 15,000ft above Alkmaar, Northern Holland when attacked by a twin-engine fighter at 02:28hrs. The first engagement was inconclusive, with the enemy fighter breaking off and disappearing. A few minutes later, the fighter reappeared astern. The rear gunner and fighter opened fire simultaneously at around 400 yards range, Sergeant Ladislav Kadlec claiming to have damaged the fighter. The German pilot's aim was equally accurate. The Wellington's turret was hit, and the young air gunner was struck by a shell just below the knee. The crew managed to evade any further trouble and made a rather bumpy landing back at R.A.F East Wretham at 03:30hrs. Sadly, Sergeant Ladislav Kadlec would, have his right leg amputated below the knee after emergency surgery. Born July 14th 1917, in Brankovice, the young Czech was determined to return to operations. Having passed a Medical Selection Board, he finally got his wish and rejoined the squadron on October 1st 1942.[36] Damage to Wellington R1451 must have been minor, on repair by No.43 Group it was returned to the squadron on July 27th.

Members of the squadron, including Squadron Leaders Batchelor and Šnajdr travelled down to Devizes for the funeral of Sergeant O Helma and crew on the 7th. While they were paying their respects the squadron completed a comprehensive day of training including Blind Flying, Cross Country flights and night circuit and bumps. While this was going on, the ground crews were bombing-up and preparing ten Wellingtons for a Cologne. 3 Group would split its forces on this night, 59 Wellingtons would bomb Cologne while 51 would tackle Munster.

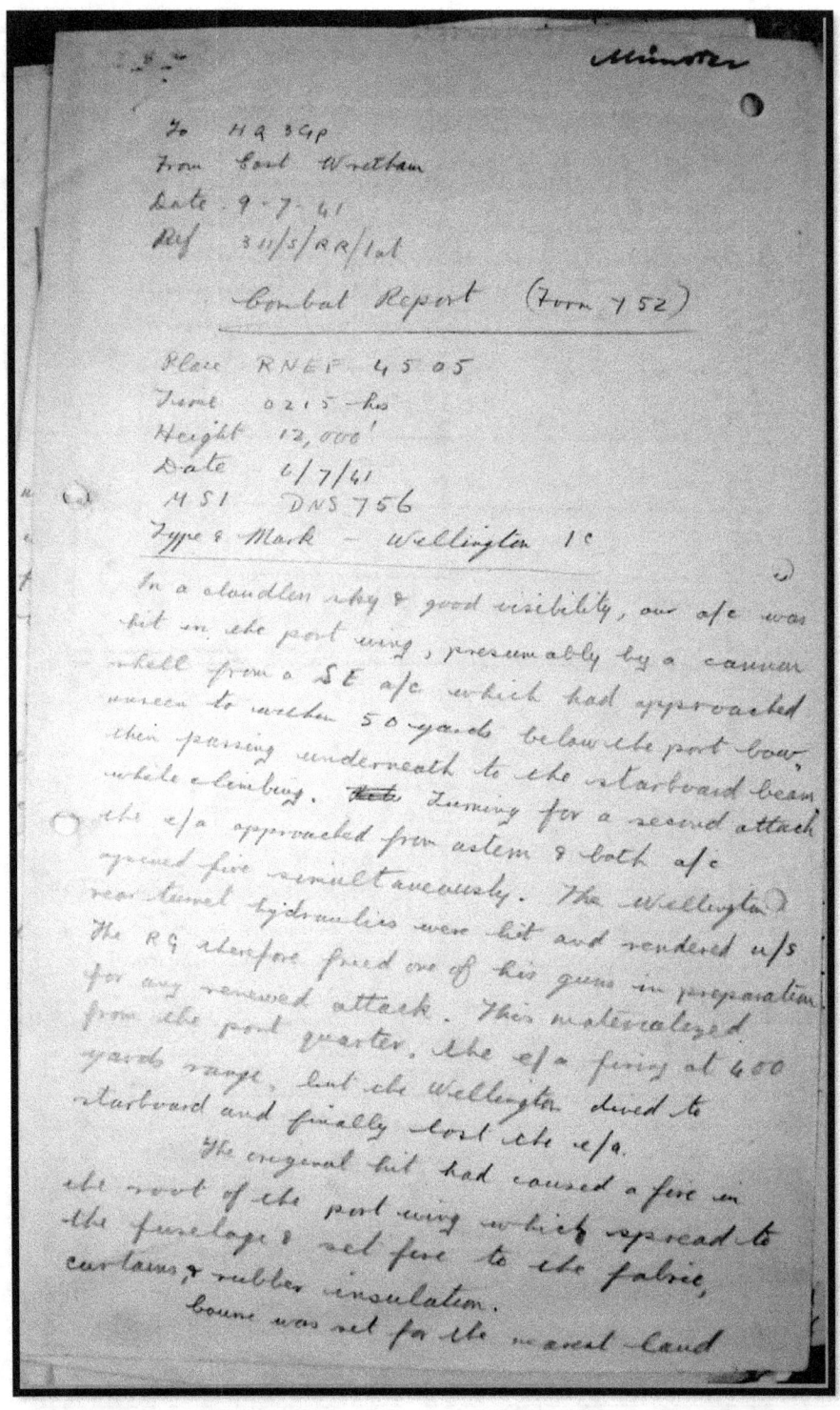

*Flight Sergeant K Schoř submitted the second Combat Report. (Pavel Vančata)*

---

[36] *Sadly, he would go missing on a patrol over the Bay of Biscay March 13th 1944 aboard Liberator VI BZ995 J.*

## July 7th 1941 : 'Trout C' – Cologne

| Pilot | 2nd Pilot | Serial | Code | Bomb Load | Result |
|---|---|---|---|---|---|
| P/O V Korda | Sgt J Horáček | T2972 | KX-G | 3x500lb+1x250lb+360x4lb | Duty Carried Out |
| F/Sgt F Fencl | Sgt V Netík | R1804 | KX-D | 3x500lb+1x250lb+360x4lb | Duty Carried Out |
| F/O F Pohlodek | Sgt O Jambor | R1458 | KX-Z | 3x500lb+1x250lb+360x4lb | Duty Carried Out |
| Sgt A Musálek | Sgt V Ryba | R1718 | KX-N | 3x500lb+1x250lb+360x4lb | Duty Carried Out |
| F/Lt J Šejbl | Sgt M Šebela | R1598 | KX-C | 1x1000lb+5x500lb | Duty Carried Out |
| F/Sgt K Schoř | Sgt M Jindra | R3166 | KX-K | 1x1000lb+5x500lb | Early Return |
| Sgt B Blatný | P/O J Nejezchleba | W5711 | KX-H | 1x1000lb+5x500lb | Duty Carried Out |
| F/Sgt J Čapka | F/O J Stránský | R1371 | KX-F | 1x1000lb+5x500lb | Duty Carried Out |
| F/O K Vildomec | Sgt V Procházka | R1015 | KX-L | 1x1000lb+5x500lb | Duty Carried Out |
| F/Lt J Breitcetl | Sgt A Mžourek | W5682 | KX-Y | 7x500lb | Duty Carried Out |

The squadron soon lost the services of Flight Sergeant K Schoř and crew with starboard engine problems, they landed back at R.A.F East Wretham within 90 minutes having jettisoned their bombs into the sea. This was one of a total of 13 crews who returned early, adding to the six that failed to take-off. Once again, it was not a particularly good night for the Group. Other than slight haze over Cologne the weather conditions were perfect. The Czechs would stay longer than usual over the target area. The first crew to bomb at 01:55hrs was Pilot Officer V Korda, who dropped his mixed load from 13,500ft onto the centre of the target, close to existing fires seen ablaze near the central train station. Searchlights were subdued due to the haze but still troublesome. Cologne's flak defences produced a withering array of multi-colour chasers and heavy flak over the cathedral city. Halfway through the raid, the crew of Sergeant A Musálek made their bomb run from 11,500ft. All the bombs were dropped at 02:20hrs across the centre of the target. Accurate observation of the results was impossible owing to the dazzle of the searchlights. A number of massive explosions were observed between aiming points A and C, which included the marshalling yards to the southwest, Cathedral Square, the central railway station and the Hohenzollern railway bridge in the city centre. One explosion was reported as *'tremendous'* and took place in the centre of the target, illuminating vast areas of the city. The last to bomb at 02:38hrs was Flight Lieutenant J Breitcetl. They dropped their all HE load from 11,000ft but could not see the results due to the crew having to take violent evasive action due to flak. On leaving Cologne, black smoke was seen rising from areas around the railway stations and what was believed to be a gas holder situated on the River Rhine's west bank. Crews reported fires still visible from over 50 miles on the return journey. The squadron dropped a total of 30,260lb of HE and incendiaries. Two Wellingtons returned damaged by flak. Flight Sergeant J Čapka at the controls of Wellington R1371 was the worst hit. Also hit, but to a lesser extent was Flying Officer F Pohlodek's Wellington R1458. On this night, a young Kiwi, Sergeant James Allen Ward of the R.A.F Feltwell-based 75(NZ) Squadron would earn the Victoria Cross for climbing out onto the wing of his Wellington to douse an engine fire during the Münster operation.

Taking full advantage of the good weather H.Q Bomber Command once again turned its attention towards Munster on the 8th. It would be an all 3 Group show with 55 Wellington crews detailed and briefed of which 311 Squadron would contribute seven.

## July 8th 1941 : 'Rudd B' – Munster

| Pilot | 2nd Pilot | Serial | Code | Bomb Load | Result |
|---|---|---|---|---|---|
| P/O V Korda | Sgt J Horáček | T2972 | KX-G | 1x1000lb+5x500lb | Duty Carried Out |
| F/Lt J Šejbl | Sgt M Šebela | R1598 | KX-C | 3x500lb+1x250lb+360x4lb | Duty Carried Out |
| F/Sgt J Čapka | F/O J Stránský | R1046 | KX-E | 3x500lb+1x250lb+360x4lb | Duty Carried Out |
| F/O K Vildomec | Sgt V Procházka | R1161 | KX-X | 3x500lb+1x250lb+360x4lb | Duty Carried Out |
| Sgt B Blatný | P/O J Nejezchleba | R1718 | KX-N | 3x500lb+1x250lb+360x4lb | Duty Carried Out |
| F/Sgt J Nyč | Sgt K Šťastný | R1804 | KX-D | 3x500lb+1x250lb+360x4lb | Duty Carried Out |
| F/Sgt J Hájek | Sgt V Soukup | W5682 | KX-Y | 7x500lb+1x250lb | Duty Carried Out |

Gremlins once again reduced the Group's contribution. Six Wellingtons failed to take-off, and four returned early. Operating on two consecutive nights put a tremendous strain on the aircrews, ground crews and aircraft. It was especially difficult for 311 Squadron with its limited pool of trained crews to call upon. Five were obliged to operate twice within 24 hours, as were five Wellingtons. The crews departed into a bright moonlit night, and this continued to the target. The usual ground haze made accurate navigation and identification difficult over North Rhine-Westphalia. The Czechs were some of the first over Munster between 01:29hrs and 01:53hrs. Flares were already slowly descending over the city when they arrived. Fires were reported in the north, south and western suburbs. Flight Lieutenant J Šejbl dropped his load across the centre of the target, leaving behind seven fires. Opposition was considered light, allowing some the opportunity to take their time in identifying the aiming point. It seemed to the crews that the whole target was ablaze as they turned for home. 3 Group H.Q seemed happy with the results recording the following, *'Results on Munster are very good, the town was left ablaze with one big central fire. Fires are more concentrated and more in number than last night.'* All the squadron was safely back by 04:25hrs. Not so lucky was 9 Squadron, for the second night they reported a crew missing. Thunderstorms thwarted the majority of the squadron's planned flying training on the 9th and probably resulted in the squadron sitting out the raid on Osnabruck by five squadrons of Wellingtons drawn from 3 Group. The weather had hardly improved by the 10th when the squadron was ordered to ready six crews for a return to Cologne.

A new Air Ministry directive was issued on July 9th, four months to the very day Bomber Command was released from its primary directive, which stated, *'We must take the offensive against the U-Boat and the Focke-Wulf wherever and whenever we can'*. The threat at sea had been checked for the time being. As difficult as it had been, the campaign was deemed a success at Bomber Command H.Q. German ports had been bombed, industrial plants had been disrupted, and many severely damaged. Considerable damage was also inflicted on the French ports, especially Brest, which housed the German battle cruisers. By some miracle, the cruisers *Scharnhorst* and *Gneisenau* were relatively unscathed and received only superficial damage during the whole campaign. There were some notable successes, but accurate details were sketchy, and most information came via neutral sources. From a Bomber Command perspective, it had carried out the directive to the full with every means at its disposal and had achieved some spectacular results, especially against Hamburg. However, the Navy Chiefs were somewhat vocal in their disappointment that the Cruisers, *Prinze Eugen*, *Scharnhorst* and *Gneisenau* had not been destroyed while in dock. After all, they were "sitting targets". Weather had been a significant issue throughout the campaign. Crews were hindered and frustrated by the lack of accurate navigational aids. Most crews still navigated by DR and found it almost impossible to locate a town, let alone a factory, if no ground detail could be seen. The few successes during this period resulted from the courage and tenacity of the crews. 3 Group had played a pivotal role, as did the Czechs of 311 Squadron.

Armed with a new directive, Bomber Command's attention now turned towards Germany, and its highly organised transport system. Nine targets were specified, Hamm, Osnabruck, Soest, Schwerte, Cologne (Kalk-Nord) Cologne (Gereon), Duisburg (Hochfeld Sud) Dusseldorf (Derendorf) Duisberg (Ruhrot). The switch to relatively small targets often located in the very heart of a town or city would test Bomber Command and its crews to the full.

### July 10th 1941 : 'Trout C' – Cologne

| Pilot | 2nd Pilot | Serial | Code | Bomb Load | Result |
|---|---|---|---|---|---|
| F/O K Vildomec | Sgt V Procházka | R1161 | KX-X | 3x500lb+1x250lb+360x4lb | Duty Carried Out |
| S/Ldr J Šnajdr | Sgt M Jindra | W5682 | KX-Y | 3x500lb+1x250lb+360x4lb | Alternative target |
| F/Sgt J Čapka | F/O J Stránský | R1046 | KX-E | 3x500lb+1x250lb+360x4lb | Duty Carried Out |
| P/O V Korda | Sgt J Horáček | T2972 | KX-G | 1x1000lb+5x500lb | Duty Carried Out |
| Sgt B Blatný | P/O J Nejezchleba | W5711 | KX-H | 1x1000lb+5x500lb | Duty Carried Out |
| F/Lt J Šejbl | Sgt M Šebela | R1598 | KX-C | 1x1000lb+5x500lb | Duty Carried Out |

Squadron Leader Šnajdr would be flying his first operation in nearly a month on this night. The relatively inexperienced Sergeant Miroslav Jindra with only four sorties would join him in the right-hand seat. The squadron would be part of a force of 62 crews detailed and briefed by 3 Group. The weather conditions created considerable problems on this night. The squadron was given Cologne C as its primary target, the west central station. Unable to locate the primary due to thick cloud, Flight Lieutenant J Šejbl would deliver his entire load on Bonn at 02:10hrs. The remaining crews all reported attacking Cologne between 02:01hrs and 02:30hrs. Squadron Leader J Šnajdr was the last to bomb. His report makes interesting reading, *'Aircraft attacked primary target, bomb bursts were observed in centre of town. Fires with intermittent explosions which developed into an orange glow with heavy grey black smoke which could be seen from 30 miles'*. The raid appeared moderately successful, some useful damage to Cologne was reported. All the squadron were safely home by 05:10hrs. Over at R.A.F Honington, 9 Squadron reported another missing crew, the third in three days. Regardless of the favourable reports filtering into 3 Group H.Q at Exning, senior staff officers were aware that the raid was far from accurate, *'Thick haze over the target. Pilots are, for the most part, uncertain of the exact locations of their bombs'*. This unusually honest assessment was written in the Group summary, it was the first inclination of trouble. Cologne reported only three bombs, 300 incendiaries, and minor damage. Bombs were reported scattered across a wide area, including Bonn and Koblenz.

Air-to-Air firing practice was the main occupation on the 11th over Berners Heath. That night there was some cause for celebration, with the announcement of the award of the DFM to Sergeant Benedikt Blatný for his actions on May 11th. 3 Group would be joined by the Lincolnshire based 5 Group on the 12th for a raid directed against Bremen. Twenty-nine Wellingtons from 9, 99 and 311 (Czech) Squadrons would be detailed and briefed. From the moment the crews took off the weather was against them and for once the meteorological experts were correct in their forecast.

### July 12th 1941 : 'Salmon B' – Bremen

| Pilot | 2nd Pilot | Serial | Code | Bomb Load | Result |
|---|---|---|---|---|---|
| F/Sgt J Nyč | Sgt K Šťastný | R1718 | KX-N | 3x500lb+360x4lb | Duty Carried Out |
| F/Sgt J Bernát | Sgt S Linka | R1371 | KX-F | 3x500lb+360x4lb | Duty Carried Out |
| F/Sgt J Hájek | Sgt V Soukup | R1804 | KX-D | 1x1000lb+4x500lb | Duty Carried Out |
| P/O V Korda | Sgt J Horáček | T2972 | KX-G | 1x1000lb+4x500lb | Duty Carried Out |
| Sgt B Blatný | P/O J Nejezchleba | W5711 | KX-H | 1x1000lb+4x500lb | Duty Carried Out |
| F/Lt J Šejbl | Sgt M Šebela | R1598 | KX-C | 1x1000lb+4x500lb+250lb | Alternative Target |
| F/O K Vildomec | Sgt V Procházka | R1161 | KX-X | 7x500lb | Duty Carried Out |
| F/Sgt J Čapka | F/O J Stránský | R1046 | KX-E | 7x500lb | Duty Carried Out |

As the Wellingtons crossed the North Sea they encountered a sinister-looking weather front. Thunderstorms, with thunderclouds reaching well above the Wellington's operational height were encountered. Somehow the squadron managed to clear these, only to encounter another electrical storm over the Zuiderzee. The crew of Flight Lieutenant J Šejbl got as far as Emden before severe icing resulted in the Wellington losing valuable height. Given the conditions, the crew jettisoned their entire bombload in the general area of Emden and turned for home. Once over Germany, the weather improved slightly, but on nearing Bremen, a vicious barrage of flak was encountered, equal to anything over the Ruhr or Berlin. Bombing between 11,000ft and 15,000ft, the squadron tried to make out the primary objective east of the old town in the haze. Sergeant J Bernát bombed at 01:50hrs, his mixed load starting two fires near the aiming point. Pilot Officer V Korda reported that their high-explosive bombload fell near the railway, southeast of the aiming point, but observation of results was impossible. They were badly knocked about by flak for their troubles. Flying Officer K Vildomec bombed at 02:10hrs, creating seven small fires. One of the last to bomb was Flight Sergeant J Nyč at 02:40hrs. Upon return, the crew worryingly reported that just one fire was seen 1 ½ miles east of the aiming point on his arrival. Group H.Q reported, *'Very few crews bombed near the aiming point. Results were not very satisfactory'*. There was one positive from an otherwise disappointing night, all of the Groups crews returned. It was no surprise that Bremen was the target the following night, the Czechs gratefully sitting the raid out. A total of 47 3 Group Wellingtons encountered, rain and icing plus 10/10th cloud over the target, bombing results were more disappointing than the raid on the 10th. To the disbelief of many on the squadron they learnt on the morning of the 14th that Bremen was once again the target for that night. 3 Group would detail 59 Wellingtons on Bremen, eight on Hanover and a further six on the oil tanks at Rotterdam. The Czechs at R.A.F East Wretham would provide eight crews against Bremen's goods station.

"The Magnificent Seven" demonstrate variety of pipes in summer 1941. L-R: Air Gunner F/O Jan Fürbach, pilot F/Lt Josef Šejbl (coincidentally later the seventh squadron Commanding Officer), Wireless Operator P/O Josef Šimet, Navigator P/O Jiří Engel, pilot P/O Václav Korda, Wireless Operator P/O Ondřej Kacíř and pilot F/O František Macháček. (Pavel Vančata)

## July 14th 1941 : 'Salmon B' – Bremen

| Pilot | 2nd Pilot | Serial | Code | Bomb Load | Result |
|---|---|---|---|---|---|
| F/Lt J Šejbl | Sgt M Šebela | R1598 | KX-C | 1x1000lb+4x500lb+1x250lb | Duty Carried Out |
| Sgt B Blatný | P/O J Nejezchleba | W5711 | KX-H | 1x1000lb+4x500lb+1x250lb | Early Return |
| F/O K Vildomec | Sgt V Procházka | R1161 | KX-X | 3x500lb+360x4lb | Duty Carried Out |
| P/O V Korda | Sgt J Horáček | T2972 | KX-G | 1x1000lb+4x500lb+1x250lb | Duty Carried Out |
| F/Lt J Breitcetl | F/O J Stránský | R1804 | KX-D | 3x500lb+360x4lb | Early Return |
| S/Ldr J Ocelka | Sgt F Radina | W5682 | KX-Y | 1x1000lb+4x500lb+1x250lb | Duty Carried Out |
| F/Sgt A Šedivý | Sgt A Mžourek | R1371 | KX-F | 3x500lb+360x4lb | Duty Carried Out |
| F/Sgt J Nyč | Sgt K Šťastný | R1718 | KX-N | 7x500lb | Duty Carried Out |

The usual spate of cancellations was made worse when ten Wellingtons returned early, eight due to the atrocious weather conditions. Two of those were from 311 Squadron. Flight Lieutenant J Šejbl turned for home for the second operation in succession. He again reported severe icing en route and jettisoned his bombload in the Zuiderzee. The crew were back at East Wretham within three hours. Sergeant B Blatný also reported icing issues. They dropped all their high-explosive load on the railway crossing at Papenburg, Lower Saxony, at 01:32hrs. Bursts were seen, and a large fire started. The rest of the crews managed to avoid the worst of the thunderstorm and icing and continued to Bremen. By sheer luck, Bremen was cloud-free, something the crews were not expecting as clouds had been encountered almost the entire route.

The raid, which was flown in collaboration with 1 and 4 Groups, began around 01:20hrs. Each Group had its own target, 1 Group Aiming Point 'C', 4 Group Aiming Point 'A' while 3 Group would tackle Aiming Point 'B'. 311 Squadron was over their target, the Goods Station, from 02:10hrs. Exceptionally heavy flak was encountered and the searchlights were particularly effective. Slight haze did not prevent Pilot Officer V Korda from bombing at 02:13hrs from 15,000ft. The crew were enthusiastic about the raid on return to East Wretham, *'Upon arrival whole town appeared to be alight. Bursts seen, practically whole town ablaze with at least five large and seven smaller scattered fires'*. Squadron Leader J Ocelka was flying at 13,800ft when his high-explosive load was dropped into the centre of the target area at 02:28hrs. He gleefully reported at de-briefing, *'Bombs in target area. Fires visible from 20 miles beyond Wilhelmshaven'*. One of the last over the blazing target from the squadron was Flight Sergeant J Nyč, *'Bursts seen, one very large fire in south of town, may have been an aerodrome, or aircraft works'*. It was almost daylight when the last crew landed safely back at R.A.F East Wretham at 05:50hrs. 3 Group had dropped at total of 178,400lb of bombs that night. 311 Squadron had contributed 22,070lb for no loss. Much to the relief of H.Q Bomber Command the raid appeared to be an outstanding success.

On the 15th, the squadron welcomed several new crews from R.A.F Honington, their arrival coincided with drizzle and low clouds which kept the squadron grounded throughout the day. The squadron ground crews were up early on the 16th, nine Wellingtons had to be tested and made ready for that night. The target was Hamburg. 3 Group would detail just 32 Wellingtons on Hamburg, plus a further six Wellingtons would visit Boulogne.

## July 16th 1941 : Dace 'A' – Hamburg

| Pilot | 2nd Pilot | Serial | Code | Bomb Load | Result |
|---|---|---|---|---|---|
| S/Ldr J Šnajdr | F/O J Stránský | R1804 | KX-D | 2x500lb+360x4lb | Alternative Target |
| Sgt B Blatný | P/O J Nejezchleba | W5711 | KX-H | 2x500lb+360x4lb | Duty Carried Out |
| P/O V Korda | Sgt J Horáček | T2972 | KX-G | 1x1000lb+3x500lb | Duty Carried Out |
| S/Ldr J Ocelka | Sgt F Radina | W5682 | KX-Y | 1x1000lb+3x500lb | Duty Carried Out |
| F/Sgt A Šedivý | Sgt A Mžourek | R1371 | KX-F | 1x1000lb+3x500lb | Duty Carried Out |
| F/O K Vildomec | Sgt V Procházka | R1161 | KX-X | 5x500lb+1x250lb | Duty Carried Out |
| Sgt J Bernát | Sgt S Linka | Z8784 | KX-U | 5x500lb+1x250lb | Duty Carried Out |
| F/Sgt J Nyč | Sgt K Šťastný | R1718 | KX-N | 5x500lb+1x250lb | MISSING |

Flight Lieutenant Šejbl's R1598 KX-C was withdrawn for unknown reason just before take-off. Only eight crews eventually departed R.A.F East Wretham at 23:09hrs. Soon after crossing the North Sea, the crew of Flight Sergeant J Nyč were intercepted and shot down by Lt. Rudolf Schoenert of 4./NJG1 at 00:40hrs, 3 miles west of Lemmer after being stalked for 17 minutes. Attacked from below and the rear, the Wellington burst into flames. It must have been quickly apparent to the crew that the Wellington was doomed as all the crew managed to take to their parachutes. Five of the crew were captured. Flight Sergeant Nyč was captured on the 19th, while Sergeant Šťastný was on the run until the 21st. On the 17th, the body of the rear gunner, Sergeant Jiří Mareš, washed ashore near Lemmer. It is presumed the young gunner drowned. He was buried on July 21st in Grave 261 of the Lemsterland Municipal Cemetery, Lemmer.[37]

<u>Vickers Wellington Mk.IC R1718 KX-N</u>

| Manufacturer | Vickers Armstrong | |
|---|---|---|
| Contract | 992424/39 | |
| Taken on Charge | 28/05/1941 via No. 37 MU. | |
| Cat E Missing | 16/07/1941 | |
| Struck Of Charge | 17/07/1941 | |
| Total Flying Hours | N/K | |
| Take-Off Time | 23:07hrs | |
| Bomb Load | 5x500lb+1x250lb | |
| | | |
| | **CREW** | |
| Captain | Flight Sergeant Jaroslav **Nyč** 787166 RAFVR | PoW |
| Second Pilot | Sergeant Karel **Šťastný** 787170 RAFVR | PoW |
| Navigator | Pilot Officer Jaroslav **Zafouk** 85217 RAFVR | PoW |
| Wireless Operator | Pilot Officer Otakar **Černý** 82590 RAFVR | PoW |
| Front Gunner | Sergeant František **Knap** 787396 RAFVR | PoW |
| Rear Gunner | Sergeant Jiří **Mareš** 787393 RAFVR. Age 25. | Plot C. Row 10. Grave 261 |
| | | |
| Posting History | F/Sgt J Nyč posted via 11 O.T.U, 06/09/1940. Sgt K Šťastný NFD. | |
| Operations Flown | F/Sgt J Nyč : 18 / Sergeant K Šťastný : 20 | |
| Buried | LEMSTERLAND (LEMMER) GENERAL CEMETERY | |

Solid cloud was encountered along the entire route. Heavy flak was experienced from around Bremen and almost the whole region surrounding Hamburg. Searchlights illuminated the almost solid cloud base as the squadron made their run into the target. 3 Group was given Aiming Point 'B', the West Altona Station, along with the Wellingtons of 1 Group. The Hampdens of 5 Group would attack the East Berliner Tor Station along with the Whitleys of 4 Group. Hamburg reported the attack commencing at 01:35hrs. The first Czech crew over the target is recorded at 01:45hrs. This was the Wellington flown by Pilot Officer V Korda, who seemed unsure of the exact aiming point due to cloud and bombed flak and searchlight positions from 15,000ft. Three crews bombed within minutes of each other, Flight Sergeants J Bernát, A Šedivý and Flying Officer K Vildomec, each bombed in one stick from heights varying from 12,000ft to 16,000ft. Only Flight Sergeant J Bernát claimed that at least one of his bombs burst in the target area. Squadron Leader J Ocelka was over Hamburg at 02:30hrs, reporting, *'Aircraft attacked primary. Dropped all bombs near Alster. All bursts seen but owing to 10/10th cloud no results seen'*. The cloud prevented the crew of Squadron Leader J Šnajdr from attacking the primary. Therefore, they turned their attention to flak and searchlight concentrations near Kiel bombing at 02:40hrs. The last of the crews landed in daylight at 06:21hrs. At briefing, a report of a Wellington shot down over the target area was noted. The operation was not a success and Group H.Q conceded that *'crews bombed somewhere in the target area'*. For the Czechs, the loss of Jaroslav Nyč and crew, was a heavy price to pay for so little in return. The

---

[37] *Another report records the body washed ashore on August 9th 1941, and was buried in Grave 259.*

Wireless Operator, Pilot Officer Otakar Černý was one of a growing number of Czechs who married English girls. He had married Miss Doris Mayes in Bury St Edmunds on December 21st 1940.

While the squadron was over Germany, a lone raider bombed R.A.F East Wretham at 00:40hrs. The bombs landed along the east side of the airfield, reducing the runway by 300 yards, thus making the station temporarily non-operational. No buildings or aircraft were damaged and thankfully no casualties reported. By mid-morning, the station was again operational allowing training over Berners Heath for the rest of the day. The squadron sat out a raid on Cologne that night. On the 18th, squadron personnel visited R.A.F Honington to join 9 Squadron for a formal inspection by Group Captain Richard Harrison DFC, AFC. While this was taking place, a number of crews were over Weybourne Ranges on air-to-air gunnery practice. At 00:21hrs, R.A.F East Wretham was again attacked by solitary raiders. One dispersal taxiway was hit and cratered, while eight other bomb craters were made in the southeast dispersal area. One 100lb bomb, painted red with yellow fins, was also found. Two machine-gunning attacks were made in a north-to-south direction. No damage to buildings or personnel was reported. Both attacks were most likely carried out by Ju88C's of NJG2 based at Gilze-Rijen. The squadron was again over Weybourne firing range during the day. Those not involved were busy making ready eight Wellingtons for operations that night against Hanover.

There was a tragic accident at East Wretham on the 19th, which resulted in the death of a civilian contractor. Flying Officer Josef Stránský was at the controls of Wellington R1804 KX-D and about to take-off for a training flight. The pilot was reported to have started excessively revving up his engines just before releasing the breaks. The Wellington lurched forward at some speed at 11:45hrs and swung violently as the speed quickly increased. The pilot attempted to correct the swing, but it was too late. The aircraft struck a stationary Steam Roller, killing the 62-year-old driver, Mr George Potter, a civilian employed by E.J.Doe Contractors. The badly burnt body of Mr Potter was taken to Thetford Cottage Hospital Mortuary. Wellington R1804 was burnt out and struck off charge on the 19th, miraculously none of the crew were injured.

| Date | 19/07/1941 |
|---|---|
| Mark | Mk.IC |
| Serial | R1804 |
| Code | KX-D |
| Taken On Charge | 18/05/1941 via No. 45 MU. |
| Manufacturer | Vickers (Chester) |
| Contract | 992424/39 |
| Pilot (s) | Flying Officer Josef Stránský |
| Flight | Daylight Training |
| Time | 11:45hrs |
| Cause | Swung on take-off |

Only two 3 Group squadrons would be involved in the raid on Hanover, 9 and 311, the effort being reduced owing to weather conditions.

## July 19th 1941 : Eel 'A' – Hanover

| Pilot | 2nd Pilot | Serial | Code | Bomb Load | Result |
|---|---|---|---|---|---|
| Sgt B Blatný | P/O J Nejezchleba | W5711 | KX-H | 3x500lb+360x4lb | Duty Carried Out |
| F/Sgt J Bernát | Sgt S Linka | Z8784 | KX-U | 3x500lb+360x4lb | Alternative Target |
| P/O V Korda | Sgt J Horáček | T2972 | KX-G | 1x1000lb+4x500lb | Duty Carried Out |
| S/Ldr J Ocelka | Sgt F Radina | W5682 | KX-Y | 1x1000lb+4x500lb | Duty Carried Out |
| F/Lt J Šejbl | Sgt M Šebela | R1598 | KX-C | 1x1000lb+4x500lb | Duty Carried Out |
| F/Sgt J Hájek | Sgt V Soukup | R1015 | KX-L | 1x1000lb+4x500lb | Early Return |
| Sgt V Netík | Sgt M Jindra | R1371 | KX-F | 1x1000lb+4x500lb | **MISSING** |
| F/O F Pohlodek | Sgt O Jambor | T2561 | KX-A | 6x500lb+1x250lb | Duty Carried Out |

*Sergeant Václav Netík failed to return from Hannover with Wellington Mk.IC R1371/KX-F on the night of 19th/20th July 1941. This was would have been his fifteenth sortie and the first as crew skipper. (Jaroslava Rozumová)*

Sergeant Václav Netík would be flying his first operation as captain on this night on what turned out to be another frustrating and costly operation. The weather was surprisingly better than forecast by the Met Officer as the small force crossed the Dutch coast and headed towards Hanover. There was just one reported early return. Flight Sergeant J Hájek suffered severe icing near the Dutch Coast. The crew jettisoned their entire bombload in the sea off the Frisian Islands. Flight Sergeant J Bernát could not climb to his bombing height due to icing. Unwilling to continue on to the primary, he bombed Emden at 01:04hrs from 10,000ft, starting some good fires.[38] On nearing the target, the crews could see that the Whitleys and new four-engined H.P Halifaxes of 4 Group had yet to start their attacks as no fires were visible. Positioned in the centre of the Steinhufer-Meer-Hildesheim-Celle area, there was no easy way of approaching the target without running the gauntlet of numerous flak and searchlight positions. German reports suggest the raid started at 01:18hrs, which is the same time that Sergeant B Blatný and crew delivered their mixed bombload, starting one medium-sized red fire. There was a gap of sixteen minutes until the next crew, skippered by Squadron Leader J Ocelka bombed. They dropped their high-explosive load from 14,000ft in the target area, their load being seen to burst in the town. Two minutes later, Pilot Officer V Korda bombed. They could not identify the aiming point but dropped in the approximate target area from 14,000ft, starting one medium-sized red fire. Flak hit the Wellington of Flight Lieutenant J Šejbl over the target area. Damage was confined to the starboard wing and wheel area, thankfully missing the fuel tanks and engine. The last to bomb at 01:55hrs was the crew of Flying Officer F Pohlodek. Surprisingly, they reported on return, *'No fires seen on arrival. Bombs seen to burst in the target area. Strong flak opposition and searchlight activity made observation of results difficult'*. The weather over England on return made finding East Wretham much more challenging for the tired

---

[38] *The ORB is at odds with 3 Group records that show the crews bombing Emden and not Hanover.*

crews. Solid clouds, in places almost to ground level, put more strain on the crews. At 05:48hrs, the crew of Flying Officer F Pohlodek touched down. One crew was unaccounted for, but given the weather conditions, no one was overly worried as rain started to fall. Wellington R1371 KX-F had been loaded with 634 gallons of fuel, giving it almost eight hours of flying time. By 07:00hrs, the usual round of telephone calls had drawn a blank on the whereabouts of the crew of Sergeant V Netík. They were presumed 'missing' by mid-morning.

## Vickers Wellington Mk.IC R1371 KX-F

| Manufacturer | Vickers Wellington | |
|---|---|---|
| Contract | 992424/39 | |
| Taken on Charge | 02/01/1941 via No. 9 MU. | |
| Cat E Missing | 19/07/1941 | |
| Struck Of Charge | 20/07/1941 | |
| Total Flying Hours | N/K | |
| Take-Off Time | 22:52hrs | |
| Bomb Load | 1x1000lb+4x500lb | |
| | **CREW** | |
| Captain | Sergeant Václav **Netík** 787211 RAFVR. Age 30. | Grave 18 |
| Second Pilot | Sergeant Miroslav **Jindra** 788021 RAFVR. Age 25. | Row 16. Grave 27. |
| Navigator | Pilot Officer Jaroslav **Partyk** 82627 RAFVR. Age 24. | 7. F. 4. |
| Wireless Operator | Sergeant Jan **Čtvrtlík** 787433 RAFVR. Age 23. | Runnymede Panel 42. |
| Front Gunner | Sergeant Václav **Valeš** 787417 RAFVR. Age 26. | Runnymede Panel 54. |
| Rear Gunner | Sergeant Pavel **Babáček** 787148 RAFVR. Age 26. | Runnymede Panel 39. |
| | | |
| Posting History | Sgt V Netík via Wilmslow Depot, 18/02/1941. Sgt M Jindra via Cosford Depot, 18/01/1941. | |
| Operations Flown | Sgt V Netík : 14 / Sgt M Jindra : 5 | |
| Buried | OLDEBROEK GENERAL CEMETERY | |
| | UITHUIZERMEEDEN GENERAL CEMETERY | |
| | SAGE WAR CEMETERY | |
| Remembered | RUNNYMEDE MEMORIAL | |

The Vickers Wellington crashed into the Waddenzee, killing all the crew. There is no known definite cause of loss. The body of Sergeant Václav Netík was washed ashore near Rottumer on August 7th 1941. He was identified and buried the same day in the Rottum Cemetery, Grave 18. Pilot Officer Jaroslav Partyk was washed ashore on August 8th near Borkum. He was buried in the Lutheran Cemetery Grave 12 on Borkum on August 12th. The body of Sergeant Miroslav Jindra was recovered on September 16th 1941, 2 miles northwest of the windmill at Goliath Molen, west of Eemshaven, on the north-Groningen coast and initially buried in the Uithuizermeeden Cemetery, Row 16, Grave 27. The remaining crew members were never found. The Wireless Operator, Sergeant Jan Čtvrtlík had a brother, Miroslav Čtvrtlík, a front gunner in the crew of Pilot Officer V Korda. There was a real sense of shock on the squadron with the loss of yet another crew just 48 hours after the disappearance of Flight Sergeant J Nyč. The squadron's excellent loss-free run seemed to be wavering.

An accident involving a lorry Bedford RAF3269 resulted in the front turret of Wellington R1598 being damaged when hit of the stationary Wellington at its dispersal. The driver 1079657 A.C.2 Tipping had been asked by the Czech ground staff to pull along a little more as they wanted to unload a spare wheel for the aircraft. While doing so, the canvas hooding snagged the turret resulting in the gun cradle and both guns being bent, plus two perspex panels broken. The unfortunate driver who later stated, *'as this is my first experience with the Czechs, it is possible that I miss-took their directions'* was disciplined.

| Date | 20/07/1941 |
|---|---|
| Mark | Mk.IC |
| Serial | R1598 |
| Code | KX-C |
| Taken On Charge | 19/03/1941 via No. 48 MU. |
| Manufacturer | Vickers (Chester) |
| Contract | 992424/39 |
| Pilot (s) | - |
| Flight | - |
| Time | Not recorded |
| Cause | Hit by lorry at dispersal |

On the 21st, almost the entire squadron was paraded at R.A.F East Wretham with the arrival of Air Vice Marshal Baldwin CB, DSO, OBE there to award the DFM to Sergeant Benedikt Blatný. Flight Sergeant Josef Bernát received a congratulatory message from the AOC R.A.F Bomber Command for his excellent effort bringing home the damaged Wellington on July 5th. The award, and the recognition went a little way to ease the gloom of losing good friends. The 22nd was primarily a day of training which invariably led to an accident. Coming into land at R.A.F Honington, Sergeant František Dostál was a little too fast and heavy handed, resulting in the port undercarriage of Wellington T2468 KX-K collapsing on touching down.

| Date | 22/07/1941 |
|---|---|
| Mark | Mk.IC |
| Serial | T2468 |
| Code | KX-K |
| Taken On Charge | 03/06/1941 via 40 Squadron. |
| Manufacturer | Vickers (Weybridge) |
| Contract | 38600/39 |
| Pilot (s) | Sergeant František Petr / Sergeant František Dostál |
| Flight | Training Flight |
| Time | Not recorded |
| Cause | Pilot error |

Sergeant František Dostál was still under instruction at this time, he only had 17 hours solo flying on the Wellington. The Commanding Officer rather casually remarked, *'Type of accident is to be expected'*. On the 23rd, General Inspector of the Polish Air Force, Air Vice Marshal Stanislav Ujejski, visited the station, and bestowed the award of the Polish War Cross to eight officers and six NCO's of the squadron. Also in attendance were Air Vice Marshal Karel Janoušek, Czechoslovak Inspector General and Group Captain Kubita. After the ceremony, the party returned to East Wretham Hall for tea, which was frightfully British! Operations had been planned for that night against Mannheim. The squadron had sat out attacks on Cologne on the 20th, and Mannheim on the 21st.

### July 23rd 1941 : Chub 'A' – Mannheim

| Pilot | 2nd Pilot | Serial | Code | Bomb Load | Result |
|---|---|---|---|---|---|
| S/Ldr J Šnajdr | Sgt A Mžourek | R1046 | KX-E | 3x500lb+360x4lb | Duty Carried Out |
| F/Lt J Breitcetl | Sgt S Linka | W5668 | KX-T | 3x500lb+360x4lb | Duty Carried Out |
| Sgt A Musálek | Sgt M Plecitý | R1777 | KX-M | 3x500lb+360x4lb | Duty Carried Out |
| P/O V Korda | Sgt M Šebela | T2972 | KX-G | 1x1000lb+4x500lb | Duty Carried Out |
| S/Ldr J Ocelka | Sgt F Radina | W5682 | KX-Y | 1x1000lb+4x500lb | Duty Carried Out |
| F/Sgt K Schoř | Sgt V Soukup | R1015 | KX-L | 1x1000lb+4x500lb | Duty Carried Out |
| Sgt B Blatný DFM | P/O J Nejezchleba | W5711 | KX-H | 1x1000lb+4x500lb | Duty Carried Out |
| F/O F Pohlodek | Sgt O Jambor | T2561 | KX-A | 6x500lb+1x250lb | Duty Carried Out |
| F/O K Vildomec | Sgt V Procházka | R1161 | KX-X | 6x500lb+1x250lb | Duty Carried Out |

There was one new name on the crew list for this night. Sergeant Miroslav Plecitý, a former air gunner would join the crew of Sergeant A Musálek for his first operation. A total of 39 Wellington crews would be detailed for Mannheim, while a further six would be given Le Havre or Ostend. The Group reported just a solitary cancellation before take-off and five early returns, none of which were 311 Squadron. Hazy conditions over Germany created the usual navigation problems, but thankfully, a reasonably clear area over the Rhine allowed the crews to glimpse the landscape below and confirm their location before they run into Mannheim. The customary ring of flak and searchlights confirmed to the crews they were over the target. A few isolated fires were seen as the first squadron arrived at 01:32hrs. Pilot Officer V Korda and crew were flying at 15,000ft and reported a vivid flash that lit up blocks of buildings at 01:32hrs. Numerous bursts could be observed below, many in or close to the aiming point, the South Square of the main town. Flak and searchlights prevented the crew of Flying Officer K Vildomec from observing where their bombs landed at 01:34hrs. Sergeant B Blatný DFM and crew dropped all their HE load from 15,000ft and reported, *'Bursts near centre of town'*. Despite the intense flak, Squadron Leader J Šnajdr bombed from 14,000ft at 01:45hrs and was confident his entire bombload hit the mark, he reported, *'Bomb bursts in centre of town. Incendiary bombs started seven fires. One large grey/red fire was seen 15 minutes after leaving'*. This fire was also seen by the crew of Squadron Leader J Ocelka, who reported it visible from 25 miles. The squadron began landing back at base at 04:43hrs. There was a tense wait of 77 minutes before Flight Lieutenant J Breitcetl finally touched down. 3 Group H.Q confirmed the hazy conditions and rather indifferently reported, *'One or two bombs near the aiming point and a few bombs in the target area'*.

*Wellington R1161 KX-X. Note the smaller individual code letter, a common application by the squadron. (John Costin)*

*The film poster for Target for Tonight.*

311 (Czech) Squadron sat out the daylight attack on Brest on the 24th. 3 Group would once again supply the bulk of the bombers taking part. A total of 99 bombers would be involved of which 36 Wellingtons were provided by 40, 57, 75(NZ), 99, 101, 115, and 218 Squadrons. The target was the cruiser *Gneisenau*. Flak and Bf109 fighters tore into the squadrons, shooting down four Wellingtons and damaging many others for little if any worthwhile results. Bomber Command H.Q had recently reintroduced, if but on a limited scale, the unpopular daylight cloud-covered operations over Germany. These almost suicidal operations were flown by solitary crews or small numbers of bombers using cloud cover for protection. They had begun in March on 3 Group and since then had been flown semi-regularly when weather permitted. Thankfully, 311 Squadron had been fortunate not to be involved. *'Target for Tonight'* was released on July 24th and would catapult Percy Pickard DSO DFC into matinee idol status, much to his embarrassment. The film was an instant success. It was back to Hamburg on the 25th, 3 Group would supply 35 Wellingtons of which 311 Squadron would contribute eight. The target was the Hamburg Berliner Bahnhof railway station. The Groups two Short Stirling squadrons, 7 and 15, would struggle on to Berlin.

### July 25th 1941 : Dace 'B' – Hamburg

| Pilot | 2nd Pilot | Serial | Code | Bomb Load | Result |
|---|---|---|---|---|---|
| P/O V Korda | P/O J Nejezchleba | T2972 | KX-G | 2x500lb+1x250lb+360x4lb | Duty Carried Out |
| S/Ldr J Ocelka | Sgt F Radina | W5682 | KX-Y | 2x500lb+1x250lb+360x4lb | Duty Carried Out |
| F/Sgt F Fencl | Sgt A Musálek | R1777 | KX-M | 2x500lb+1x250lb+360x4lb | Duty Carried Out |
| F/Sgt J Hájek | Sgt V Soukup | R1015 | KX-L | 1x1000lb+3x500lb+1x250lb | Duty Carried Out |
| F/O F Pohlodek | Sgt O Jambor | T2561 | KX-A | 1x1000lb+3x500lb+1x250lb | Duty Carried Out |
| F/Lt J Šejbl | Sgt M Šebela | Z8784 | KX-U | 1x1000lb+3x500lb+1x250lb | Duty Carried Out |
| S/Ldr J Šnajdr | Sgt A Mžourek | R1046 | KX-E | 6x500lb | Duty Carried Out |
| F/O K Vildomec | Sgt V Procházka | X9741 | KX-D | 6x500lb | Alternative Target |

The now familiar long haul over the North Sea and northern Holland was completed, and the usual unfriendly welcome of flak and searchlights were encountered. Flying conditions were almost perfect, apart from ground haze over the entire route. Flying Officer K Vildomec was unable to reach the target. Despite his best efforts, he could not coax Wellington X9741 KX-D above 9,000ft, and wisely decided that this was not a safe altitude to bomb Hamburg. The crew bombed the town of Rendsburg, situated on the River Eider, at 01:15hrs. The remaining crews were over the target between 01:38hrs and 01:47hrs. Visibility was good but hampered by haze. Opposition over Hamburg was initially quiet, but then it suddenly appeared to increase in volume and accuracy as the first bombs landed. Flight Sergeant J Hájek was at 15,000ft when he dropped his entire load at 01:40hrs in the dock area. He reported back at East Wretham, *'Large red fire developed between Binner Alster and Docks turning violet in colours'*. Flak and searchlight activity had increased dramatically as the raid progressed, resulting in some crews being unable to confirm the accuracy of their bombing. However, all were confident it was in the target area. As the Czechs turned for home, two areas of Hamburg, one in the north and the other in the west of the city, were seen ablaze. These fires were reported visible by Flight Lieutenant J Šejbl from 35 miles on the return flight. Both 3 Group and Bomber Command H.Q seemed satisfied with the results, especially as some damage had been inflicted on the Blohm and Voss shipyards. On the 25th, the squadron recorded yet another ground accident. At 23:00hrs Sergeant Karel Weiss was taking off with his instructor for a night training flight when his Wellington R3286 KX-W struck a MT vehicle. Confusion between the pilot and ground control over flare path lighting was the cause.

| Date | 25/07/1941 |
|---|---|
| Mark | Mk.IC |
| Serial | R3286 |
| Code | KX-W |
| Taken On Charge | 07/07/1941 via 9 Squadron. |
| Manufacturer | Vickers (Weybridge) |
| Contract | B3913/39 |
| Pilot (s) | Sergeant Karel Weiss |
| Flight | Training Flight (Czech Training Flight) |
| Time | 23:00hrs |
| Cause | Collison with MT Van. |

Cologne was the intended target on the 28th, but this was cancelled at 21:40hrs for fear of the bases being unusable on return. The weather was better on the 30th when 41 Wellingtons were detailed and briefed from 3 Group to attack Cologne, 311 Squadron was not required on this occasion. The weather took a hand in operation for the last few days of July. Low clouds and heavy rain meant the squadron was stood down.

July had seen the squadron fly 101 sorties, for the loss of two crews 'missing' and one crew shot down by friendly fire. The losses were no higher than most within the Group, and given the targets and weather conditions, it was an excellent effort by all concerned on the squadron. More worrying was the departure of some experienced crews who had flown 200 operational hours. These included Flight Sergeants J Čapka, F Fencl, J Bernát, A Šedivý and Sergeant B Blatný DFM. These pilots' contribution to the early success of 311 Squadron cannot be overstated, they had endured the good times and the bad.

Their departure drastically reduced the availability of crews on the squadron from twelve down to seven. It was hoped that the arrival of the Czech Initial Training Flight from R.A.F Honington to East Wretham mid-month would counter their departure. Five crews had almost completed their training and were just weeks away from becoming operational, while another five were well advanced. A number of promotions also took place, one of which was the promotion of Squadron Leader Kenneth Batchelor to Wing Commander. The squadron was flooded with awards in July. The previously mentioned DFM and Polish medals were joined by twenty-six Czechoslovak War Crosses, 34 Bars to the Czechoslovak War Cross, two second Bars and a further 34 recipients were awarded the Czechoslovak War Medal.

# August 1941 : English Summer - Rain!

The month started with some positive news from the International Red Cross in Geneva that Flight Sergeant J Nyč, his second pilot Sergeant K Šťastný and front gunner Sergeant Knap František were safe and PoWs in Germany. Rain interrupted training throughout the day and would continue to cause problems for the squadron and the Group throughout the month. Hamburg would start the month off, and Group H.Q hoped the success of July 25th would be replicated. 3 Group would attack three targets on this night, a force of 45 Wellingtons would be joined by 1 and 4 Groups over Hamburg. A smaller force of just seven Wellingtons and five Stirlings would, along with Halifaxes and Wellingtons of 4 Group, keep the residents of Berlin awake. Finally, four 'Freshmen' crews would bomb the docks at Cherbourg. 3 Group would commit eleven of its squadrons. The only squadron not operating was 115 Squadron based at R.A.F Marham. Despite losing so many of its experienced crews, 311 Squadron contributed eight Wellingtons on Hamburg.

### August 2nd 1941 : Dace 'B' – Hamburg

| Pilot | 2nd Pilot | Serial | Code | Bomb Load | Result |
|---|---|---|---|---|---|
| P/O V Korda | Sgt J Filler | T2972 | KX-G | 2x500lb+1x250lb+360x4lb | Duty Carried Out |
| F/Sgt F Radina | Sgt A Mžourek | W5682 | KX-Y | 2x500lb+1x250lb+360x4lb | Duty Carried Out |
| F/Sgt K Schoř | Sgt S Linka | W5668 | KX-T | 2x500lb+1x250lb+360x4lb | Duty Carried Out |
| F/Lt J Šejbl | Sgt M Šebela | Z8784 | KX-U | 1x1000lb+3x500lb+1x250lb | Duty Carried Out |
| F/O F Pohlodek | Sgt O Jambor | X3221 | KX-O | 1x1000lb+3x500lb+1x250lb | Alternative Target |
| S/Ldr J Šnajdr | P/O J Nejezchleba | W5711 | KX-H | 1x1000lb+3x500lb+1x250lb | Duty Carried Out |
| Sgt A Musálek | Sgt V Ryba | R1777 | KX-M | 6x500lb | Duty Carried Out |
| F/O K Vildomec | Sgt V Procházka | X9741 | KX-D | 6x500lb | Duty Carried Out |

Sergeant J Filler had not flown on the squadron since June 7th. An unfortunate incident on June 11th resulted in a serious altercation with the squadron commander, Wing Commander Josef Schejbal. The squadron commander wanted Filler Court Martialled and posted off the squadron and transferred to the army. Now, seven weeks on, the Court Martial dropped, Josef Filler was back. Experienced Czech pilots were hard to come by, but it was an uneasy truce. Flying Officer F Pohlodek and crew got as far as Wilhelmshaven before deciding to continue was too dangerous. Engine problems aboard Wellington X3221 KX-O, on its first raid since the taxying accident on July 3rd after being repaired and returned to the squadron on July 23rd, meant the crew dropped their bombload on Wilhelmshaven at 01:50hrs in one stick. The sirens of Hamburg began wailing at 01:20hrs, although it was not until 02:04hrs that the first crew of 311 Squadron was over the target. The Czechs reported that only one medium fire was visible on approach to Hamburg but plenty of flak. Pilot Officer V Korda bombed from 15,500ft. The crew watched their bombs burst just south of the Alster, producing four fires, one on the south side and four on the northern bank. Cloud varied between 5/10th and 9/10th over the city. However, it did not prove overly troublesome. Flight Sergeant Schoř reported all his bombs landed on the Central railway station, while Flight Lieutenant J Šejbl confidently reported his high explosive load burst in the dock area. A large fire was reported in the Altona area by Flying Officer K Vildomec as he turned for home. All the squadron were safely down by 06:32hrs. Flight Sergeant Schoř for some unrecorded reason landed at R.A.F Coltishall. One crew had a narrow and rather embarrassing escape. The navigator to Flying Officer F Pohlodek had requested a QDM when south of Texel. This was duly received, but due to the distance, it was inaccurate. Having received the QDM, the navigator gave the pilot a new course, but had yet to realise the mistake and that they were heading south over Holland instead of west over the North Sea and home. Believing they were over East Anglia, they found an aerodrome near a lake with a lit flare path. The navigator, confident of his whereabouts, informed the pilot it was R.A.F Honington. The Wellington's undercarriage was lowered in preparation for landing. Thankfully, a burst of machine gun fire quickly made them realise that they were not over Honington but still over enemy territory. The undercarriage was quickly retracted, and the crew frantically climbed away. Finally, having identified their position, they set a new course for home, landing at 06:10hrs. They had been airborne for 7 ½ hours. It had been a lucky escape! The squadron was not required on the 3rd when 3 Group bombed Hanover, and the operation planned for the 4th to Frankfurt was cancelled before take-off. On the 5th, the squadron complete its previously arranged, if limited flying training programme, while nine Wellingtons were prepared for operations. Number 3 Group would make available 11 of its front line squadrons for operations on this night. Sixty-nine Wellingtons, including nine from 311 Squadron would attack Mannheim. Nine Stirlings and two Wellingtons would attack Karlsruhe while six Wellingtons would bomb the docks at Boulogne. For the Czechs, a trip to southern Germany was a welcome change from the northern ports.

### August 5th 1941 : Chubb 'A' – Mannheim

| Pilot | 2nd Pilot | Serial | Code | Bomb Load | Result |
|---|---|---|---|---|---|
| P/O V Korda | Sgt J Filler | T2972 | KX-G | 3x500lb+360x4lb | Duty Carried Out |
| F/Sgt F Radina | Sgt A Mžourek | W5682 | KX-Y | 3x500lb+360x4lb | Duty Carried Out |
| F/O K Vildomec | Sgt V Procházka | X9741 | KX-D | 3x500lb+360x4lb | Duty Carried Out |
| F/O F Pohlodek | Sgt O Jambor | X3221 | KX-O | 1x1000lb+3x500lb+1x250lb | Duty Carried Out |
| P/O J Nejezchleba | Sgt S Linka | W5711 | KX-H | 1x1000lb+3x500lb+1x250lb | Duty Carried Out |
| F/Lt J Šejbl | Sgt M Šebela | R1598 | KX-C | 1x1000lb+4x500lb | Duty Carried Out |
| F/Sgt J Hájek | Sgt J Horáček | R1015 | KX-L | 6x500lb | Duty Carried Out |
| Sgt A Musálek | Sgt V Ryba | R1777 | KX-M | 6x500lb+1x250lb | Duty Carried Out |
| F/Sgt K Schoř | F/O J Stránský | X9742 | KX-F | 6x500lb+1x250lb | Alternative Target |

*John Gelliner. His skill and diplomacy were used on numerous occasions throughout his time on the squadron. (John Costin)*

Pilot Officer Josef Nejezchleba was deemed ready to captain his own crew on this night. He had previously completed 13 operations as a second pilot, almost all flown with the departed Sergeant Benedikt Blatný DFM. The squadrons of 3 Group would be joined on this night by 33 Hampdens of 5 Group which would attack Aiming Point 'C' while the Wellington force would be divided over three Aiming Points, 'A', 'C' and 'D'. It was a bright moonlit night as the Wellingtons climbed for altitude. A troublesome layer of cloud was encountered between 12,000ft and 14,000ft which gave the crews two options, fly below it, or above it, staying in it meant encountering icing. Despite the bright moon most Czechs opted to fly above the white cotton like clouds. One crew that did not have to worry was Flight Sergeant K Schoř, they suffered port engine trouble while en route. The crew reached just south of Charleroi, Belgium, before they turned for home. Deciding to bomb Oostende, they dropped their bombs at 01:08hrs in the general area of the docks, 10/10th cloud preventing observation of the results. The run into Mannheim resulted in the usual cauldron of flak which intensified over the target area. The crews bombed from between 14,000ft and 17,000ft. The glow of fires was just visible through the clouds but no features were observed. Pilot Officer V Korda delivered his entire load in one stick at 00:50hrs near the aiming point, his incendiaries starting a fire. Only two crews reported actually seeing the target. The rear gunner aboard Wellington X9741 KX-D flown by Flying Officer K Vildomec reported that he saw buildings in the glow of fires. Flight Sergeant J Hájek was flying at 17,000ft, they bombed at 01:10hrs and submitted the following on return, *'Aircraft attacked primary objective. All bombs burst in one stick and seen to burst in dock area in north part of town. One large fire started which spread rapidly'.*

The weather over England on return was atrocious, with low clouds, rain, and icing. Despite this, the crews began landing back at East Wretham at 04:37hrs, and by 05:32hrs, they were all safely back. There was reason to celebrate. Four of Pilot Officer V Korda's crew had returned from their last operation. Václav Korda had flown a remarkable 41 operations, John Gellner, Navigator, 37 operations, Vladimír Slánský, Wireless Operator, 44 and front gunner Miroslav Čtvrtlík had flown 41. They had defied the odds and completed a hectic tour. Each would be keenly missed. This was especially true of John Gelliner, whose diplomacy had been crucial at key times when feelings amongst the squadron were running high. Group H.Q seemed generally pleased with the results reporting *'Most crews bombed their target and some large fires were seen, one very near or on a large cross roads in the centre of Mannheim'*. The now customary training flights followed an operation. On the 6th, a wet and depressing day of heavy showers, would see a number of extensive cross-country flights flown, followed in the evening by night flying circuit and bumps. Training continued the following day by the Training Flight which completed a series of long cross country flights. Operations were on for that night, and the ground crews made ready nine Wellingtons. The target was the Marshalling Yards at Hamm, which would be an all 3 Group effort.

### August 7th 1941 : Poodle 'B' – Marshalling Yards – Hamm

| Pilot | 2nd Pilot | Serial | Code | Bomb Load | Result |
|---|---|---|---|---|---|
| F/O F Pohlodek | Sgt O Jambor | X9741 | KX-D | 6x500lb+1x250lb+120x4lb | Duty Carried Out |
| Sgt A Musálek | Sgt A Šiška | R1777 | KX-M | 6x500lb+1x250lb+120x4lb | Duty Carried Out |
| F/Sgt K Schoř | Sgt J Filler | X9742 | KX-F | 6x500lb+1x250lb+120x4lb | Duty Carried Out |
| F/O J Stránský | Sgt V Procházka | Z8784 | KX-U | 6x500lb+1x250lb+120x4lb | Duty Carried Out |
| F/Sgt F Radina | Sgt A Mžourek | W5682 | KX-Y | 6x500lb+1x250lb+120x4lb | Duty Carried Out |
| P/O J Nejezchleba | Sgt S Linka | W5711 | KX-H | 6x500lb+1x250lb+120x4lb | Duty Carried Out |
| F/Lt J Breitcetl | Sgt J Svoboda | Z8805 | KX-V | 6x500lb+1x250lb+120x4lb | Duty Carried Out |
| F/O K Vildomec | F/O L Němec | T2972 | KX-G | 6x500lb+1x250lb+120x4lb | Duty Carried Out |
| Sgt V Ryba | Sgt M Plecitý | R1046 | KX-E | 6x500lb+1x250lb+120x4lb | Duty Carried Out |

The crews were a mixed bunch on this night, with two pilots appearing for the first time. Sergeant Alois Šiška and Sergeant Jindřich Svoboda would accompany a veteran captain for operational experience. At the same time, two pilots, Flying Officer Josef Stránský and Sergeant Václav Ryba, would be given the captain's role. The night would also see the welcome reappearance of Flying Officer Ludvík Němec. He had not flown on operations since the tragic Anson crash in October 1940 when he had suffered a fracture of the neck and spine.

A force of 47 Wellingtons, plus a solitary Short Stirling of 15 Squadron would be detailed and briefed for Hamm. However, the unwelcomed reduction in numbers once again blighted the Groups contribution. Five Wellingtons failed to take-off, and a further six aircraft returned early with various technical issues. Over the North Sea, the small force encountered icing and electrical storms, which did not bode well for the remainder of the flight to Hamm. Crossing the coast between Knocke and The Hague, they slipped south of Munster and its formidable flak defences. It was here the crews encountered varying clouds ranging between 6/10th and 8/10th as they made their final run into the target. Between 02:15hrs and 02:35hrs the Czechs pressed home their attacks, resulting in a devastating series of explosions and fires in the Marshalling Yards. All nine crews claimed to have bombed the aiming point from between 12,000ft and 15,000ft, encountering vicious flak throughout. Flying Officer F Pohlodek visually identified the aiming point and reported a large red fire already ablaze in the centre of the yard. A few bombs, mostly overshoots were seen to fall in the town, but the majority appeared to burst in the target area. Flight Sergeant K Schoř was at 13,500ft. He reported back at East Wretham, *'HE fell across the yards, one large fire started close to the station, visible 10-15 minutes after leaving'*. The ever-dependable Flight Lieutenant J Breitcetl also reported his entire bombload falling across the Marshalling Yards, starting a small fire. Flak very nearly claimed Flight Sergeant F Radina and crew. They were at 15,000ft when a shell exploded close enough to puncture holes in the Wellington's starboard tailplane, rudder trimmer, fuselage and astrodome, thankfully without injury to the occupants. The crew of Flying Officer J Stránský were the last to leave Hamm at 02:35hrs. They reported, *'Four large fires, two in the centre of the Marshalling Yards, two on its northern extremity'*. The squadron turned for home, having dropped 33,070 lbs of bombs on or very near the aiming

*Josef Stránský, seen here with the DFC, would captain a crew for the first time against Hamm. (John Costin)*

point. One colossal explosion was seen from which smoke rose to 11,000ft and drifted for 10 miles. The glow from the fires remained visible on the horizon for up to 30 minutes. All nine crews were back at East Wretham by 05:47hrs. It was later confirmed by Bomber Command H.Q that only 32 Wellingtons carried out this devastating raid, it was a tremendous effort. Unbeknown to the crews attacking Hamm, R.A.F East Wretham was on the receiving end of an intruder attack by a Ju88 of I./NJG2. Ten bombs were dropped creating small craters in a semi-circle in the south-west part of the aerodrome. No injuries or damage was reported.

Bad weather on the 8th kept the squadron grounded, allowing some essential maintenance on the squadron's ageing Wellingtons. The Krupps works at Essen were the intended target on the 9th, but the weather again intervened. Conditions improved slightly on the 10th, with only essential flying undertaken. Group Captain Josef Berounský, 3 Group Czech Liaison Officer, left the squadron on this date. He was sent to Moscow as Deputy Chief of the Czechoslovak Military Mission to the Soviet Union. Finally on the 11th, the weather had improved sufficiently for two 'Freshmen' crews to operate.

### August 11th 1941 : Rotterdam Docks

| Pilot | 2nd Pilot | Serial | Code | Bomb Load | Result |
|---|---|---|---|---|---|
| F/O K Vildomec | F/O L Němec | T2972 | KX-G | 6x500lb+1x250lb+120x4lb | Duty Carried Out |
| Sgt V Ryba | Sgt M Plecitý | R1046 | KX-E | 6x500lb+1x250lb+120x4lb | Duty Carried Out |

It was a later-than-normal take-off, with both crews away by 02:04hrs. 3 Group's efforts on this night were restricted to just 24 aircraft, of which 21 Wellingtons would be provided by R.A.F Marham's resident squadrons, 115 and 218. These were briefed to attack Munchen-Gladbach. 311 Squadron would be joined over Rotterdam by a single crew from 9 Squadron. Joining them would be Hampdens of 5 Group. Sergeant Ryba was the first to bomb at 03:15hrs, dropping his mixed load in two sticks across Dock 28 from 12,000ft. It was not until 03:45hrs that Flying Officer K Vildomec commenced his attack. They bombed from 15,000ft in one stick. Clouds prevented the crew from accurately confirming the accuracy, but one bomb burst was observed, which was believed to be on Dock 21.[39] Both were back at R.A.F East Wretham just after daybreak.

---

[39] *The ORB reports Dock 21. 3 Group Records and the Form E record Dock 28.*

n the 12th, 3 Group committed all 13 of its front line bomber squadrons for the first time. They would bomb no less than five separate targets, it was going to be a busy night. The chart below shows the contribution that 3 Group made on this night.

### No.3 Group Operations 12/13th August 1941

| Squadron | Hanover | Essen | Berlin | Le Havre | Bielefield | Totals |
|---|---|---|---|---|---|---|
| 7 | - | 2 Stirling | 4 Stirling | - | - | 6 |
| 9 | 12 Wellington | - | - | 1 Wellington | - | 12 |
| 15 | 5 Stirlings | 1 Stirling | 1 Stirling | - | 1 Stirling | 8 |
| 40 | 10 Wellingtons | - | - | - | - | 10 |
| 57 | 11 Wellingtons | - | 1 Wellington | 2 Wellingtons | - | 14 |
| 75(NZ) | 9 Wellingtons | - | - | 3 Wellingtons | - | 12 |
| 99 | 8 Wellingtons | - | 1 Wellington | 3 Wellingtons | - | 12 |
| 101 | 7 Wellingtons | - | | - | - | 7 |
| 115 | 2 Wellington | 7 Wellingtons | - | - | - | 9 |
| 149 | 7 Wellingtons | - | - | 3 Wellingtons | - | 10 |
| 214 | 10 Wellingtons | - | 1 Wellington | 2 Wellingtons | - | 13 |
| 218 | 1 Wellington | 6 Wellingtons | 2 Wellingtons | - | - | 9 |
| 311 (Czech) | 7 Wellingtons | - | - | 3 Wellingtons | - | 10 |
| Grand Total | 89 | 16 | 10 | 17 | 1 | 133 |

Seven aircraft would fail to take-off, which was offset by the relatively low figure of just four early returns. Three more recently trained pilots would be making an appearance on this night. Sergeants Václav Pánek, Jindřich Svoboda and Jiří Fína had proven themselves ready for operations. Sadly, none would survive the war to return to their homeland. The first Wellington lifted off the grass runway at 21:20hrs into an overcast sky. The squadron recorded the early return of just one crew. Flying Officer F Pohlodek had just cleared the English coast when starboard engine trouble aboard Wellington X3221 KX-O resulted in their operation being abandoned and the bombs jettisoned.

### August 12th 1941 : Eel 'A' – Hanover

| Pilot | 2nd Pilot | Serial | Code | Bomb Load | Result |
|---|---|---|---|---|---|
| F/Sgt F Radina | Sgt V Pánek | R1598 | KX-C | 3x500lb+360x4lb | Duty Carried Out |
| Sgt A Musálek | Sgt A Šiška | R1777 | KX-M | 3x500lb+360x4lb | Duty Carried Out |
| F/O F Pohlodek | Sgt O Jambor | X3221 | KX-O | 6x500lb+1x250lb | Early Return |
| F/Sgt J Hájek | Sgt V Soukup | R1015 | KX-L | 6x500lb+1x250lb | Duty Carried Out |
| F/Lt J Breitcetl | Sgt S Linka | W5711 | KX-H | 6x500lb+1x250lb | Duty Carried Out |
| F/O J Stránský | Sgt V Procházka | Z8784 | KX-U | 6x500lb+1x250lb | Duty Carried Out |
| F/Sgt K Schoř | Sgt J Filler | X9742 | KX-F | 6x500lb+1x250lb | Duty Carried Out |

The first of the squadron arrived over Hanover at 23:45hrs to find the defences dormant. Sergeant A Musálek dropped his mixed load from 11,500ft in one stick. Bursts were seen, and a small fire started. No other results were observed due to haze. Five minutes later, Flight Sergeant K Schoř delivered his bombload across the target. The city's defences were by now in full swing. Flak and searchlights swept the sky above Hanover. Three crews bombed simultaneously at 00:15hrs, Flight Sergeant F Radina, Flight Lieutenant J Breitcetl and Flying Officer J Stránský each claimed to have bombed the target. However, results were unobserved due to haze. 3 Group H.Q were not totally convinced about the raid's success, their report stated, *'attack seems quite successful'*.

## August 12<sup>th</sup> 1941 : Le Havre Docks

| Pilot | 2<sup>nd</sup> Pilot | Serial | Code | Bomb Load | Result |
|---|---|---|---|---|---|
| F/O K Vildomec | F/O L Němec | T2972 | KX-G | 6x500lb+1x250lb+120x4lb | Duty Carried Out |
| Sgt V Ryba | Sgt J Svoboda | R1046 | KX-E | 6x500lb+1x250lb+120x4lb | Duty Carried Out |
| Sgt A Mžourek | Sgt J Fína | Z8805 | KX-V | 6x500lb+1x250lb+120x4lb | Duty Carried Out |

Flying Officer K Vildomec bombed at midnight from 16,000ft. His bombs were seen to explode across Dry Dock No.1. Following closely behind was Sergeant A Mžourek, who was unable to establish precisely where his bombs landed in the docks due to cloud. At 00:03hrs, Sergeant V Ryba dropped his entire load in one stick on Dock No.4. Only one burst was observed, but the crews reported good fires at Docks 4 and 10 on leaving. The appearance of a Junkers Ju88 of I./NJG2 created problems for the returning crews. At 03:50hrs, the Intruder attacked three Wellingtons while in the East Wretham circuit. One Wellington flown by Sergeant A Musálek was badly mauled while flying at 1,000ft. Despite the unwelcomed arrival of the Intruder, all the Wellingtons landed safely. The following morning Wellington R1777 KX-M was inspected, the crew had been fortunate. Cannon fire had damaged the port inner mainplane, the trailing edge and port flaps. Several hydraulic pipes were punctured and an elevator badly holed. The Wellington would be repaired-on-site and would eventually return to front-line service on September 20<sup>th</sup>.

Bomber Command H.Q continued the pressure on Germany's northern cities on the 14th. Three would be visited, Hanover, Braunschweig (Brunswick) and Magdeburg. 3 Group would commit 88 crews from ten of its squadrons on this night, ten provided by 311 Squadron. Joining the Group were 30 Wellingtons of 1 Group, all were given Aiming Point 'A'. Also over Hanover would be 55 Whitleys of 4 Group who had been given Aiming Point 'B' to attack.

## August 14<sup>th</sup> 1941 : Eel 'A' – Hanover

| Pilot | 2<sup>nd</sup> Pilot | Serial | Code | Bomb Load | Result |
|---|---|---|---|---|---|
| F/O K Vildomec | F/O L Němec | T2972 | KX-G | 3x500lb+360x4lb | Duty Carried Out |
| Sgt V Ryba | Sgt M Plecitý | R1046 | KX-E | 3x500lb+360x4lb | Duty Carried Out |
| P/O J Nejezchleba | Sgt S Linka | W5711 | KX-H | 3x500lb+360x4lb | Duty Carried Out |
| F/Sgt J Hájek | Sgt V Soukup | R1015 | KX-L | 1x1000lb+4x500lb | Duty Carried Out |
| F/Lt J Šejbl | Sgt M Šebela | R1598 | KX-C | 1x1000lb+4x500lb | Early Return |
| Sgt A Musálek | Sgt A Šiška | X9741 | KX-D | 1x1000lb+4x500lb | Duty Carried Out |
| F/Sgt K Schoř | Sgt V Pánek | X9742 | KX-F | 6x500lb+1x250lb | Duty Carried Out |
| F/O J Stránský | Sgt V Procházka | Z8784 | KX-U | 6x500lb+1x250lb | Duty Carried Out |
| F/Lt J Breitcetl | Sgt J Svoboda | Z8805 | KX-V | 6x500lb+1x250lb | Duty Carried Out |
| Sgt A Mžourek | Sgt J Fína | R1021 | KX-W | 6x500lb+1x250lb | Alternative Target |

Flight Lieutenant J Šejbl aborted the operation soon after take-off due to an overheating engine. The bombs were jettisoned in the sea 40 miles off Ipswich. Hanover was spared another bombload when the crew of Sergeant A Musálek experienced persistent engine problems. They decided to bomb Munster from 15,000ft and nurse the engine back to East Wretham. The squadron was in the vanguard over the target. Bombing commenced at 23:00hrs, with the Czechs starting their bomb runs from 23:56hrs. There were already some isolated fires visible through the 6/10th cloud on arrival. Flak and searchlights were fully engaged as Flight Lieutenant J Breitcetl and Flight Sergeant K Schoř dropped their High-Explosive loads at 23:56hrs on the aiming point. Flight Lieutenant J Breitcetl was unable to see the results of his bombs, he was busy taking violent evasive action, having been caught and held by searchlights. Sergeant K Schoř reported his bombs landing 1 ½ miles southwest of the aiming point. At 00:06 hrs, Flight Sergeant J Hájek reported his bombs dropping in the main railway and station area, resulting in a large bright flash. Another crew confident that they hit the target was Flying Officer J Stránský, who bombed at 00:10hrs from 15,000ft. They report at de-briefing, *'All bombs burst across railway station'*. By the time the squadron turned for home, isolated fires ranging in size from small to medium were reported almost circling the aiming point, with the fiercest in the northeast, east, south and southwest of the railway. The bombing would continue until 02:47hrs. All the Czech crews were back by 04:35hrs, having dropped

21,570lbs of bombs on Hanover. The Czech Training Flight recorded an accident involving the recently promoted Warrant Officer Karel Weiss on the 14th. Returning to R.A.F East Wretham at 22:55hrs after a night cross country training flight, the young pilot overshot his landing and ended up wrapping a barbed wire fence around Wellington T2624.

| Date | 14/08/1941 |
| --- | --- |
| Mark | Mk.IC |
| Serial | T2624 |
| Code | KX-? |
| Taken On Charge | 02/08/1941 via 15 Squadron after being repaired by No.43 Group. |
| Manufacturer | Vickers (Weybridge) |
| Contract | B38600/39 |
| Pilot (s) | Warrant Officer Karel Weiss |
| Flight | Night Training Flight (Czech Training Flight) |
| Time | 22:55hrs |
| Cause | Overshot Landing |

None of the crew were injured. Damage to T2624 was extensive, considering the report just mentions tangling with barbed wire. The port wheel was damaged, bomb doors and port flaps were unserviceable, airscrew and aileron were also damaged, and the geodetic was torn away in places. Repairs were not completed until August 28th when it returned to Czech Training Flight.

A brief period of good weather on the morning of the 15th allowed a number of cross-country exercises to be flown. By mid-afternoon, low cloud and drizzle arrived bringing training and test flights to a premature conclusion. The Ruhr city of Duisburg was selected by H.Q Bomber Command for attention on the 16th, which would be an all 3 Group operation. Group detailed and briefed 58 crews from six of its front line squadrons. 311 Squadron were back into double figures with ten crews detailed.

August 16th 1941 : Cob 'B' – Duisburg

| Pilot | 2nd Pilot | Serial | Code | Bomb Load | Result |
| --- | --- | --- | --- | --- | --- |
| F/Lt J Breitcetl | Sgt J Svoboda | Z8805 | KX-V | 3x500lb+1x250lb+360x4lb | Duty Carried Out |
| Sgt V Ryba | Sgt M Plecitý | R1046 | KX-E | 3x500lb+1x250lb+360x4lb | Duty Carried Out |
| Sgt A Musálek | Sgt A Šiška | X9741 | KX-D | 1x1000lb+5x500lb+1x250lb | Duty Carried Out |
| F/Sgt K Schoř | Sgt J Filler | R1451 | KX-P | 1x1000lb+5x500lb+1x250lb | Duty Carried Out |
| P/O J Nejezchleba | Sgt S Linka | W5711 | KX-H | 1x1000lb+5x500lb+1x250lb | Duty Carried Out |
| F/O J Stránský | Sgt V Procházka | Z8784 | KX-U | 8x500lb | Duty Carried Out |
| Sgt A Mžourek | Sgt J Fína | R1161 | KX-X | 8x500lb | Duty Carried Out |
| F/Lt J Šejbl | Sgt V Pánek | R1598 | KX-C | 8x500lb | Early Return |
| F/O K Vildomec | F/O L Němec | T2972 | KX-G | Withdrawn | Cancelled |
| F/Sgt J Hájek | Sgt V Soukup | X9742 | KX-F | Withdrawn | Cancelled |

Two crews were withdrawn before take-off,[40] and a third crew skippered by Flight Lieutenant J Šejbl, was obliged to jettison their bombload when Wellington R1598 KX-C experienced engine trouble. The crew were back at their dispersal within 60 minutes. Josef Šejbl had every right to be cautious. He was one of the most experienced pilots on the squadron at the time, having flown 34 operations and was very close to being withdrawn from operations and rested. Pressing on to the target at this stage in his operational career would have been foolhardy. The squadron was over the target between 01:53hrs and 02:50hrs in reasonable weather conditions but with wisps of cloud and the usual industrial

---

[40] *The second pilots names are not confirmed.*

*Part of Korda's crew with ground staff in front of Wellington Mk IC T2972/KX-G. L-R: 3. Pilot Officer V. Korda, 4. Flight Sergeant M. Čtvrtlík, 5. Pilot Officer K. Janšta. Note both blackened out and covered in windows. (V. Kolesa)*

haze. Fortunately, the River Rhine and its distinctive Ruhrort complex, the largest inland docks in Germany, provided a reference for some. In contrast, others were guided by searchlight and flak activity, reported by returning crews as the most intense yet encountered as the Germans bolstered their defences around the Ruhr region. Some useful damage was inflicted. Seven large fires were reported, one of the larger conflagrations was observed from 90 miles on the return flight. Only 33 Wellingtons claimed to have actually bombed Aiming Point 'B', while three others bombed what they believed to be Duisburg.

Local thundery showers on the 17th restricted training, but cross-country flights were completed by crews close to finishing their time on the Training Flight. One of those however made a rather heavy landing at R.A.F Langham at midday resulting in the undercarriage collapsing. The crew, captained by Sergeant Karel Danihelka, were uninjured.

| Date | 17/08/1941 |
|---|---|
| Mark | Mk.IC |
| Serial | R1523 |
| Code | KX-B |
| Taken On Charge | 02/08/1941 |
| Manufacturer | Vickers (Chester) |
| Contract | B.992424/39 |
| Pilot (s) | Sergeant Karel Danihelka |
| Flight | Daylight Cross-Country (Czech Training Flight) |
| Time | 12:00hrs |
| Cause | Heavy Landing u/c collapsed. |

The damage to the almost new Wellington was beyond the capacity of the groundcrews at Langham. The Wellington would be repaired by No.43 Maintenance Unit, it never returned to squadron service. Once repaired, it would have a lengthy career with various Operational Training Units. The old war-horse was finally scrapped in 1946!

Weather conditions were still causing issues on the 19th. Rain and low clouds throughout the day again restricted flying. A few training flights did take place, one of which resulted in yet another accident on the Training Flight. Sergeant Vladimír Šponar was taxiing Wellington R3166 KX-K at East Wretham a little too fast when a tyre burst resulting in the undercarriage collapsing. The crew unlike the Wellington were unharmed.

| Date | 19/08/1941 |
|---|---|
| Mark | Mk.IC |
| Serial | R3166 |
| Code | KX-K |
| Taken On Charge | 12/06/1941 via 75 Squadron. |
| Manufacturer | Vickers (Weybridge) |
| Contract | B3913/39 |
| Pilot (s) | Sergeant Vladimír Šponar / Flying Officer Oldřich Hořejší |
| Flight | Daylight Cross-Country (Czech Training Flight) |
| Time | 12:00hrs |
| Cause | Heavy Landing tyre-burst, undercarriage collapsed. |

On the 18th of August 1941, the Butt Report was published. Within hours of its publication, shock waves were felt throughout Bomber Command H.Q and the Air Ministry. Reaction from Bomber Command H.Q was instant, and almost immediately, they dismissed its findings. The report was initiated by Lord Cherwell, a friend of Churchill and chief scientific advisor to the Cabinet. David Bensusan-Butt, a civil servant in the War Cabinet Secretariat and an assistant of Cherwell, was tasked with assessing 633 target photos and comparing them with crews' claims. Dismiss it they might, but it had the backing of some influential people and many Bomber Command detractors. The following is a small part of the report but highlights the apparent failure of the crews given the rudimentary navigation equipment then available.

> *Dear Wing-Commander Lewin Bowring Duggan,*
>
> *I attach the report on the work I have been doing in the past fortnight. It is, I am afraid, very lengthy, but the summary on top will perhaps be of assistance.*

*You will see that I have not attempted any study of the day photographs or of comparative damage in German and British towns. This was partly because the time at my disposal was not long enough, but mainly because to be done properly the work requires skill at interpreting photographs and technical knowledge of bomb damage, neither of which I possess. But as mentioned in the last part of the report I very much hope that this job will get done by those suitably qualified. I can, however, record a definite impression – not more – that such a study would give results consistent with those I reached.*

*If there are points about this report which you would like me to amplify perhaps you would be kind enough to give me a ring here. I can be obtained via the Air Ministry, Whitehall, switchboard. In the evenings I am usually at Marlow 487 and could easily get over to see you.*

*I cannot end without thanking you and all the other officers at Bomber Command whom I saw, and particularly those of the Photographic Interpretation Section, for their kindness and patience. You made my stay most enjoyable and I only hope in return that I have been of some use.*

*Yours sincerely,*

*David Miles Bensusan-Butt*

*MOST SECRET*

## SUMMARY

### STATISTICAL CONCLUSIONS.

*An examination of night photographs taken during night bombing in June and July points to the following conclusions:*

1) *Of those aircraft recorded as attacking their target, only one in three got within five miles.*
2) *Over the French ports, the proportion was two in three; over Germany as a whole, the proportion was one in four; over the Ruhr, it was only one in ten.*
3) *In the Full Moon, the proportion was two in five; in the new moon it was only one in fifteen.*
4) *In the absence of haze, the proportion is over one half, whereas over thick haze it is only one in fifteen.*
5) *An increase in the intensity of [AAA] fire reduces the number of aircraft getting within 5 miles of their target in the ratio three to two.*
6) *All these figures relate only to aircraft recorded as attacking the target; the proportion of the total sorties which reached within five miles is less by one third.*

*Thus, for example, of the total sorties only one in five get within five miles of the target, i.e. within the 75 square miles surrounding the target.*

### RECOMMENDATIONS

*These results though fairly reliable should be checked by a thorough expert study of the day photographs, and by a comparative study of photographs of German and British towns.*

*In order to keep these figures up to date, and to obtain continuous records of the success of air our navigation, staff should be set up to maintain statistical records of night photographs and any other evidence that may be available.*

*This staff should consist of at least one trained statistician, with sufficient clerical staff. He should have authority to modify forms and questionnaires in order to make sample [inquiries], e.g. to replace some existing questions for a certain period by others designed to elucidate some particular point.*

There was already some doubt about the weather even before the Wellingtons of 3 Group took off on the evening of August 19th for a raid on Kiel. 3 Group H.Q committed 44 of its crews, plus four 'Freshmen' on this night despite concerns about the weather over northern Germany.

### August 19th 1941 : Minnow 'C' – Kiel

| Pilot | 2nd Pilot | Serial | Code | Bomb Load | Result |
|---|---|---|---|---|---|
| F/Lt J Šejbl | Sgt M Šebela | Z8784 | KX-U | 2x500lb+360x4lb | Duty Carried Out |
| F/O K Vildomec | F/O L Němec | T2972 | KX-G | 2x500lb+360x4lb | Duty Carried Out |
| P/O J Nejezchleba | Sgt S Linka | W5711 | KX-H | 1x1000lb+3x500lb | Duty Carried Out |
| F/Sgt K Schoř | Sgt J Filler | R1451 | KX-P | 1x1000lb+3x500lb | Duty Carried Out |
| Sgt A Musálek | Sgt A Šiška | X9741 | KX-D | 1x1000lb+3x500lb | Duty Carried Out |
| F/Lt J Breitcetl | Sgt J Svoboda | Z8805 | KX-V | 5x500lb+1x250lb | Duty Carried Out |
| Sgt V Ryba | Sgt M Plecitý | R1046 | KX-E | 5x500lb+1x250lb | Duty Carried Out |
| Sgt A Mžourek | Sgt J Fína | W5668 | KX-T | 5x500lb+1x250lb | Alternative Target |
| Sgt V Soukup | Sgt V Pánek | R1015 | KX-L | 5x500lb+1x250lb | Duty Carried Out |

Having crossed the North Sea, the squadron found the Schleswig-Holstein peninsula covered in an impregnable layer of cloud that stretched as far as Kiel and beyond. It was here that the crew of Sergeant A Mžourek abandoned the operation. A faulty port engine prevented the crew from continuing to target, so they opted to drop their bombs on a concentration of searchlights located on the western edge of the Island of Terschelling at 00:08hrs. Flashes were seen, and some searchlights flickered off. With no landmarks visible and attempting to bomb below the clouds suicidal, the squadron's navigators had an almost impossible job. Arriving over the target area at 23:31hrs, the crews bombed on either E.T.A, DR or the faint flashes of flak or searchlights. The last to bomb was Sergeant V Soukup and crew at 00:40hrs. This was Vilém Soukup's first operation as captain. There was an anxious wait for Flight Lieutenant J Šejbl, who finally landed at 05:13hrs. He and his crew had been airborne for over 8 hours. H.Q Exning summed up the conditions that night as *'Rotten'*, the raid was a failure. Bomber Command H.Q seemed to put their usual spin on the raid, reporting, *'Some big fires started'*. However, exactly where was not stated.

The accidents continued the following day when fellow trainee pilot Sergeant František Naxera landed heavily at East Wretham on return from a cross-country exercise at 16:15hrs. The tail wheel fractured and then collapsed, but the pilot unwisely decided to taxi for ½ mile!

| Date | 20/08/1941 |
|---|---|
| Mark | Mk.IC |
| Serial | N2772 |
| Code | KX-K |
| Taken On Charge | 28/09/1940 via No. 18 MU. |
| Manufacturer | Vickers (Chester) |
| Contract | B124362/40 |
| Pilot (s) | Sergeant František Naxera |
| Flight | Daylight Cross-Country (Czech Training Flight) |
| Time | 16:15hrs |
| Cause | Tailwheel collapsed heavy landing. |

Weather restricted the squadron's activities on the 21st and meant the cancellation of the planned attack on Duisburg organised by Bomber Command H.Q. There was some excitement, however. Officials from the B.B.C paid a visit to the airfield and interviewed and recorded several officers and airmen who spoke about their experiences. They would, over the coming weeks be used by the B.B.C Foreign and Home Service Broadcasts. The weather had improved slightly the following day, allowing a number of training flights over Berners Heath for gunnery practice. Mannheim would be

*Josef Bernát DFM*

*Vilém Bufka*

*František Sixta*

*Bohumil Landa*

attacked that night, but 311 Squadron would not be involved. Instead, it would send a solitary 'Freshmen' crew to Le Havre.

### August 22nd 1941 : CC.24(A) – Le Havre

| Pilot | 2nd Pilot | Serial | Code | Bomb Load | Result |
|---|---|---|---|---|---|
| Sgt J Horáček | Sgt K Knaifl | X9742 | KX-F | 7x500lb | Early Return |

This would be Sergeant Josef Horáček's first operation as captain, having previously flown 17 operations in the 2nd pilot's seat, most of which were besides experienced pre-war airline pilot Václav Korda. Sergeant Karel Knaifl would be flying his first operation. He began his military service in 1936 at the Military Aviation School in Prostějov. Like so many of his comrades, he escaped the territory of the Protectorate and ended up in France before arriving in England. Taking off at 01:53hrs, the crew headed south and then out over the English Channel. The operation was brought to an early conclusion when an overheating engine resulted in the operation being abandoned and the bombs jettisoned 30 miles south of Bognor Regis on the Sussex coast. The crew landed safely back at East Wretham at 05:52hrs. An Intruder, probably a Ju88C of 2./NJG2 dropped a number of 'incendiaries' on the eastern side of the aerodrome without causing any damage or injuries. Bomber Command remained grounded on the 23rd and 24th. At East Wretham, a Night Flying Programme was cut short when Sergeant J Filler landed his Wellington with its undercarriage retracted at 21:15hrs in the centre of the flare path. It was a careless mistake by a pilot with over 200 flying hours on the Vickers Wellington.

| Date | 24/08/1941 |
|---|---|
| Mark | Mk.IC |
| Serial | L7847 |
| Code | KX-? |
| Taken On Charge | 23/05/1941 via 99 Squadron. |
| Manufacturer | Vickers (Chester) |
| Contract | B.992424/39 |
| Pilot (s) | Sergeant Josef Filler |
| Flight | Night Flying |
| Time | 21:15hrs |
| Cause | Landed with undercarriage retracted. |

The Wellington, once repaired rejoined the squadron on September 6th. After an enforced absence, Bomber Command H.Q planned an operation against Karlsruhe on the 25th. Once again 311 Squadron was not required. On the 26th, Form B.591 arrived with instructions to prepare ten crews for a raid on Cologne plus one 'Freshmen' crew against the docks at Boulogne. During the day British Movietone News arrived at R.A.F East Wretham and took a number of photographs to illustrate the varied and challenging work of the Czech personnel.

### August 26th 1941 : Trout 'C' – Cologne

| Pilot | 2nd Pilot | Serial | Code | Bomb Load | Result |
|---|---|---|---|---|---|
| F/Lt J Šejbl | Sgt M Šebela | R1598 | KX-C | 7x500lb | Early Return |
| P/O J Nejezchleba | Sgt S Linka | W5711 | KX-H | 3x500lb+1x250lb+360x4lb | Duty Carried Out |
| F/Sgt F Radina | Sgt V Pánek | R1015 | KX-L | 1x1000lb+4x500lb+1x250lb | Duty Carried Out |
| Sgt V Ryba | Sgt M Plecitý | R1046 | KX-E | 1x1000lb+4x500lb+1x250lb | Duty Carried Out |
| F/O K Vildomec | F/O L Němec | T2972 | KX-G | 7x500lb | Duty Carried Out |
| F/Lt J Breitcetl | Sgt J Svoboda | Z8805 | KX-V | 7x500lb | Duty Carried Out |
| F/O F Pohlodek | Sgt O Jambor | X3221 | KX-O | 1x1000lb+4x500lb+1x250lb | Duty Carried Out |
| F/O J Stránský | Sgt V Procházka | Z8784 | KX-U | 3x500lb+1x250lb+360x4lb | Duty Carried Out |

Bomber Command H.Q gathered all its heavy groups for the attack on Cologne. 19 Wellingtons would be provided by 1 Group, which would bomb Aiming Point 'B'. 3 Group would provide 38 Wellingtons and attack Aiming Point 'C'.

The Whitleys of 4 Group would tackle Aiming Point 'D' who would be joined by 29 Hampdens and 1 Manchester. Although 311 Squadron was ordered to provide ten crews for Cologne, only eight would take-part. This number would be reduced when Flight Lieutenant J Šejbl and crew suffered engine issues aboard Wellington R1598 KX-C. An overheating port engine prevented the Wellington from gaining altitude. The crew jettisoned the bombs 'safe' into the sea 30 miles east of Clacton. This was one of only three early returns reported by 3 Group that night.

Weather conditions while en route varied considerably and this continued over Cologne where some crews were fortunate in finding gaps in the cloud. The squadron was over the target between 00:50hrs and 01:15hrs operating between 13,000ft and 17,000ft. All the crews claimed to have bombed the primary. Flight Lieutenant J Breitcetl and Flight Sergeant F Radina both reported their loads bursting close to the Aiming Point, Radina even reporting one of his bombs may have been a direct hit on the railway station. Unable to see the Aiming Point, Pilot Officer F Pohlodek and crew dropped their mixed load at 01:13hrs on a concentrated mass of searchlights starting one large red fire, and a few smaller ones. By the time the crews cleared the target a few fires were observed, one of which was reported as large. Local sources in Cologne claimed that most of the bombs had missed the eastern side of the city and no significant damage had been inflicted by the few bombs which had landed within its boundaries.

August 26th 1941 : CC.29(A) – Boulogne

| Pilot | 2nd Pilot | Serial | Code | Bomb Load | Result |
|---|---|---|---|---|---|
| Sgt J Horáček | Sgt K Knaifl | X9741 | KX-D | 7x500lb+120x4lb | Duty Carried Out |

Conditions were considerably better for the 16 crews allocated Boulogne Docks. Sergeant J Horáček and crew dropped their mixed load at 02:07hrs from 13,000ft and were satisfied to see them straddle Dock No.3 and 4. The incendiaries started a fire, which was seen to develop on departure. Crews reported that fires could be seen from 40 miles away on the return journey. The squadron notched up yet another accident on the 26th, when Wellington R1021 KX-W captained by Sergeant Alois Mžourek taxied into a recently unmarked trench at 22:25hrs.

| Date | 26/08/1941 |
|---|---|
| Mark | Mk.IC |
| Serial | R1021 |
| Code | KX-W |
| Taken On Charge | 28/09/1940 via No.22 MU. |
| Manufacturer | Vickers (Chester) |
| Contract | B.992424/39 |
| Pilot (s) | Sergeant Alois Mžourek |
| Flight | Night Flying |
| Time | 22:25hrs |
| Cause | Taxied into ditch. |

No blame was attributed to the pilot, but steps were taken to ensure that any work undertaken by civilian workmen, or RAF personnel would in future be reported and duly marked. The Wellington would not return to service until October.

Training in between heavy rain and clouds was safely undertaken on the 27th. A film was shown in the N.A.A.F.I. for the first time since the squadron arrived at East Wretham on this date. Until then, personnel would have to drive to R.A.F Honington for any shows or films. Now, for the first time, the East Wretham N.A.A.F.I was to be utilised. After a few short documentaries about the R.A.F, the first film shown, was not surprisingly, *'Target For Tonight'* starring Percy 'Pick' Pickard DSO DFC. The evening was an enormous success and confirmed that the N.A.A.F.I was an ideal venue, and everyone hoped that it would be used again for future entertainment. Thunderstorms, low clouds and rain resulted in a restricted flying schedule for the squadron the following day but this did not prevent 3 Group H.Q putting

*The men behind the scenes. With the hard work, skill and dedication of the ground crews, Bomber Command achieved the success it did. Here, a smiling Czech LAC Armourer poses beside the rear turret.*

together a force of 58 bombers to attack Duisburg with reasonable results. The following night, Group detailed 73 crews to bomb Mannheim, eight of which would be supplied by the Czechs.

### August 29th 1941 : Chub 'D' – Mannheim

| Pilot | 2nd Pilot | Serial | Code | Bomb Load | Result |
|---|---|---|---|---|---|
| F/O K Vildomec | F/O L Němec | T2972 | KX-G | 3x500lb+360x4lb | Duty Carried Out |
| F/Lt J Breitcetl | Sgt J Svoboda | Z8805 | KX-V | 3x500lb+360x4lb | Duty Carried Out |
| Sgt V Ryba | Sgt M Plecitý | R1046 | KX-E | 3x500lb+360x4lb | Duty Carried Out |
| F/Sgt F Radina | Sgt V Pánek | W5682 | KX-Y | 3x500lb+360x4lb | Alternative Target |
| F/Sgt K Schoř | Sgt J Filler | R1451 | KX-P | 3x500lb+360x4lb | Duty Carried Out |
| Sgt V Soukup | Sgt J Miklošek | R1015 | KX-L | 1x1000lb+4x500lb | Duty Carried Out |
| F/O F Pohlodek | Sgt O Jambor | X3221 | KX-O | 1x1000lb+4x500lb | Duty Carried Out |
| F/O J Stránský | Sgt V Procházka | X9742 | KX-F | 6x500lb+1x250lb | Duty Carried Out |

The raid would be a predominantly 3 Group effort but supplemented by 25 Wellingtons from 1 Group. The 3 Group contribution would be depleted, firstly three crews failed to take-off and a further nine returned early, the majority of which was caused by severe icing and electrical storms.

Sergeant Ján Miklošek would be taking part in his first operation, he would join the crew of Sergeant V Soukup aboard Wellington R1015 KX-L. Conditions as previously mentioned were terrible, one Czech crew was obliged to find an alternative target. Flight Sergeant F Radina got as far as the town of Kaiserlautern located 30 miles west of Mannheim. Severe icing prevented the Wellington from gaining altitude and the crew intelligently decided that to continue would be foolish. They dropped their mixed load at 23:20hrs from 11,000ft through a gap in the cloud, from which they identified a river and railway. It was here their bombload was seen to create a fire. Those reaching the target area found it concealed beneath 9/10th cloud. This greatly hindered the identification of the Aiming Point, the Main Railway Station. Two of the early crews, Flight Sergeants V Soukup and K Schoř bombed on E.T.A. or the glow of what was believed to be fires. Soukup had his 1000-pounder hang up due to ice, which they brought back to R.A.F East Wretham. Some crews were luckier and found a gap in the cloud. Sergeant V Ryba managed to glimpse the river's loop south of the target, his bombs were dropped at 23:15hrs starting a blaze. This same gap also gave the crew of Flying Officer F Pohlodek a fleeting glance of the rivers loop, where he dropped his bombs on an estimated position at 23:30hrs. The last to bomb at 23:32hrs was Flying Officer K Vildomec from 16,000ft. He identified Mannheim through a gap in the cloud, but worryingly reported back at R.A.F East Wretham, *'No fires seen on arrival'*. It was later confirmed that bombing results were poor with only scattered damage.

### August 29th 1941 : CC24 – Le Havre

| Pilot | 2nd Pilot | Serial | Code | Bomb Load | Result |
|---|---|---|---|---|---|
| Sgt A Mžourek | Sgt K Knaifl | X9741 | KX-D | 6x500lb+1x250lb+120x4lb | Duty Carried Out |

Sergeant A Mžourek was one of just three crews from 3 Group detailed to bomb the docks at Le Havre. The crew identified the town in between the clouds and dropped their bombload from 12,000ft starting a series of fires between Dock 7 and Bassin-de-Maree. The crew landed back at base at 02:05hrs.

August was a month of exceptionally poor weather which resulted in the squadron operating on only 11 nights flying 86 sorties and in the process dropping 303,950lb of bombs without loss. The month saw the award of one Czechoslovak Medal of Valour, and four Czechoslovak War Crosses. The squadron lost the services of Flight Lieutenant Angus MacNicol, Adjutant during the month, along with another R.A.F officer, Flying Officer Gregory MacMahon. Posted to 311 Squadron was the impressively named Flight Lieutenant Owen Geoffrey Langford Pillivant Powell to fill the vacant Adjutant post.

# September 1941 : Italian Skies

The month started with the squadron involved in a small joint 3 and 5 Group effort directed against Cologne. Thirty-five Wellingtons would be provided by 3 Group, of which 311 Squadron supplied nine. They would be joined over Cologne by twenty Hampdens from 5 Group.

### September 1st 1941 : Trout 'B' – Cologne

| Pilot | 2nd Pilot | Serial | Code | Bomb Load | Result |
|---|---|---|---|---|---|
| S/Ldr J Šejbl | F/Sgt A Jedounek | Z8784 | KX-U | 3x500lb+360x4lb | Duty Carried Out |
| F/O K Vildomec | F/O L Němec | T2972 | KX-G | 3x500lb+360x4lb | Duty Carried Out |
| Sgt V Ryba | Sgt M Plecitý | R1046 | KX-E | 3x500lb+360x4lb | Duty Carried Out |
| F/Sgt F Radina | Sgt V Pánek | W5682 | KX-Y | 3x500lb+360x4lb | Duty Carried Out |
| F/Lt J Breitcetl | Sgt J Svoboda | Z8805 | KX-V | 6x500lb+1x250lb | Duty Carried Out |
| F/Sgt K Schoř | Sgt J Filler | R1451 | KX-P | 1x1000lb+4x500lb | Duty Carried Out |
| P/O J Bala | Sgt K Knaifl | X9741 | KX-D | 6x500lb+1x250lb | Duty Carried Out |
| F/O J Stránský | Sgt B Hradil | X9742 | KX-F | 6x500lb+1x250lb | Duty Carried Out |
| P/O L Anderle | W/O K Weiss | X3221 | KX-O | 6x500lb+1x250lb | Duty Carried Out |

The squadron welcomed back two pilots after a lengthy absence on this night. Flight Sergeant Arnošt Jedounek, who had last flown on March 14th and Leo Anderle, absent from operations since April. Two new pilots would be appearing on the Battle Order for the first time. French veteran 26-year-old Sergeant Bohuslav Hradil, and the accident prone Warrant Officer Karel Weiss. The recently promoted Josef Šejbl would be operating as Squadron Leader for the first time, to the delight of everyone on the squadron.

All the squadron had safely departed by 20:37hrs and climbed away into better-than-forecast conditions.
The sirens in Cologne were the first warning of impending trouble as the civilian population made their way to their basements or shelters. The first 3 Group bombs started dropping at 22:29hrs over a hazy target. The aiming point was the East Kalk Marshalling Yards on the river's eastern bank and the railway lines leading to the Hohenzollern railway bridge. The first Czech crew to bomb was Pilot Officer L Anderle at 22:46hrs from 15,000ft. His bombload landed on the east side of the river in one stick. Bursts were seen, but the haze prevented accurate observation. Flying Officer K Vildomec pin-pointed the river and Autobahn before he dropped his mixed load at 23:00hrs. His bombs were seen to burst just south of the target, but the glare from numerous searchlights prevented observation of the results. A large fire, red in colour, was seen on the east side of the river as Sergeant V Ryba and crew began their bomb run at 23:10hrs. Intense flak and searchlights resulted in the crew of Flight Sergeant K Schoř bombing on E.T.A. ten minutes later. The last over the target at 15,500ft was Flight Sergeant F Radina at 23:37hrs. They reported, *'Bombed one stick. Bursts slightly east of Aiming Point. Burst seen, fires started'*. As the crews turned away, some large fires were seen south and north-east of the city, while smaller fires were reported on the east side of the river in the area of the Marshalling Yards. There were some concerns at Bomber Command H.Q. about crews bombing a series of decoy dummy fires. Cologne reported only 35 bombs of any kind were dropped, damaging one house.

Familiar names appeared on the operations board on the 3rd, *Gneisenau* and *Scharnhorst* both relatively safe in Brest Harbour were to be attacked. 3 Group would provide the bulk of the 140 bombers detailed to take part. Sixty-seven crews would be detailed and briefed including ten from 311 Squadron.

### September 3rd 1941 : 'Toads' – Brest Harbour

| Pilot | 2nd Pilot | Serial | Code | Bomb Load | Result |
|---|---|---|---|---|---|
| S/Ldr J Šejbl | F/Sgt A Jedounek | Z8784 | KX-U | 7x500lb | Duty Complete |
| F/Lt K Vildomec | F/O L Němec | T2972 | KX-G | 1x1000lb+4x500lb+1x250lb | Duty Complete |

| Sgt A Musálek | Sgt A Šiška | X9742 | KX-F | 1x1000lb+4x500lb+1x250lb | Duty Complete |
| F/Sgt K Schoř | Sgt J Filler | R1451 | KX-P | 7x500lb | Duty Complete |
| W/Cdr J Ocelka DFC | Sgt J Svoboda | Z8805 | KX-V | 1x1000lb+2x500lb+1x250lb | Duty Complete |
| F/Sgt F Radina | Sgt V Pánek | W5682 | KX-Y | 1x2000lb+2x500lb+1x250lb | Duty Complete |
| Sgt V Ryba | Sgt M Plecitý | R1046 | KX-E | 7x500lb | Duty Complete |
| F/O F Pohlodek | Sgt O Jambor | X3221 | KX-O | 7x500lb | Early Return[41] |
| P/O J Bala | Sgt K Knaifl | X9741 | KX-D | 7x500lb | Duty Complete |
| P/O L Anderle | W/O K Weiss | R1161 | KX-X | 7x500lb | Duty Complete |

The squadron commanding officer, Wing Commander J Ocelka DFC, would be accompanying the crews this night. He would command the partially changed crew of Flight Lieutenant J Breitcetl who was withdrawn from operations to become an Operational Training Flight instructor. Ocelka was seconded by Breitcetl's former 2nd pilot Sergeant Jindřich Svoboda who had flown already 8 sorties. Jindřich Svoboda was born on 23rd May 1917. Pre-war, he was a member of the Junák association, Czech equivalent to the Scout organisation in England. In 1936, he attended the Military Aviation School in Prostějov. Deciding his future lay elsewhere after the occupation, he travelled to France via Hungary, Yugoslavia, Greece, Turkey, Syria, and North Africa. When the French threw in the towel, he arrived in England to continue training. From a 3 Group perspective, the 'Toads' had been neglected since 'Operation Sunrise' apart from the occasional 'Freshmen' raid. It had been almost two months since 311 Squadron's last visit, and little if anything, appeared to have changed. The crews were warned at briefing that there was a very good possibility that East Wretham could be fog-bound on return.

*Josef Filler was a complicated and opinionated individual. His fondness for drinking, high jinks, and issues with authority often found him in trouble with the commanding officer. However, he was an excellent pilot. (John Costin)*

It was not the news the crews wanted to hear on completion of a long raid. All ten Wellingtons were airborne between 19:00hrs and 19:09hrs. The concerns about the weather resulted in 1 Group withdrawing its contribution before take-off. However, one squadron, the Elsham Wolds based 103 was already airborne, and two of its Wellingtons did not receive the recall message. The Yorkshire-based 4 Group, with the furthest to fly, either recalled or cancelled the operation. This also applied to the Lincolnshire based 5 Group. However, a 4 Group Wellington and a 5 Group Hampden also appeared not to have heard the recall message and continued to Brest. This left 3 Group to continue alone with the four aircraft that had failed to pick up the recall signal. One 311 crew aborted, Flying Officer F Pohlodek suffered engine trouble, resulting in the entire bombload being jettisoned into the sea. They landed safely at R.A.F Cerney, Gloucestershire. The first bombs to fall on Brest were dropped around 21:35hrs. Conditions above the port were ideal, but an effective smoke screen quickly enveloped the cruisers and surrounding docks, causing considerable problems for the bomb aimers. Bombing between 9,000ft and 16,000ft, the squadron crews had varying success. Flight Lieutenant Vildomec, Flight

---

[41] *Other sources state crew bombed target.*

Sergeants Schoř, Radina and Sergeant Musálek all claimed to have bombed Dock No.1, and each reported seeing their bombs bursts. Wing Commander J Ocelka DFC bombed on the estimated position of the docks from 15,000ft. Results were unobserved, smoke and the glare from the searchlights making it virtually impossible to see any ground detail. Intense flak persuaded him not to loiter in the target area. The crew of Squadron Leader J Šejbl glimpsed a convoy leaving the harbour, so they opted to bomb this at 22:06hrs from 9,000ft. Dropping the bombs in one stick, they landed neatly between the first and second ship. Fog, smoke and withering flak made any attempt to observe results impossible. By 22:53hrs, the squadron were on their way home. As predicted, fog had settled over most of the region. Thankfully, a few airfields within the Group remained open. Five squadron crews landed at R.A.F Mildenhall, while a single crew landed at Alconbury, Wyton and Newmarket. Squadron Leader J Šejbl landed a R.A.F Wittering, a fighter airfield in Cambridgeshire. In total, 45 crews were diverted on return from Brest.

The squadron sat out the raid on Huls on the 6th. It was just as well as nine bombers failed to return. 4 Group would take the brunt, losing seven Whitley crews. 3 Group got off lightly, reporting just two missing. The following day, Bomber Command H.Q gathered a force of 197 bombers for a raid on Berlin. Once again, 3 Group would provide the majority detailing 75 bombers, including 12 four-engined Short Stirlings. They would bomb Aiming Point 'B' along with the Wellingtons of 1 Group. At the same time, a force of 54 bombers would attack Kiel, including three crews from 311 Squadron, while 47 'freshmen' crews would visit Boulogne. 3 Group would supply 22 crews against Kiel while another 22 were given Boulogne. It was a busy night at 3 Group H.Q.

The Czechs at R.A.F East Wretham provided six crews on Berlin. It had been over four months since their last visit and there was a sense of excitement mixed with foreboding throughout the squadron. Only one pilot had visited Berlin before, Flight Sergeant František Radina.

<u>September 7th 1941 : 'Whitebait'– 'B' – Berlin</u>

| Pilot | 2nd Pilot | Serial | Code | Bomb Load | Result |
|---|---|---|---|---|---|
| F/Sgt F Radina | Sgt V Pánek | X3221 | KX-O | 1x500lb+1x250lb+360x4lb | Duty Carried Out |
| Sgt V Ryba | Sgt M Plecitý | R1046 | KX-E | 1x500lb+1x250lb+360x4lb | Duty Carried Out |
| Sgt A Musálek | Sgt A Šiška | X9742 | KX-F | 1x500lb+1x250lb+360x4lb | Duty Carried Out |
| P/O J Bala | Sgt K Knaifl | X9741 | KX-D | 1x1000lb+1x500lb+250lb | Duty Carried Out |
| P/O J Nejezchleba | Sgt S Linka | W5711 | KX-H | 1x1000lb+1x500lb+250lb | Early Return |
| Sgt M Šebela | F/Sgt A Jedounek | Z8784 | KX-U | 1x1000lb+1x500lb+250lb | Early Return |

The first Wellington got the green light from the Control Van at 20:11hrs, and within 11 minutes, the Berlin crews were airborne into a clear night sky. The squadron reported one early return. Pilot Officer J Nejezchleba reported an electrical issue aboard Wellington W7511 KX-H. The crew decided to bomb Osnabruck at 22:15hrs. They landed safely back at East Wretham at 00:55hrs. Another crew in trouble was Sergeant M Šebela. Overheating port engine meant that they were not able to gain the needed height and bombed the dock area at Kiel at 23:35hrs from 10,000ft. Early returns on this night for 3 Group were at a record high, with 13 aborts, almost all with technical trouble. This added to the six that failed to take-off, was something that 3 Group H.Q took very seriously. Conditions over Berlin were ideal, with good visibility and no cloud. The first of 311 Squadron's crews over Berlin was Pilot Officer Bala, who approached the city at 16,000ft at 23:30hrs and reported no fires on his arrival. He dropped his high-explosive bombload near the Wannsee Railway Station located in the southwestern quarter of Berlin, producing a large fire. Two crews bombed within a minute of each other. Sergeant V Ryba dropped his entire bombload in one stick at 23:44hrs near Aiming Point 'B' the Alexander Platz. He was followed by Flight Sergeant F Radina, who bombed from 17,000ft, not on Aiming Point 'B' as instructed, but on 'C'. He reported, *'two medium fires near aiming point 'C' on arrival. Own burst near aiming point 'C' and fires started. Good fire developing near city centre'*. The last crew to bomb was Sergeant A Musálek at 00:01hrs. He dropped from 18,000ft and delivered his bombload between Aiming Point 'C' and Tempelhof. All his bombs were seen to burst, and his incendiaries were observed to ignite creating a small fire. The raid still had three hours to go as the squadron turned for home. 3 Group H.Q at Exning were delighted with the results reporting *'brilliant fires in the target area'*. One in particular in the centre of the town near the Alexander Platz was reported as *'very large'*.

### September 7th 1941 : 'GR3588' – Kiel Harbour – Dry Docks & Floating Docks

| Pilot | 2nd Pilot | Serial | Code | Bomb Load | Result |
|---|---|---|---|---|---|
| P/O L Anderle | W/O K Weiss | R1161 | KX-X | 1x1000lb+4x500lb | Duty Carried Out |
| Sgt J Horáček | Sgt B Hradil | R1451 | KX-P | 1x1000lb+4x500lb | Early Return |

The Kiel raid brought further inconveniences caused by technical problems. T2972 KX-G captained by Flying Officer F Pohlodek did not even take-off due to engine failure while Sergeant J Horáček and crew returning with an overheating and oil leaking port engine. They got as far as 70 miles northwest of the Frisian Islands before jettisoning their bombs. Pilot Officer L Anderle arrived over Kiel at 23:05hrs, stating that fires were already abundant in the target area. He dropped his bombload in one stick from 15,000ft in the target area. These were seen to burst and create a violent explosion. Bombing appeared excellent, with explosions seen across Dock No.3,4,6 and 7. These fires could be seen from 50 miles away. The nights successes were tempered by the three early returns, it was not a particularly good night for the squadron groundcrews.

Bad weather on the 8th and 9th restricted the squadron's activities to a few short training flights. There was some unsavoury business on the 9th to attend too. A Field Court Martial at S.H.Q Honington was held against Pilot Officer Alois Tolar, a navigator of the Operational Training Flight. He has been found guilty of stealing a quantity of service petrol on July 27th and sentenced to be dismissed from his Majesty's service.

The squadron was informed probably early on the 9th that it would be required the following night for a raid on Turin, northern Italy. If there was excitement about a raid on Berlin, the idea of attacking an Italian target had the crews clambering to have their names put forward. This would be the first operation by the squadron over the Alps. Much had to be organised, fuel and bombloads, routes and importantly, timing. 3 Group would detail and brief 45 Wellingtons who would bomb the Royal Arsenal while 14 Short Stirlings would tackle Turin's main railway station. Five crews were chosen, all experienced. Slender aluminium additional fuel tanks were lifted into the Wellington's spacious bomb bay suspended by the bomb shackles. These held an additional 120 gallons of precious fuel. The crews were given instructions by the Commanding Officer on their usage.

### September 10th 1941 : 'S.19' – Royal Arsenal – Turin

| Pilot | 2nd Pilot | Serial | Code | Bomb Load | Result |
|---|---|---|---|---|---|
| F/Lt K Vildomec | F/O L Němec | T2972 | KX-G | 1x1000lb+2x500lb | Duty Carried Out |
| Sgt V Ryba | Sgt M Plecitý | R1046 | KX-E | 1x1000lb+2x500lb | Duty Carried Out |
| F/Sgt F Radina | Sgt V Pánek | X3221 | KX-O | 1x1000lb+2x500lb | Duty Carried Out |
| F/O F Pohlodek | Sgt A Šiška | Z8805 | KX-V | 1x1000lb+2x500lb | Duty Carried Out |
| Sgt M Šebela | Sgt J Svoboda | Z8784 | KX-U | 1x1000lb+2x500lb | Early Return |

All five crews had departed by 20:23hrs for the long flight ahead. Each had been instructed to carry out a slow climb to conserve fuel which they did, with eyes transfixed to the fuel and temperature gauges. Eventually, the crews cleared a fog covered England and emerged into a cloudless sky as they headed south. There was another early return logged by the squadron. Sergeant M Šebela found that he could not climb to the briefed height, so the crew reluctantly opted to bomb Flushing from 10,000ft. It was another bad night for the Group, with 11 early returns. 40 Squadron based at R.A.F Wyton reported six of the eight Wellingtons detailed aborted. One of those who did complete the operation was an old friend of 311 Squadron, Squadron Leader Thomas Kirby-Green. Also based at R.A.F Wyton was the Stirling-equipped XV Squadron. Three of its six Stirlings also abandoned the raid. The Marham-based 218 had a crew return early when the long-range fuel tank malfunctioned.

The squadron threaded its way across France where blackouts seemed non-existent. The further south the crews flew, the more lights were seen making navigation that much easier. Flying between 16,000ft and 17,500ft, the city of Geneva

*The imposing Italian Alps. The Wellingtons of 311 Squadron had to cross these beautiful but deadly peaks to reach Turin.*

was observed in the excellent conditions, its streets and building ablaze with lights. Eventually, the snow-capped Alps loomed ahead as the crews neared the city of Chambéry, France. Once clear of the Alps, the Czechs were over the Po Valley and within minutes, flying over a mist-covered Turin. Three crews were over the target between 00:20hrs and 00:27hrs. Flight Lieutenant K Vildomec was at 17,000ft as he approached the target. He reported fires already well ablaze as his bombs were dropped ½ mile southeast of the Royal Arsenal. Also at 17,000ft was Sergeant V Ryba, who reported back at R.A.F East Wretham, *'three fires seen on arrival, not pinpointed, believed south of Arsenal. Our own bombs just north of these. Bursts seen but no result seen'.* Mist hindered Flight Sergeant F Radina's crew. They bombed from 16,000ft between two large existing fires, bursts were seen but again no results were observed. The last crew over the Royal Arsenal was Flying Officer F Pohlodek at 00:40hrs.[42] Mist and drifting smoke resulted in the crew bombing the 'target area', but they were unsure of the accuracy as no bursts were observed. The Czechs reported a number of fires on leaving the target, the larger ones visible from 50 miles into the return flight. Flak over the city was sparse and haphazard, and a solitary balloon was observed. Crews reported seeing flak and searchlight activity over Milan. Flying Officer F Pohlodek was forced to crash land due to fuel shortage on return. The Wellington came down in a ploughed field at Thornden Farm, Headcorn, Kent. Unfortunately, anti-invasion obstacles had been placed in the field, causing more damage to the Wellington. Amazingly, none of the crew were injured. Soon after the crash, local Home Guard and military personnel arrived on the scene. After confirming that they were 'friendly', the crew were taken to the nearby farm for a welcome breakfast.

| Date | 11/09/1941 |
| --- | --- |
| Mark | Mk.IC |
| Serial | Z8805 |
| Code | KX-V |
| Taken On Charge | 28/07/1940 via No.33 MU. |
| Manufacturer | Vickers (Weybridge) |
| Contract | B71441/40 |
| Pilot (s) | Flying Officer František Pohlodek / Sergeant Alois Šiška |
| Flight | Operations |
| Time | 06:15hrs |
| Cause | Fuel shortage. |

---

[42] *Times vary depending on what report is used.*

The repairs were lengthy, and it was not until March 1942 Z8805 returned to action, not with 311 Squadron but 214 (FMS) Squadron.

Unsettled weather conditions created some concerns at H.Q High Wycombe on the 12th resulting in a late change of targets for that night. Initially, 3 Group was detailed Nuremberg with Kiel as an alternative. This was however switched, and the Group would instead attack Frankfurt while 'Freshmen' crews would visit the Docks at Cherbourg. A total of 54 Wellingtons and 9 Stirlings would be briefed. At R.A.F East Wretham, the Czechs made ready ten Wellingtons but R1598 KX-C with Sergeants A Mžourek and J Fína was later withdrawn for unspecified technical defect.

### September 12th 1941 : Sole 'B' – Frankfurt

| Pilot | 2nd Pilot | Serial | Code | Bomb Load | Result |
|---|---|---|---|---|---|
| Sgt S Linka | Sgt V Procházka | W5682 | KX-Y | 3x500lb+1x250lb+360x4lb | Duty Carried Out |
| F/Lt K Vildomec | F/O L Němec | X9741 | KX-D | 3x500lb+1x250lb+360x4lb | Early Return |
| Sgt A Musálek | F/Sgt A Jedounek | T2553 | KX-B | 3x500lb+1x250lb+360x4lb | Duty Carried Out |
| F/O J Stránský | Sgt B Hradil | X9742 | KX-F | 1x1000lb+4x500lb | Duty Carried Out |
| Sgt M Šebela | Sgt J Svoboda | W5668 | KX-T | 1x1000lb+4x500lb | Duty Carried Out |
| P/O J Nejezchleba | Sgt K Knaifl | W5711 | KX-H | 1x1000lb+4x500lb | Duty Carried Out |
| P/O L Anderle | W/O K Weiss | R1161 | KX-X | 1x1000lb+4x500lb | Duty Carried Out |
| Sgt V Soukup | Sgt J Miklošek | R1015 | KX-L | 1x1000lb+4x500lb | Duty Carried Out |
| F/O F Pohlodek | Sgt O Jambor | T2561 | KX-A | 1x1000lb+4x500lb | Duty Carried Out |

Sergeant Stanislav Linka would be occupying the pilot's seat for the first time. The first crew was away at 20:00hrs. Within 15 minutes, all nine were safely airborne. Problems were soon experienced by Flight Lieutenant K Vildomec who suffered hydraulic failure aboard Wellington X9741 KX-D soon after take-off. The crew jettisoned the bombload 'Safe' in the sea 12 miles off Orfordness but were unable to close the bomb doors. Showing commendable flying skill, the pilot approached East Wretham with the undercarriage lowered, this having been completed by hand. With the flaps in the 'up position', the Wellington came into land at a considerable speed at 21:25hrs. Unfortunately, the Wellington overshot its landing and hit the station flag pole located in the northwest corner of the airfield. All the crew were shaken but uninjured apart from the second pilot, Flying Officer Ludvik Němec. He received severe fractures to his hand and a nasty head injury. The damaged hand required an operation, and a stay at Ely Hospital and convalescence at Torquay, Cornwall. His injuries were such that he did not return to the squadron.

| | |
|---|---|
| Date | 11/09/1941 |
| Mark | Mk.IC |
| Serial | X9741 |
| Code | KX-D |
| Taken On Charge | 22/07/1941 via No.7 MU. |
| Manufacturer | Vickers Armstrong |
| Contract | 124362/40 |
| Pilot (s) | Flight Lieutenant Karel Vildomec / Flying Officer Ludvík Němec |
| Flight | Operations |
| Time | 21:25hrs |
| Cause | Hydraulic failure |

Damage to the Wellington was surprisingly light. Repair to the damaged starboard aircrew and structural testing to the engine was all that was needed. There were no other early returns, and the remain crews reached the target area to find it concealed beneath thick cloud. Sergeant S Linka and crew were among the first over the target at 16,000ft. They reported 9/10th cloud and bombed on E.T.A. Bursts were seen, but no other results. It was 25 minutes before the next crew bombed at 23:30hrs. Pilot Officer L Anderle arrived to find no fires visible and a blanket of clouds. They were honest about the results, *'Bombs dropped on town, believed Frankfurt'*. Remarkably following only a minute behind was the crew of Sergeant M Šebela who reported at interrogation, *'Bombs dropped near railway sidings. Bursts were followed by 20 huge explosions followed by red fires and much smoke'*. The last crew to bomb at 22:58hrs was Sergeant V Soukup from 16,500ft. Unable to identify the target, they decided to drop on flak and searchlight concentrations, adding at de-briefing *'on what was thought to be Frankfurt'*. The Frankfurt authorities reported only seventy-five high explosive bombs hitting the city and its surrounding area, along with more than six hundred incendiaries, and although a number of fires resulted, the damage was not severe. Also reporting bombs falling was Mainz, 20 miles south-west of Frankfurt.

A minor incident occurred at 21:50hrs involving the crew of Sergeant František Naxera, who once again appears to have made a rather heavy landing at East Wretham snapping the tail wheel. This was the second occasion he had been overzealous on touchdown!

| Date | 15/09/1941 |
|---|---|
| Mark | Mk.IC |
| Serial | L7841 |
| Code | KX-S |
| Taken On Charge | 28/09/1940 via No. 22 MU. |
| Manufacturer | Vickers Armstrong |
| Contract | 992424/39 |
| Pilot (s) | Sergeant František Naxera |
| Flight | Night landing practice (Czech Training Flight) |
| Time | 21:50hrs |
| Cause | Tailwheel collapsed heavy landing. |

Bomber Command switched its attention to northern Germany on the 15th. Hamburg was the primary target with its three main railway stations the aiming points. A total of 53 Wellingtons and eight Short Stirlings would be offered by 3 Group, with 311 Squadron contributing ten crews. Their target was the West Altona Railway Station, Aiming Point 'A' located in the western quarter of the city.

September 15th 1941 : Dace 'A' – Hamburg[43]

| Pilot | 2nd Pilot | Serial | Code | Bomb Load | Result |
|---|---|---|---|---|---|
| Sgt A Musálek | Sgt A Šiška | T2553 | KX-B | 3x500lb+1x250lb+360x4lb | Alternative Target |
| Sgt M Šebela | Sgt J Svoboda | W5668 | KX-T | 1x1000lb+3x500lb+1x250lb | Alternative Target |
| P/O L Anderle | W/O K Weiss | R1161 | KX-X | 1x1000lb+3x500lb+1x250lb | Duty Carried Out |
| F/O J Stránský | Sgt B Hradil | X3221 | KX-O | 1x1000lb+3x500lb+1x250lb | Duty Carried Out |
| F/Sgt A Jedounek | Sgt V Pánek | R1451 | KX-P | 3x500lb+1x250lb+360x4lb | Duty Carried Out |
| Sgt S Linka | Sgt V Procházka | W5682 | KX-Y | 3x500lb+1x250lb+360x4lb | Duty Carried Out |
| F/O F Pohlodek | Sgt O Jambor | R1046 | KX-E | 6x500lb | Alternative Target |
| P/O J Nejezchleba | Sgt K Knaifl | W5711 | KX-H | 6x500lb | Duty Carried Out |
| Sgt A Mžourek | Sgt J Fína | R1598 | KX-C | 3x500lb+1x | Alternative Target |
| Sgt V Soukup | Sgt J Miklošek | R1015 | KX-L | 1x1000lb+3x500lb+1x250lb | MISSING |

It was a night of excellent conditions that would afford unlimited visibility making navigation that much easier. Held by searchlights at 17,500ft ten miles west of Oldenburg, the crew of Sergeant A Musálek were attacked by a Bf109, which made a head-on pass after firing a single red star. Musálek instinctively put the nose of the Wellington down and dived. A Bf109 approached on the Green beam two minutes later and opened fire. The rear gunner replied firing 200 rounds from between 300-400 yards. To avoid the fighter, the pilot once again dived but, in doing so, ran into the flak and searchlights of Oldenburg, where they dropped their bombs and turned for home. Flying Officer F Pohlodek and crew were badly knocked about by flak damaging an engine en route to Hamburg. Unable to maintain height the crew

---

[43] *For an unknown reason, the ORB records W/Cdr J Ocelka flying on this operation. This was incorrect.*

eventually jettisoned their bombs over Bremen from just 3,000ft at 22:50hrs. Sergeant A Mžourek had trouble with the port engine which was showing worrying signs of overheating. Deciding they would be pushing their luck if they continued, they too bombed Oldenburg at 23:15hrs and headed back to East Wretham.[44] Another crew who bombed Bremen was Sergeant M Šebela. Worries about having sufficient fuel to complete the operation, they dropped the entire bombload and turned for home. It was not a promising start. The Germans opposed this raid in some strength. Approximately 30 night-fighters from NJG1 and I./NJG3 were airborne and on the prowl in the clear night sky. The crew of Sergeant V Soukup were intercepted and shot down by Gefreiter Rudolf Frank of I./NJG3 at 23:30hrs. The Vickers Wellington crashed near Andervenne, Lower Saxony, Germany, killing all the crew. The Germans initially only identified three bodies, Sergeant Zdeněk Babíček, Pilot Officer Mojmír Sedláček and Sergeant Richard Husman. The three identified crew, plus the three unidentified members, were buried with Military Honours on September 18th 1941, in the Lingen-on Ems Cemetery, Row 5, Grave 1A & 1B.

Vickers Wellington Mk.IC R1015 KX-L

| Manufacturer | Vickers Armstrong | |
|---|---|---|
| Contract | 992424/39 | |
| Taken on Charge | 30/11/1940 via No.48 MU. | |
| Cat E Missing | 15/09/1941 | |
| Struck Of Charge | 16/09/1941 | |
| Total Flying Hours | N/K | |
| Take-Off Time | 19:55hrs | |
| Bomb Load | 1x1000lb+3x500lb+1x250lb | |
| | **CREW** | |
| Captain | Sergeant Vilém **Soukup** 787246 RAFVR. Age 27. | Coll. grave 27. F.12-15. |
| Second Pilot | Sergeant Ján **Miklošek** 787339 RAFVR. Age 25. | Coll. grave 27. F.12-15 |
| Navigator | Pilot Officer Mojmír **Sedláček** 82634 RAFVR. Age 27. | Coll. grave 27. F.12-15. |
| Wireless Operator | Pilot Officer Antonín **Zimmer** 82649 RAFVR. Age 27. | Coll. grave 27. F.12-15. |
| Front Gunner | Sergeant Zdeněk **Babíček** 787505 RAFVR. Age 21. | Coll. grave 27. F.12-15. |
| Rear Gunner | Sergeant Alois **Jarnot** 787394 RAFVR. Age 23. | Coll. grave 27. F.12-15. |
| Posting History | Sgt V Soukup posted via 2 S.F.T.S, 17/07/1941. Sgt J Miklošek posted via Wilmslow Depot, 11/04/1941. | |
| Operations Flown | Sgt V Soukup : 17 / Sgt J Miklošek : 2 | |
| Buried | REICHSWALD FOREST WAR CEMETERY | |

Confirmation of the crew's fate arrived at R.A.F East Wretham via the International Red Cross in early October, but the inclusion of Sergeant Richard Husman on the list of dead caused much head-scratching on the squadron. Sergeant Richard Husman was still very much alive! The mystery was finally solved when Sergeant Husman explained that on September 15th, he was in hospital but shared a locker with Alois Jarnot, a fellow air gunner. The only explanation that Richard Husman could give was that Sergeant Jarnot borrowed his flying jacket, which had his name in it and flying boots for the operation as Jarnot's may have been wet or damp. Post-war, the bodies of the crew were removed from the cemetery at Lingen-on Ems and re-buried in the Reichswald Forest War Cemetery.

Only four crews reported bombing Hamburg, which put up a spirited defence, with the searchlights being particularly bothersome. The first crew with a reported bombing time of 21:44hrs[45] was Flight Sergeant A

---

[44] *Once again, conflicting details. The ORB reports that the crew bombed Hamburg, while another reports the crew bombed Oldenburg!*
[45] *Once again reports differ on time.*

*Left: Sergeant Vilém Soukup and his co-pilot Sergeant Ján Miklošek.*

Jedounek. They watched as their bombs landed approximately 500 yards east of the aiming point, starting a good fire. There was a delay of 36 minutes before Pilot Officer L Anderle arrived over Hamburg at 16,000ft. They reported their all HE load landing just south of Alster creating a small fire. Sergeant S Linka was the next to bomb at 23:25hrs. The last reported crew over Hamburg was Pilot Officer J Nejezchleba at 22:30hrs. They reported on return, *'Burst seen on docks east of aiming point. Two small fires started, many fires seen on arrival'.*

The last of the crews landed at 02:35hrs, and the realisation that the squadron had suffered its first missing crew in eight weeks of operations soon swept through the squadron. The returning crews reported a number of aircraft engaged by both flak and night-fighters, with at least three shot down in flames. All the squadron crews mentioned the ferocious flak along the entire route especially from Emden, Oldenburg, Bremerhaven, and Hamburg. It had been a tremendous loss-free period for the squadron, making the news of missing friends much harder. Results were classed as *'satisfactory'* by 3 Group H.Q. On this night, 75(NZ) Squadron reported the loss of two crews, one of which was skippered by Sergeant J.A Ward **VC** RNZAF.

On return from Hamburg the squadron spent the next four days relatively idle apart from training flights flown in between some poor summer weather. On the 20th, a small force of 3 Group Wellingtons bombed Berlin, 311 sat the attack out. On the 22nd, the squadron was informed it would be required that night for an attack on Stettin, this was then changed to Berlin and then finally cancelled due to adverse weather. The Training Flight notched up another avoidable accident on the 22nd when Sergeant Jaroslav Rolenc landed at East Wretham on completion of a photographic flight but failed to lower the undercarriage. Wing Commander J Ocelka put it down to 'Negligence' and recommended disciplinary action.

| Date | 22/09/1941 |
|---|---|
| Mark | Mk.IC |
| Serial | R3206 |
| Code | KX-B |
| Taken On Charge | 11/05/1941 via No.20 MU. |
| Manufacturer | Vickers Armstrong |
| Contract | 3913/39 |
| Pilot (s) | Sergeant Jaroslav Rolenc |
| Flight | Photographic Flight (Czech Training Flight) |
| Time | 14:50hrs |
| Cause | Failed to lower undercarriage. |

Two days of fog halted any planned operations and restricted flying to a minimum. On the 26th, instructions were received from Bomber Command H.Q detailing 53 crews to attack the important port of Genoa the following night if weather permitted. Crews would also be visiting Cologne and Emden. 311 Squadron would detail only four crews on Genoa and a further three on Cologne.

### September 26th 1941 : Shark 'A' & 'B' – Genoa

| Pilot | 2nd Pilot | Serial | Code | Bomb Load | Result |
|---|---|---|---|---|---|
| Sgt S Linka | Sgt V Procházka | R1598 | KX-C | 1x500lb+180x4lb | Recalled |
| Sgt A Musálek | Sgt A Šiška | T2553 | KX-B | 1x500lb+180x4lb | Duty Carried Out |
| Sgt V Ryba | Sgt M Plecitý | R1046 | KX-E | 1x1000lb+1x250lb | Recalled |
| Sgt M Šebela | Sgt V Pánek | X3221 | KX-O | 1x1000lb+1x250lb | Recalled |

Surprisingly, the crews chosen were all captained by NCO pilots for the long haul to Genoa on the coast of the Ligurian Sea. The largest port in Italy, it was also home to the Ansaldo Shipyards and the influential Gio. Ansaldo & C, one of Italy's most prominent engineering companies. Given the range, the installation and connection of the auxiliary fuel tanks were completed and, importantly, tested by the groundcrews before take-off. The Armourers had less to do, the bombloads having been reduced in favour of the fuel. Between 19:25hrs and 19:36hrs the four Wellingtons were airborne and slowly climbing away for the long flight ahead. Before taking off, the crews had been warned that fog may result in diversions on return. Little did they realise that Group H.Q would send a general recall signal to all its aircraft within an hour of departure. By this time, the small force was over France and quickly turned for home with most jettisoning their bombloads over the sea. A few, the more daring, sought out targets of opportunity. Three Czech crews did precisely that.

Both Sergeant V Ryba and S Linka bombed Dunkirk. Sergeant Linka dropped his mixed load on searchlight and flak concentrations around the docks at 20:40hrs, followed five minutes later by Sergeant Ryba. The docks at Calais were attacked by Sergeant M Šebela before heading home. One crew did not hear the recall signal. The Wireless Operator aboard Wellington T2553 KX-B was Pilot Officer Josef Ščerba. He missed the signal, and the crew continued to Genoa, oblivious to the recall. They arrived over a cloud-covered Genoa at 00:45hrs. The defences were already active, with flak busy firing at non-existent aircraft. From 14,000ft, the bombs were dropped on the estimated position of the port in amongst light tracer flak. No results were seen. The crew turned for home, eventually landing at the Fighter Command airfield at Manston, Kent. They had been fortunate, airborne for 9 hours and 25 minutes, much of which was over enemy territory, they had somehow survived. Why H.Q had agreed for the raid to take place given the unpredictable weather is unknown. The crews briefed to attack Cologne departed between 19:29hrs and 19:46hrs. They like those on Genoa were ordered to abort and return to base. Only the vigilant Wireless Operator Flight Sergeant Viktor Tégel in the crew of Pilot Officer J Nejezchleba heard the recall message. They landed at R.A.F Mildenhall on return. The two other crews continued on, totally unaware.

### September 26th 1941 : 'Trout' – Cologne

| Pilot | 2nd Pilot | Serial | Code | Bomb Load | Result |
|---|---|---|---|---|---|
| P/O J Nejezchleba | Sgt K Knaifl | W5711 | KX-H | 1x1000lb+5x500lb | Recalled |
| F/O F Pohlodek | Sgt J Lenc | T2561 | KX-A | 1x1000lb+5x500lb | Duty Carried Out |
| F/O J Stránský | Sgt B Hradil | X9742 | KX-F | 3x500lb+1x250lb+360x4lb | Duty Carried Out |

On this raid, Sergeant Jan Lenc would experience his first operation sitting beside Flying Officer F Pohlodek. Jan Lenc was born on April 27th 1916, in Komařice, South Bohemia. He was drafted into the army on June 23rd, 1937. In the first half of 1938, Jan Lenc underwent his basic training before progressing onto two-seater aircraft at the 1st Air Regiment. From then until his discharge from the army, he served at Milovice airfield. On August 7th, 1939, he left his native Komařice, never to return. He first travelled to Poland and then arrived in France on August 21st of the same year. Like many of his fellow Czechs, he signed a pledge to the Foreign Legion on September 21st 1939. In October, with the rank of Corporal, he was transferred to a base in the southern French town of Pau. Here, he and his colleagues underwent flight training. After the fall of France in June 1940, he embarked from Bordeaux on the vessel *Robur III*. Four days later, on June 22<sup>nd</sup> 1940, he set foot on English soil at Falmouth for the first time.

Both crews bombed Cologne within five minutes of each other. Flying Officer J Stránský dropped his bombload at 22:30hrs, having identified the river Rhine. Their mixed load was seen to burst near the aiming point, starting a fire. Flying Officer F Pohlodek followed flying at 14,000ft. They reported back at interrogation that their bombs started two fires. Flak was as expected fierce and in abundance. However, luck was on the side of the two crews, who both landed back at R.A.F East Wretham. Sadly, an error of judgement by Flying Officer F Pohlodek would result in an avoidable accident. He decided that his second pilot, the inexperienced Sergeant J Lenc, should make the landing despite the airfield being blanketed by fog. At 02:05hrs, Wellington T2561 KX-A dropped a wing while on its approach and hit the ground too hard, collapsing the port undercarriage. The crew emerged unscathed, but the Wellington was less fortunate. The squadron commander, Wing Commander J Ocelka, reported it was an unwise decision by the more experienced Flying Officer F Pohlodek to allow his novice second pilot to land in the poor visibility. It was probably not the best start to Sergeant Lenc's operational career.

| Date | 27/09/1941 |
|---|---|
| Mark | Mk.IC |
| Serial | T2561 |
| Code | KX-A |
| Taken On Charge | 09/08/1940 via No.10 MU. |
| Manufacturer | Vickers Armstrong |
| Contract | 38600/39 |
| Pilot (s) | Flying Officer František Pohlodek / Sergeant Jan Lenc |
| Flight | Operations |
| Time | 02:05hrs |
| Cause | Heavy Landing |

3 Group H.Q were not overly pleased that a number of its crews had either ignored the order to return to base and chose instead to bomb a target of opportunity, putting them, and their aircraft at risk. Or simply failed to hear it, indicating a lack of diligence. They rather sanctimoniously reported that 'troublesome' crews failed to hear the recall!

Genoa was again on the Battle Order on the 28<sup>th</sup>, along with Frankfurt. 311 Squadron would brief just four crews for the trip to Italy, they would join a force of forty-seven aircraft including six Short Stirlings for the long trip to Genoa. A further three were detailed for a raid against Frankfurt.

September 28<sup>th</sup> 1941 : 'Sole' 'B' – Frankfurt

| Pilot | 2<sup>nd</sup> Pilot | Serial | Code | Bomb Load | Result |
|---|---|---|---|---|---|
| F/O J Stránský | Sgt B Hradil | X9742 | KX-F | 6x500lb+1x250lb | Duty Carried Out |
| Sgt A Mžourek | Sgt J Fína | R1451 | KX-P | 1x1000lb+4x500lb+1x250lb | Early Return |
| Sgt M Šebela | Sgt J Janek | X3221 | KX-O | 1x1000lb+4x500lb+1x250lb | Duty Carried Out |

Bomber Command allocated just 31 bombers to Frankfurt. 3 Group would provide 14 Wellington crews, the bulk of the attacking force would be provided by the Hampdens of 5 Group. Sergeant Jan Janek would undertake his first operation sitting beside Sergeant Šebela. The Frankfurt force was the first away, in what was described as a *'very dark night'*. Sergeant A Mžourek experienced fuel problems[46] over Givet, northern France, which ultimately resulted in the crew abandoning the operation. On the return trip they dropped their mixed load on a flashing beacon near Dunkirk at 22:37hrs. Conditions over Frankfurt were clear. However, there was thick industrial haze. Sergeant M Šebela bombed

---

[46] *Other reports suggest engine issues.*

*A wonderful photograph of Wellington Mk.IC R1532/KX-R which completed 17 sorties (4 abortive) between 31st March and 6th July 1941 when it was damaged by German night fighter. After repairs the aircraft was returned to 311 Squadron on 9th November 1941 and flew a further 11 sorties (3 abortive) as KX-B between 1st January and 12th April 1942 being transferred to 27 Operational Training Unit on 24th April 1942. Two members of the ground personnel posing with the aircraft are: AC1 Jakub Hochberger (Instrument Repairer) leaning on the elevator and Cpl Josef Blažek (Fitter I) standing on the wing. (Zdeněk Hurt)*

first at 22:15hrs. He and his crew were unable to determine the accuracy due to the haze but were confident their bombs landed in the south of the town. They had done well to get to the target having experienced wireless and instrument issues en route.

Flying Officer J Stránský found himself over the northern suburbs of Frankfurt at 16,500ft. They dropped at 22:43hrs and confirmed at de-briefing that their bombs landed in the town amongst buildings, creating a large fire which was visible from 50 miles away. Bad weather over East Wretham got the better of the crew of Sergeant M Šebela, who were diverted to R.A.F Stradishall. The crew had already suffered instrument and wireless malfunctions, which had resulted in navigational problems. To add to the crew's difficulties, heavy rain and clouds down to 300ft were also encountered as the crew headed towards R.A.F Stradishall. Now, hopelessly lost, and with a sick wireless operator the crew were in desperate trouble. Eventually, their Wellington force landed at Knowles Green[47] at 04:00hrs. All but two of the crew, who stayed with the aircraft were driven by a civilian driver to R.A.F Stradishall. Here the crew telephoned both Honington and East Wretham explaining what had happened. Preferring to be interrogated back at their own base, R.A.F Stradishall arranged transport, collecting the two airmen at the crash site on the way.

---

[47] *The location for the crash landing is vague, Knowles Green, Hargrave, 5 miles Stradishall. Times also vary, Stradishall Ops Records Book stated 03:30hrs.*

Damage to the Wellington was relatively minor, with a badly bent bomb beam and wing tip. It was an excellent piece of flying by the pilot. However, there was criticism from the squadron commander for not carrying out a proper lost aircraft procedure. Thankfully, none of the crew, including the novice Sergeant J Janek, were injured.

| Date | 29/09/1941 |
| --- | --- |
| Mark | Mk.IC |
| Serial | X3221 |
| Code | KX-O |
| Taken On Charge | Not Recorded on the AM Form 78 |
| Manufacturer | Vickers Armstrong |
| Contract | 92439/40 |
| Pilot (s) | Sergeant Metoděj Šebela / Sergeant Jan Janek |
| Flight | Operations |
| Time | 04:00hrs |
| Cause | Forced Landing |

The ground crews once again had the arduous task of installing the cigar-shaped additional fuel tanks into and then testing them prior to take-off. This done, the bombs and ammunition could be loaded.

### September 28<sup>th</sup> 1941 : Shark 'A' – Genoa Docks

| Pilot | 2nd Pilot | Serial | Code | Bomb Load | Result |
| --- | --- | --- | --- | --- | --- |
| Sgt A Musálek | Sgt A Šiška | T2553 | KX-B | 2x500lb+180x4lb | Duty Carried Out |
| F/Lt K Vildomec | Sgt J Filler | R1161 | KX-X | 2x500lb+180x4lb | Alternative |
| P/O J Nejezchleba | Sgt K Knaifl | W5711 | KX-H | 1x1000lb+1x250lb | Duty Carried Out |
| Sgt V Ryba | Sgt M Plecitý | R1046 | KX-E | 1x1000lb+1x250lb | Duty Carried Out |

The first of the Genoa-bound crews was away at 19:27hrs. Once airborne, they headed to Orfordness on the Suffolk coast. Once passed, they began their slow climb across the North Sea until landfall was made at Nieuwpoort, Belgium. From here, the crews headed south, weaving their way past Paris and Reims until they skirted Dijon. Once again, they were astounded by the lack of blackout but glad of it as it made navigation less demanding. Once again, the city of Geneva shone like a beacon in the blackness as the crews approached the city of Chambéry and its glittering lights. Here, the crews would have to decide if they had the altitude and fuel to cross the Alps and on to Genoa. Having crossed the Alps and entering Fascist Italy, the crew of Flight Lieutenant K Vildomec made the decision that their remaining fuel would not allow them to continue onto Genoa, so they opted to bomb Turin. This they did through a gap in the cloud at 00:45hrs. On their arrival over Genoa, Sergeant A Musálek and crew were fortunate to find a gap in the cloud allowing them to drop their bombload from 14,000ft and observe two explosions in the target area at 01:10hrs followed by a fire. Sergeant V Ryba was not so lucky. They had to bomb between the glow of two large fires as clouds obscured the target area. Five minutes later at 01:26hrs, the crew of Pilot Officer J Nejezchleba were over the city at 17,000ft. They estimated the location of the dock area by the flak bursts over what they presumed was the target. With the bombs dropped, the crews turned for home. They now had to re-trace their route across France without interference. Two landed away from home, Sergeant A Musálek landed at R.A.F Abington while Pilot Officer J Nejezchleba landed at R.A.F Honington, after being airborne for 10 hours 23 minutes.

The squadron prepared for another long-range operation on the 29<sup>th</sup>, this time to Stettin, the largest city of the Western Pomerania. However, for reasons unknown the squadron was withdrawn late in the day. The operation was flown, and nine squadrons of 3 Group carried out a successful raid. There was a flurry of activity at R.A.F East Wretham on the 30<sup>th</sup>. 311 Squadron would provide six crews for a return visit to Stettin, while a solitary crew would attack the docks at Cherbourg. The Stettin raid would be a predominantly 3 Group affair, 36 Wellingtons would be joined by a handful of Wellingtons from 1 Group.

## September 30th 1941 : Tarpon 'A' – Stettin

| Pilot | 2nd Pilot | Serial | Code | Bomb Load | Result |
|---|---|---|---|---|---|
| Sgt S Linka | Sgt V Procházka | R1777 | KX-M | 3x500lb+9x50lb I | Duty Carried Out |
| Sgt J Svoboda | Sgt V Pára | R1451 | KX-P | 3x500lb+9x50lb | Alternative |
| Sgt V Pánek | Sgt K Danihelka | R1161 | KX-X | 1x1000lb+1x500lb+1x250lb | Duty Carried Out |
| F/O F Pohlodek | Sgt J Lenc | W5668 | KX-T | 1x1000lb+1x500lb+1x250lb | Duty Carried Out |
| Sgt A Mžourek | Sgt J Fína | R1598 | KX-C | 4x500lb | Duty Carried Out |
| F/O J Stránský | Sgt B Hradil | X9742 | KX-F | 4x500lb | Alternative |

Two new names appeared on the crew list for the night's activities, Sergeants Vladimír Pára and Karel Danihelka. They would be joining Sergeant J Svoboda and Sergeant V Pánek respectively, who both would be captaining their own crew for the first time. Weather conditions over northern Germany were ideal as the crews picked their way past the formidable northern coastal flak defences. Kiel unsurprisingly put up a tremendous heavy flak barrage mixed with numerous searchlights. The severity of the defences resulted in the crew of Flying Officer J Stránský being driven badly off course. They ended up attacking Berlin. Arriving over Berlin at 23:10hrs, the crew were immediately engaged by flak, which burst all around the Wellington. Wasting no time, they quickly dropped their 500-pounders from 16,000ft. These were seen to burst south of the Charlottenburger Chaussee, a paved road leading from Berlin's old town to the Tiergarten Zoo. With flak still accurate, Flying Officer J Stránský throttled back hard and dived to 13,500ft to evade the flak. This initially worked. For 60 seconds, the Wellington flew on unmolested until flak again exploded all around the aircraft. One burst was close enough to damage the front turret, wounding the gunner Sergeant Jan Peprníček in the face. Finally, they cleared Berlin, shaken but alive. The remaining crews skirted Lubeck and Rostock, both easily identified by the cauldron of flak and searchlights. Sergeant J Svoboda lost his starboard engine en route, forcing the crew to dispose of their bombs near Politz at 23:40hrs. They returned to base on one engine. Arriving over the Baltic coast between 23:05hrs and 23:40hrs, the squadron found Stettin under clear skies but shrouded by ground haze. Over the target at 23:05hrs, Sergeant S Linka dropped his bombload from just 13,000ft north of the target area. He was closely followed by Flying Officer F Pohlodek, who reported on return, *'Very large fire, 700 yards long and horseshoe shaped, 1,500 yards north of Aiming Point 'A' on arrival. Very red with much smoke. Also a large fire 1000 yards north of Aiming Point 'C'. Own bursts near first fire'*. Conditions varied for Sergeants A Mžourek and V Pánek. The latter dropped his load from 17,000ft in one stick from south to north across the eastern side of Stettin. While Sergeant A Mžourek could not pinpoint the Aiming Point and bombed on flashes of flak below the clouds at 23:25hrs. All the crew returned to R.A.F East Wretham, tired and cold but confident that Stettin had been hit hard. Both 3 Group H.Q and Bomber Command H.Q seemed happy with the results achieved by the modest force.

## September 30th 1941 : CC16 – Cherbourg

| Pilot | 2nd Pilot | Serial | Code | Bomb Load | Result |
|---|---|---|---|---|---|
| Sgt A Musálek | Sgt J Janek | T2553 | KX-B | 7x500lb+120x4lb | Duty Carried Out |

The crew departed at 18:23hrs. They were part of a group of just 41 bombers, six of which were provided by 3 Group. The small force made their way south to Cherbourg, located on the northern end of the Cotentin peninsula in clear conditions. The Wellingtons of 3 Group were over the docks between 20:37hrs and 21:40hrs. Sergeant A Musálek was one of the first at 20:40hrs. The crew dropped the entire bombload from 16,000ft, which was observed to straddle Dock No.4 and No.5, the incendiaries creating a fire on Dock 5. Returning crews reported the visibility over the target as 'exceptional'.

September had been a difficult month operationally. Fog and bad weather had reduced the squadron's operational commitment, and numbers were well down on the previous month. Thankfully, the weather began to improve towards the end of the month, and the longer nights allowed Bomber Command to flex its growing muscle to distant targets, most notably against Fascist Italy, Stettin and Berlin. A total of 71 sorties were flown on 13 separate raids, dropping in the process 78,170lbs of bombs for the loss of just one crew. The squadron had continuously pressed home their attacks despite what the weather and defences threw at them. Unfortunately, 3 Group's end of month figures show the squadron well behind the other Wellington-equipped squadrons in sorties flown, bombs dropped and worryingly, serviceability. Only the four-engine Short Stirlings had a lower serviceability rate in the Group.

There was however some good news, former commanding officer, Wing Commander Toman, was awarded the DFC for his time in command. Two DFMs were also announced, Flight Sergeant Josef Bernát and Flight Sergeant Rudolf Haering, for their actions on the night of July 5th. Flight Lieutenant Fantl, Squadron Operations Officer, was awarded the MBE. There were also a number of Czech awards, five recipients of the Czechoslovak Medal of Valour and five Czechoslovak War Medals. The squadron parted ways with pilot, Pilot Officer Václav Korda, who had flown a creditable 41 operations. The Czech Training Flight was producing crews on a regular basis despite the frequent accidents being reported. With one of the Group's lowest loss rates, the demand for trained crews was thankfully not causing any undue problems on the squadron.

*311 Squadron OTF commanding staff photographed on 22nd September 1941. L-R: Air Gunner-Instructor Flight Lieutenant Thomas James Desmond Baber (New Zealander), ATF Commanding Officer Flight Lieutenant Karel Vildomec, OTF Commanding Officer Squadron Leader Josef Šejbl, an adviser to the OTF Commanding Officer Squadron Leader Murrey Vernon Peters-Smith (British). (Jitka Špatná)*

*Incomplete five crews undergoing operational training at 311 Squadron OTF were photographed on the transfer from the ITF to the ATF at East Wretham on 22nd September 1941. Back row L-R: Sgt Ondřej Špaček, Sgt František Šipula, Sgt Zdeněk Sichrovský, Sgt Karel Hurt, Sgt Karel Mazurek, Sgt Vratislav Žežulka, Sgt Jaroslav Klvaňa, Sgt Josef Holub, Sgt Otakar Janůj, Sgt Jaroslav Poledník, 787881 Sgt Josef Svoboda. Middle row L-R: Sgt František Horký, Sgt Přibyslav Strachoň, Sgt František Binder, Sgt Rudolf Poledník, Wireless Operator-Instructor Sgt Jaroslav Mareš, Sgt Jaromír Bajer, Sgt Bedřich Gissübel, Sgt Alois Tolar, Sgt Jaroslav Rolenc, Sgt Jaroslav Jebáček, Air Gunner-Instructor F/Sgt ? (British). Seating L-R: F/O Antonín Hruška, F/O Oldřich Hořejší, ATF Pilot-Instructor P/O Anthony Herbert Foord (British), ATF Commanding Officer F/Lt Karel Vildomec, OTF Commanding Officer S/Ldr Josef Šejbl, an adviser to the OTF Commanding Officer S/Ldr Murrey Vernon Peters-Smith (British), Air Gunner-Instructor F/Lt Thomas James Desmond Baber (New Zealander), P/O Jaromír Brož, P/O Karel Sláma. (Pavel Vančata)*

# October 1941 : The Weather Sets In

There was limited flying for the first two days of October, not that anyone was complaining, fog being the culprit. Over at Honington, General Sir Alan Brooke, Commander-in-Chief Home Forces paid a visit. It was on the 3rd that a sizeable force of mostly 'Freshmen' crews would blitz the docks and facilities at Rotterdam, Antwerpen, Dunkirk and Brest. It would be mostly a 3 Group show with 311 Squadron supplying just five crews.

<u>October 3rd 1941 : CC25A – Dunkirk</u>

| Pilot | 2nd Pilot | Serial | Code | Bomb Load | Result |
|---|---|---|---|---|---|
| F/O J Stránský | Sgt F Naxera | Z8784 | KX-U | 7x500lb+120x4lb | Duty Carried Out |
| Sgt J Svoboda | Sgt V Pára | W5711 | KX-H | 7x500lb+120x4lb | Duty Carried Out |
| Sgt V Pánek | Sgt J Janek | T2553 | KX-B | 7x500lb+120x4lb | Duty Carried Out |
| Sgt V Procházka | Sgt F Petr | R1046 | KX-E | 7x500lb+120x4lb | Duty Carried Out |
| Sgt O Jambor | Sgt K Danihelka | T2971 | KX-J | 7x500lb+120x4lb | Duty Carried Out |

The raid was perfect for two novice pilots to experience an operation first-hand. Sergeants František Petr and František Naxera would finally get to put their long training into practice. Sergeant Naxera's route to operations had been anything but smooth, with two prangs to his credit while on the Training Flight. Sergeants V Procházka and O Jambor had completed their initiation in the 2nd pilot's seats and were given the chance to captain their own crews on this night. Flying Officer J Stránský and crew were the first away at 19:10hrs. Conditions once airborne were perfect as the crews cleared the Suffolk coast. The Czechs were over Dunkirk between 20:30hrs and 21:35hrs. Three crews made their bomb runs across the target from west to east, each dropping one stick of bombs. Flying Officer J Stránský was flying at 15,000ft, and had the satisfaction of seeing his bombs explode across Dock No.5 and Dock No.8. Sergeant J Svoboda bombed north to south and reported his bombs bursting across Dock No.6. Only the crew of Sergeant V Procházka chose to fly inland and then attack Dunkirk from east to west. His bombs were observed exploding along the railway station just south of the dock area. On leaving Dunkirk, fires were seen ablaze from 20 miles distant. All the crews were back by 20:21hrs apart from Sergeant O Jambor, who landed at R.A.F Marham. The squadron dropped a total of 19,900lb of bombs.

Widespread fog descended like a blanket on the 4th and would bring operations to a standstill. Operations were planned and then quickly cancelled. Conditions improved slightly on the afternoon of the 6th but resulted in the first accident of the month when Sergeant Jan Kotrch tried to lift Wellington R1771 off the runway with insufficient speed. The Wellington dropped heavily back to earth on its port wheel, bringing the dual training flight to a speedy conclusion.

| Date | 06/10/1941 |
|---|---|
| Mark | Mk.IC |
| Serial | R1771 |
| Code | KX-P |
| Taken On Charge | 23/09/1941 via 75 Squadron after being repaired by No.43 Group |
| Manufacturer | Vickers Armstrong |
| Contract | 992424/39 |
| Pilot (s) | Sergeant J Kotrch / ??? |
| Flight | Dual Training (Czech Training Flight) |
| Time | 16:00hrs |
| Cause | Insufficient speed on take-off. |

The AOC No.3 Group AVM J.E.A Baldwin CB, OBE, DSO, visited Honington on October 6th. The same day 311 Squadron has been informed that AOC No.3 Group confirmed the finding of Field Court Martial held on September 9th against the Pilot Officer Alois Tolar but commuted the sentence of dismissal to one of severe reprimand. The culprit

*AVM J. E. A. Baldwin CB, DSO, OBE exits Miles Mentor Mk.I L4422 belonging to No. 3 Group Communication Flight at East Wretham during his visit in October. The AVM rank pennant (two thick dark blue stripes with two thin red stripes between them on a light blue background) painted behind the cockpit means this was Baldwin's personal aircraft. Approaching the aircraft is RAF Station Honington Commanding Officer A/G/Cpt John Astley Gray DFC, GM. (Jaroslav Popelka)*

was sent to the Czech Depot at Wilmslow straight away. There he was reduced in rank to Sergeant and on October 10th he was back with the unit and continued in operational training

For the next three days, dense fog kept the squadron grounded. It was not until the morning of the 10th that the weather improved enough for the resumption of training and operations. The 'Freshmen' crews of 3 Group were given a relatively easy target to cut their teeth, the Docks at Rotterdam. Five crews from 311 Squadron would join a modest force of just 13 Wellingtons.

October 10th 1941 : CC42A – Rotterdam

| Pilot | 2nd Pilot | Serial | Code | Bomb Load | Result |
|---|---|---|---|---|---|
| F/O J Stránský | Sgt F Naxera | X9742 | KX-F | 7x500lb+120x4lb | Duty Carried Out |
| Sgt O Jambor | Sgt K Danihelka | R1451 | KX-P | 7x500lb+120x4lb | Duty Carried Out |
| Sgt K Knaifl | Sgt V Pára | W5711 | KX-H | 7x500lb+120x4lb | Duty Carried Out |
| Sgt V Pánek | Sgt J Janek | R1046 | KX-E | 7x500lb+120x4lb | Duty Carried Out |
| Sgt V Procházka | Sgt F Petr | Z8784 | KX-U | 7x500lb+120x4lb | Duty Carried Out |

The crews were all away by 23:07hrs into a decidedly unfriendly-looking sky. Clouds over the target created problems in identifying the Aiming Point. The first of the crews arrived over the docks at 00:15hrs. Flying Officer J Stránský had the briefest of views of the target, allowing his bomb aimer to drop their bombs close to the floating dock, starting a fire. Sergeant O Jambor dropped his entire load in one stick from 11,000ft. These landed south of the target, creating a large fire that produced a large volume of black smoke. The last crew over the target was Sergeant K Knaifl at 01:08hrs. They could not identify the aiming point, so they elected to bomb searchlights just south of the target. It was another satisfying and quick operation completed. The squadron was ordered to prepare for an operation to Emden on the 11th, but the order was cancelled at 13:25hrs. This order was itself cancelled at 14:50hrs when the squadron was ordered to bomb-up and stand-by. Finally at 17:00hrs after some frantic work by the groundcrews yet another order arrived cancelling the planned operation.

There was no such confusion on the 12th. The weather had improved enough for Bomber Command H.Q to send its bombers all over Germany, from Bremen in the north to Nuremberg in the south and Huls in the Ruhr. 311 (Czech) Squadron would detail five crews against Nuremburg while three 'Freshmen' crews would bomb the docks at Boulogne. A sizeable force of 152 aircraft were detailed and briefed for Nuremburg. 3 Group providing the largest contribution with 68 Wellingtons and seven Stirlings. The Yorkshire based 4 Group would supply 54 Whitleys, 4 Wellingtons and nine of its Halifaxes, while ten Wellingtons of 1 Group would also be involved.

### October 12th 1941 : Grayling 'A' – Nuremburg – Train Station /Marshalling Yards

| Pilot | 2nd Pilot | Serial | Code | Bomb Load | Result |
|---|---|---|---|---|---|
| Sgt A Šiška | Sgt J Filler | W5668 | KX-T | 1x500lb+1x250lb+410x4lb | Duty Carried Out |
| Sgt A Musálek | Sgt F Dostál | T2553 | KX-B | 1x500lb+1x250lb+410x4lb | Duty Carried Out |
| Sgt A Mžourek | Sgt J Fína | R1598 | KX-C | 1x500lb+1x250lb+410x4lb | Duty Carried Out |
| P/O J Nejezchleba | Sgt J Lenc | R1777 | KX-M | 1x1000lb+3x500lb | Alternative Target |
| Sgt M Šebela | Sgt B Hradil | R1161 | KX-X | 5x500lb+1x250lb | Duty Carried Out |

It was a mixed bag of crews on this night with new pilots and 2nd pilot combinations, some flying together for the first time. It was dark when the first crews started to taxi for take-off on what would be a long flight to southern Germany. 3 Group committed 13 of its squadrons on this maximum effort. Fog descended on R.A.F Oakington just before take-off removing the contribution of 7 and 101 Squadrons. A further four aircraft failed to take-off due to technical issues. Once airborne, the squadron headed south crossing the coast of Essex and out over the sea. There were no early returns from 311 Squadron although a further six from other squadrons returned early with various mechanical issues and one due to an encounter with a night fighter.

The crew of Sergeant M Šebela were the first over Nuremberg at 22:35hrs. Two medium-sized fires had already appeared to have taken hold north of the target. The crew dropped their bombload from 16,000ft. Scrutiny of the accuracy was not possible due to clouds. Within five minutes, Sergeant A Musálek was over the target at the controls of Wellington T2553 KX-B. He reported there were no fires visible on his arrival. Having established their position and the aiming point, they dropped their bombload close to the main train station and marshalling yards from 16,000ft. The incendiaries started an intense white fire with columns of smoke, which was reported to be visible from 40 miles away. There was a delay of twenty minutes before Sergeant A Šiška and crew arrived. Ominously, they, too, reported no visible fires on arrival. They bombed the marshalling yards, where their mixed load started a large fire with clouds of white smoke. The last to bomb was Sergeant A Mžourek at 23:10hrs. The crew reported at interrogation back at East Wretham, *'Fires seen in town and near marshalling yards on arrival. Bombs started fires south-east of station, followed by explosions'*. Pilot Officer J Nejezchleba got to within 120 miles of the target before issues with a fluctuating port engine forced him to find an alternative target. The target they chose was the small town of Weinheim, which, up until the rise of National Socialism, was popular with German Jews. Strangely, they did not bomb until 23:30hrs. Bursts were seen, but no other results. The raid appeared to be a success, at least at 3 Group H.Q. They reported *'the whole centre of the town appeared to be ablaze'*. This confidence was, however, dashed when the target photographs were examined by the

*The decoration of 311 Squadron members held at East Wretham on 11th October 1941. L-R: G/Cpt Karel Mareš-Toman, AVM J. E. A. Baldwin CB, DSO, OBE, the 3 Group Commanding Officer pinning the DFC to G/Cpt Mareš-Toman, Intelligence Officer F/Lt A. Fantl (hidden) awaiting the MBE, Pilot F/Sgt J. Bernát and Wireless Operator F/Sgt R. Haering both ready to be decorated by the DFM. (Pavel Vančata)*

boffins at Bomber Command H.Q. Unfortunately, the conclusion was that the weight of the attack fell upon outlying villages 8-10 miles from the centre of town. Further investigation later established that bombs were dropped on the town of Lauingen, located on the left bank of the Danube and 65 miles from Nuremberg. Worse still were reports that bombs had been dropped on the small town of Lauffen, on the River Neckar and a regular decoy site for Stuttgart, located 95 miles from Nuremberg.

### October 12th 1941 : CC.29A - Boulogne Docks

| Pilot | 2nd Pilot | Serial | Code | Bomb Load | Result |
|---|---|---|---|---|---|
| Sgt O Jambor | Sgt K Danihelka | R1451 | KX-P | 7x500lb+120x4lb | Duty Carried Out |
| Sgt K Knaifl | Sgt V Pára | W5711 | KX-H | 7x500lb+120x4lb | Duty Carried Out |
| Sgt V Procházka | Sgt F Petr | R1046 | KX-E | 7x500lb+120x4lb | Duty Carried Out |

The crews arrived over an already blazing dock area, flying between 11,000ft and 13,000ft. A large fire was reported between Docks 4 and 5, which attracted the bombloads of all three.

Obviously disappointed with the results of the previous attack, Bomber Command H.Q ordered another raid on Nuremburg on the 14th in what turned out to be terrible weather conditions. Group would supply 51 Wellingtons from seven squadrons with the Czechs of 311 providing six. Seven Stirlings of XV Squadron would also be detailed, although the usual mechanical troubles reduced this to four. Constant drizzle throughout the morning and early afternoon did not bode well for Op's that night. Thankfully, the rain stopped late afternoon as the first of the Wellingtons were bombed-up.

### October 14th 1941 : Grayling 'A' – Nuremburg – Train Station/Marshalling Yards

| Pilot | 2nd Pilot | Serial | Code | Bomb Load | Result |
|---|---|---|---|---|---|
| W/Cdr J Ocelka DFC | Sgt K Danihelka | T2971 | KX-J | 1x500lb+1x250lb+390x4lb | Alternative Target |
| Sgt V Ryba | Sgt M Plecitý | X9741 | KX-D | 1x500lb+1x250lb+390x4lb | Duty Carried Out |
| Sgt A Šiška | Sgt J Filler | W5711 | KX-H | 1x500lb+1x250lb+390x4lb | Duty Carried Out |
| Sgt S Linka | Sgt J Fína | R1598 | KX-C | 1x500lb+1x250lb+390x4lb | Duty Carried Out |
| Sgt M Šebela | Sgt B Hradil | R1161 | KX-X | 1x1000lb+3x500lb | Alternative Target |

Only five crews were eventually briefed for the operation as Z8784 KX-U assigned to Pilot Officer J Nejezchleba's crew had been found unserviceable. One was captained by Wing Commander J Ocelka DFC, accompanying Sergeant K Danihelka, who would be flying only his fifth operation. It was the mark of an excellent commanding officer to fly with an inexperienced crew.

The Group's contribution would be reduced by the failure of six aircraft to take-off and a further six which returned early either due to mechanical trouble or icing. An almost unbroken blanket of solid cloud over Germany made navigation particularly difficult on this night. Conditions were made worse when the bombers encountered a severe weather front as they neared Mannheim. Sergeant A Šiška was over a cloud covered Nuremburg at 02:11hrs having estimated his position. They dropped their mixed load from 14,000ft. Three minutes after bombing a fire was observed followed by an explosion. Sergeants V Ryba and S Linka also estimated their arrival over the target, Sergeant Ryba bombed at 02:30hrs and reported a glow beneath the clouds. Two crews never bombed Nuremburg. Sergeant M Šebela and crew encountered severe icing, causing the Wellington to become unstable. They selected to bomb Mannheim, dropping their all high-explosive load from 16,000ft at 01:35hrs. No results were seen. Wing Commander J Ocelka DFC reached Nuremberg. However, he retained his bombs as cloud obscured the target. Having turned for home, the crew chose to bomb Mannheim, which they had pin pointed on their way to Nuremberg. They bombed at 02:55hrs, flashes were seen, but no details were observed. All the crews safely returned to R.A.F East Wretham.

It had been a frustrating and arduous trip for little or no reward. Once again, poor weather made navigation extremely difficult for the participating crews. Nuremberg reported only three groups of bombs landing within the city's boundaries. It was yet another failure.

On the evening of the 14th, the squadron notched up another accident. However, no blame could be attributed to the pilot, Sergeant Jaroslav Hájek, Operational Training Flight Instructor. Scheduled for circuit and bumps training, the experienced pilot with over 400 flying hours on Wellingtons began his take-off at 19:15hrs when, without warning, a tyre burst when almost airborne. The Wellington's undercarriage immediately collapsed, bringing the aircraft to an undignified halt. New tyres had only just been fitted to the Wellington, and further investigation revealed that shrapnel had punctured the tyre. The squadron commander ordered a thorough runway check for further shrapnel or sharp objects.

| Date | 14/10/1941 |
|---|---|
| Mark | Mk.IC |
| Serial | L7847 |
| Code | KX-? |
| Taken On Charge | 25/05/1941 via 99 Squadron. |
| Manufacturer | Vickers (Chester) |
| Contract | 992424/39 |
| Pilot (s) | Sergeant Jaroslav Hájek |
| Flight | Training |
| Time | 19:15hrs |
| Cause | Tyre burst on take-off |

The veteran Wellington had previously seen service with 214 and 99 Squadron and had only just returned to the Czechs following repairs after Sergeant J Filler landed with a retracted undercarriage on August 24th.

Training was undertaken for the next two days, culminating in bombing practice over the Lakenheath bombing range. A planned raid on Emden on the 18th was cancelled at 16:03hrs after all the hard work of fuelling and bombing up had been completed. That night, the R.A.F Orchestra, under the direction of Wing Commander R.P O'Donnell, played in the recently opened Airmen's Dining Hall. Wing Commander R.P O'Donnell was a Londoner, his previous military experiences included duty with the infantry, cavalry, artillery, Royal Marines, and the R.A.F. During the First World War, he served in the 21st Lancers in India. The orchestra played works by both Czech and English composers and was warmly received by a large appreciative audience, especially the Czechs, who had a particular fondness for orchestral music. Air gunnery training was undertaken on the 19th over Langham, four crews taking part. The following day, the squadron prepared for yet another visit to Bremen, the squadron was to provide ten crews, nine on Bremen, one against Antwerp.

October 20th 1941 : Salmon 'A' – Bremen

| Pilot | 2nd Pilot | Serial | Code | Bomb Load | Result |
|---|---|---|---|---|---|
| Sgt A Mžourek | W/O K Weiss | R1598 | KX-C | 2x500lb+1x250lb+420x4lb | Duty Carried Out |
| Sgt O Jambor | Sgt K Danihelka | W5711 | KX-H | 2x500lb+1x250lb+420x4lb | Jettisoned |
| Sgt S Linka | Sgt J Janek | T2971 | KX-J | 3x500lb+1x250lb+18x50lb | Jettisoned |
| Sgt V Ryba | Sgt M Plecitý | X9741 | KX-D | 1x1000lb+4x500lb | Duty Carried Out |
| Sgt J Svoboda | Sgt J Filler | R1777 | KX-M | 1x1000lb+4x500lb | Early Return |
| Sgt V Pánek | Sgt J Lenc | R1161 | KX-X | 1x1000lb+4x500lb | Duty Carried Out |
| F/O F Pohlodek | Sgt F Dostál | T2553 | KX-B | 6x500lb+1x250lb | Duty Carried Out |
| Sgt V Procházka | Sgt F Petr | R1046 | KX-E | 6x500lb+1x250lb | MISSING |
| Sgt J Fína | Sgt F Bulis | R1451 | KX-P | 6x500lb+1x250lb | Duty Not Carried Out |

Just one new name would be chalked up on the crew board this night, Sergeant František Bulis. He would join the crew of Sergeant J Fína on what would be a difficult night and one which would see the squadron lose a crew. 3 Group would commit eight squadrons against Bremen and three on Antwerp. Bremen would be attacked by 153 bombers, and for once, the majority was not supplied by 3 Group. On this occasion, 5 Group put up 82 Hampdens and eight Manchesters. The first crew airborne was Sergeant O Jambor at 18:30hrs, and the last away was Sergeant V Procházka at 18:55hrs. The crew had aborted an earlier take-off when the port engine started misfiring. The following is taken from the book *Vocelovej* by author and historian Pavel Vančata based on Procházka's post-war report:

> *"Originally, we were going to take-off at 19:30 [actually first start at 18:44, re-start at 18:55] but as the port engine was running erratically and shaking, I stopped the start about halfway down the runway, afterwards we stopped at the last red light and returned back to the start. I reported to the squadron commander (škpt.[48] Ocelka) that the aircraft was unfit for operational flight and that I had no confidence in the aircraft. Škpt. Ocelka ordered the mechanic to check the plane. After the examination, škpt. Ocelka and the mechanic stated that the engine is fine. I disagreed with this opinion and insisted that the aircraft was not capable of operational flight. The commander then declared, if I recall in the presence of a mechanic, that I was a 'funk' and to go about my task, which I did."*

Conditions over the North Sea and Holland were good, but a little too good. German night fighters from NJG1 were airborne over the Low Countries and north-west Germany. One crew aborted, Sergeant J Svoboda reported engine trouble and jettisoned their entire load into the sea 10 miles north of the island of Ameland, West Frisians. Two crews had encounters with prowling fighters. The first involved Sergeant O Jambor at 20:30hrs while flying at 12,000ft, 10 miles from Wesermünde. The crew reported red flare-type shells being fired from the ground on their red beam up to about 10,000ft, apparently in an attempt to track the bomber. These seemed to be working in conjunction with searchlights, the master beam having a peculiar reddish tint. Two Me110s were observed flying in loose formation, followed by a pair of single engine Bf109s[49] on the green beam. One of the Me110s opened fire at 400 yards, on conclusion of his attack and pulling away past the Wellington, the front gunner managed to engage with two short bursts. Sergeant O Jambor took avoiding action by banking 30 degrees to port and starboard while at the same time diving and climbing 1000ft at a time. It was during these actions the bombload was jettisoned. Despite the manoeuvres, two more attacks were made, one from below and stern and the other on the green beam. All the while two fighters approached without firing, trying to confuse the gunners. One fighter made a head-on attack, passing just 10 feet above the Wellington. The front and rear gunners kept up a steady stream of short, controlled bursts at anything that came into range. The encounter lasted 29 minutes. During that time, the fighters were lost from view on several occasions.

Sergeant S Linka was also involved with a fighter east of Wesermünde at 21:07hrs. Flying at 16,500ft, the Wellington was being held in searchlights when a previously unseen Me110 was observed slightly above and on the green beam. The Me110 opened fire with cannons. Sergeant S Linka instantly turned sharply into his attacker while at the same time jettisoning the bombload. The manoeuvre worked, both the fighter and the searchlights were lost. Neither Wellington gunners opened fire, and there was no damage to the Wellington. Heavy, intense flak working in conjunction with numerous searchlights made the run into the target uncomfortable. This was particularly true between Wesermund and Bremen. Five crews claimed to have bombed the target. The first, at 20:45hrs, were the crew of Sergeant V Procházka, who reported a red fire in the town centre on arrival. They bombed from 15,000ft, and bursts were seen close to the aiming point, east of the old town. Sergeant V Ryba followed at 21:13hrs, and he dropped from 15,000ft. Two small fires were seen, and the crew added their bombload near the target area. Two crews arrived almost simultaneously at 21:28hrs. Sergeant V Pánek was at 17,000 feet, while Sergeant A Mžourek was three thousand feet lower. Neither crew were able to identify the aiming point and could only estimate the accuracy of their bomb bursts. Sergeant J Fína jettisoned his bombload 10 miles east of the target due to engine problems. The raid was not a success. Only half of the force reported bombing the target and of those very few identified the aiming point due to haze and the vicious flak.

---

[48] *Wing Commander*
[49] *The Combat report records two single engine Me110, this probably is a typo.*

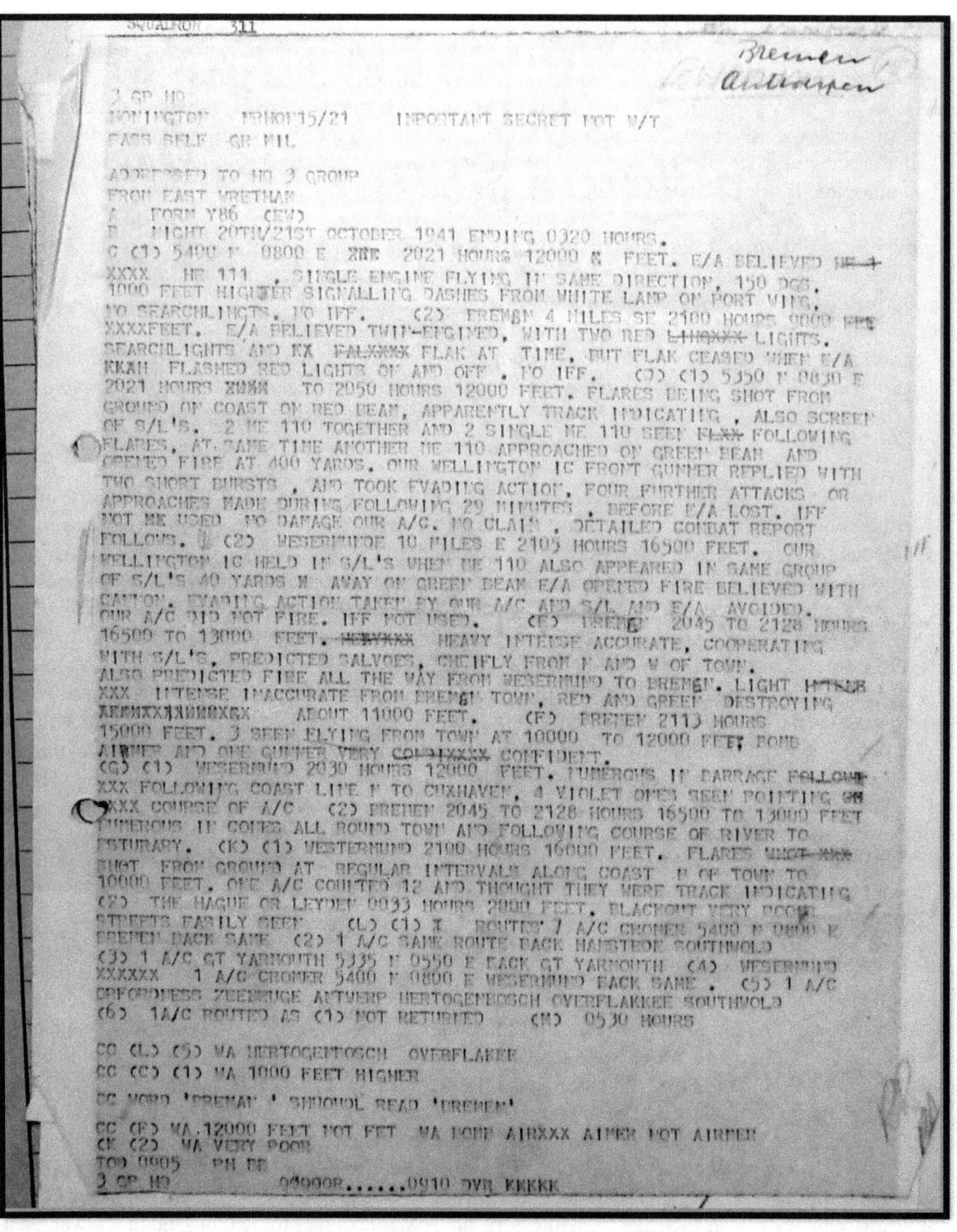

*It was a busy night for 311 Squadron on October 20th, and this report records the various night fighter encounters. (Pavel Vančata)*

The operation would claim another crew from 311 (Czech) Squadron skippered by Sergeant V Procházka aboard Wellington R1046 KX-E.

There are conflicting details regarding where the crew bombed and cause of loss. A NAP message was sent at 21:26hrs confirming bombs had been dropped. This is often quoted as Bremen and the cause of loss on the return journey due to Marine flak. However, contrary to this, the following details record the circumstances surrounding the eventual loss of Wellington R1046 KX-E. Again taken from Pavel Vančata's excellent book, '*Vocelovej*'.

*"Immediately after take-off the left engine ran erratically (shudders). About 30 minutes into the flight, the wireless operator reported significant radio malfunctions, and the internal connection failed. The second pilot (Sgt Petr) was at the controls from the start until the arrival over the auxiliary target. The auxiliary target was Emden, I made the decision to bomb this target because of a fault on the port engine, which continued to worsen as the flight continued. Flames shot from the engine. I was flying on the lowest boost. After the bombing (successfully hitting the given target) I set the course for the return. On the way back we were shelled by enemy flak – but we were not hit. Even above the Dutch coast, the port engine was losing revs, the oil temperature was already over 100°. I therefore switched off the port engine and balanced the aircraft on one engine. The port engine was still spinning due to spontaneous combustion. The control instruments of the port engine stopped showing. At the time there was a strong westerly wind (against the flight) at around 85 mph and I was therefore forced to switch to gliding flight. We were losing altitude from the original 13,000 feet, passing through cloud cover twice, and in the process some icing had formed and I could see that it was out of the question to get to England on one engine under the circumstances. I therefore decided to make an emergency landing at sea. Around 8.30 pm [actually at 10.30 pm] I landed on the water about 8 miles off the Dutch coast in the area north east of Leeuwarden.*

*On landing I found that I had landed on quicksand just as the tide was returning. Even before landing, the port engine flew away from the plane (the second day it could be seen after the water receded about 40 metres in front of the plane). We found ourselves in the water with the tide. I left the cabin and stood on the fuselage with the rest of the crew. The dinghy had to be pulled out using a cable, but since it was not attached to the plane, the dinghy floated away after being pulled out. The rear gunner, Zvolenský swam to her, got inside, but he did not manage to swim with her back to the plane."*

The remaining crew sat helplessly on the semi-submerged Wellington waiting for rescue. It was a cold, wet and unpleasant night exposed to the elements. The following morning the Wireless Operator, Sergeant Bedřich Valner in desperation acted.

*"In the morning hours, I managed to break into the plane and take out the signal cartridges, which, although completely soaked, could be used. I fired them at hourly intervals. ... I then managed to connect the Aldis lamp to the accumulators and send an SOS light signal to the shore, which was now looming on the horizon, to which I received an answer with a light mark. At 11:30 a.m. on October 21$^{st}$, 1941, when the situation seemed hopeless due to the rising tide, a German lifeboat arrived and pulled us, all five crew members, out of the thirteen-hour bath."*

Sergeant Jozef Zvolenský, the rear gunner was carried by the tide to the Dutch coast, where he landed on October 21$^{st}$ at 21:50hrs. He tried to escape but he was eventually captured.

## Vickers Wellington Mk.IC R1046 KX-E

| Manufacturer | Vickers Armstrong | |
|---|---|---|
| Contract | 992424/39 | |
| Taken on Charge | 14/04/1941 via 214 Squadron after being repaired by No.43 Group | |
| Cat E Missing | 20/10/1941 | |
| Struck Of Charge | 21/10/1941 | |
| Total Flying Hours | N/K | |
| Take-Off Time | 18:55hrs | |
| Bomb Load | 6x500lb+1x250lb | |
| | **CREW** | |
| Captain | Sergeant Václav **Procházka** 787193 RAFVR | PoW Stalag 4C Wistritz bei Teplitz |
| Second Pilot | Sergeant František **Petr** 787175 RAFVR | PoW Stalag 4B Muhlberg (Elbe) |
| Navigator | Pilot Officer Erazim **Veselý** 82534 RAFVR | PoW Oflag O4C Saalhaus **Colditz** |
| Wireless Operator | Sergeant Bedřich **Valner** 787899 RAFVR | PoW Stalag 4B Muhlberg (Elbe) |
| Front Gunner | Sergeant Josef **Sůsa** 787036 RAFVR | PoW Stalag Luft L3 Sagan |
| Rear Gunner | Sergeant Josef **Zvolenský** 787902 RAFVR | PoW Stalag Luft L1 Barth |
| | | |
| Posting History | Sgt V Procházka via RAF Hendon, 14/03/1941<br>Sgt F Petr via RAF Church Fenton, 12/05/1941 | |
| Operations Flown | Sgt V Procházka : 26 / Sgt F Petr : 3 | |
| Buried | N/A | |

Bomber Command detailed 35 crews for the attack on Antwerp, half of which were supplied by 3 Group. A solid layer of cloud covered the target. Most crews aborted, and a few bombed on E.T.A. Flying Officer J Stránský, had other ideas.

### October 20th 1941 : CC47 – Antwerp Docks

| Pilot | 2nd Pilot | Serial | Code | Bomb Load | Result |
|---|---|---|---|---|---|
| F/O J Stránský | Sgt F Naxera | X9742 | KX-F | 7x500lb+120x4lb | Alternative |

The crew, unable to bomb the primary, decided to head for Bremen. Unfortunately, they encountered a severe weather front where the Wellington began to ice up. Realising that to continue was reckless, they bombed a flak position near Hertogenbosch at 20:15hrs from 10,000ft. Bursts were seen through gaps in the clouds, followed by fires. The last of the crews landed at 03:16hrs. It had been a gruelling night with little if anything to cheer about. The loss of the experienced Sergeant Procházka was a cruel blow, especially so close to completing his operational tour. It could have been a lot worse if it had not been for the excellent coordination between pilot and gunners in fending off the night fighters over northern Germany.

On the 21st, those who did not operate the previous night were training. Bombing practice was flown over Lakenheath while others carried out long-range cross-country exercises. On the return from one of these exercises, Sergeant Oldřich Soukup landed too fast back at East Wretham and struck the boundary fence ridge, the Wellington's undercarriage buckled and then collapsed.

| Date | 21/10/1941 |
|---|---|
| Mark | Mk.IC |
| Serial | P9299 |
| Code | KX-A |
| Taken On Charge | 16/06/1941 (Czech Training Flight) via 218 Squadron. |
| Manufacturer | Vickers (Weybridge) |
| Contract | 549268/36 |
| Pilot (s) | Sergeant Oldřich Soukup |
| Flight | Training |
| Time | 16:30hrs |
| Cause | Landed too fast. |

Rain and drizzle on the 22nd kept the Training and Operations Flight grounded. During the day, there was one unexpected incident involving a Hawker Hurricane. Frenchman Sous Lt M.H.I. Daligot landed Hurricane IIB Z3029 of 257 Squadron on the grass flarepath, having become lost. While taxing slightly too fast in gusty conditions, the usually robust undercarriage collapsed in the soft ground. The embarrassed pilot was unhurt and immediately taken to the Operations Room, where the Czechs tested their French learnt while serving in France with the unfortunate Frenchman.

The bad weather continued into the following day with intermittent thunder, drizzle, and hail. However, on this occasion, seven crews managed to get airborne for cross-country training. Sergeant Karel Hurt skippered one of the crews. Airborne at 11:40hrs aboard Wellington T2624, they were routed from base to Llangollen, located on the River Dee, Wales, then out into the Irish Sea and onto Maughold Head on the Isle of Man. From here, they were routed to make landfall at Aberystwyth on the coast of Cardigan Bay, then to Banbury, Oxfordshire and finally back to base. Unfortunately, the crew never returned to R.A.F East Wretham, it is presumed the Wellington crashed in Cardigan Bay killing all on board. The cause has never been determined, the AM Form 1180 Accident Card records the cause could have been due to one of four reasons: *"1) Technical failure, 2) Enemy Aircraft, 3) Weather, 4) Error of judgement by the pilot."*

*Unexpected visitor. Frenchman Maurice Henri Daligot. He would be posted to 340 Free French Squadron based in Scotland and equipped with Spitfire Mk II's. Maurice did not return from a training flight on December 20th, 1941, his body was never found.*

At East Wretham, there had been no communication via W/T indicating trouble and no clue as to what had happened. Crews were expected to go missing over enemy territory but not on a routine training flight.

## Vickers Wellington Mk.IC T2624 KX-?

| Manufacturer | Vicker Armstrong | |
|---|---|---|
| Contract | B38600/39 | |
| Taken on Charge | 02/08/1941 (Czech Training Flight) via 15 Squadron after being repaired by No.43 Group. | |
| Cat E Missing | 23/10/1941 | |
| Struck Of Charge | 24/10/1941 | |
| Total Flying Hours | N/K | |
| Take-Off Time | 11:04hrs | |
| Bomb Load | N/A | |
| | **CREW** | |
| Captain | Sergeant Karel **Hurt** 787557 RAFVR. Age 24. | Runnymede Panel 45 |
| Second Pilot | Sergeant Jaroslav **Rolenc** 787686 RAFVR. Age 26. | Runnymede Panel 51 |
| Navigator | Pilot Officer František Karel **Dittrich** 87616 RAFVR. Age 24. | Grave D.17[50] |
| Wireless Operator | Sergeant Otakar **Janůj** 787846 RAFVR. Age 28. | Runnymede Panel 46 |
| Front Gunner | - | |
| Rear Gunner | Sergeant Jaroslav **Poledník** 787598 RAFVR. Age 21. | Runnymede Panel 50 |
| | | |
| Posting History | Sgt K Hurt posted to 311 Sqdn, 13/07/1941. Sgt J Rolenc posted to 311 Sqdn, 18/07/1941. | |
| Operations Flown | - | |
| Buried | CARDIGAN CEMETERY | |
| Remembered | RUNNYMEDE MEMORIAL | |

There was little time to ponder the fate of Sergeant K Hurt and crew as preparations were in full swing for an operation against Kiel that night. Only five squadrons would be provided by 3 Group, R.A.F Marham's pairing of 115 and 218 Squadron, plus eight crews from 311, six from 9 Squadron, and eight Wellingtons from 214 Squadron. They would be joined by a mixed bag of Hampdens, Whitleys, Wellingtons and Manchester's from 1, 4 and 5 Groups. The Wellingtons of 3 Group were given Kiel's important shipbuilders, the Deutsche Werkes famous for the heavy cruiser the *Lützow*, the Battleship *Gneisenau* and building the German's only aircraft carrier, the *Graf Zeppelin*.

### October 23rd 1941 : 'Ant' – Kiel Harbour – Deutsche Werkes

| Pilot | 2nd Pilot | Serial | Code | Bomb Load | Result |
|---|---|---|---|---|---|
| Sgt M Plecitý | Sgt J Tománek | X9741 | KX-D | 1x1000lb+420x4lb | Duty Carried Out |
| Sgt S Linka | Sgt J Janek | T2971 | KX-J | 1x500lb+420x4lb | Duty Carried Out |
| F/O F Pohlodek | Sgt F Dostál | T2553 | KX-B | 1x500lb+420x4lb | Duty Carried Out |
| Sgt V Pánek | Sgt J Lenc | R1161 | KX-X | 1x500lb+420x4lb | Duty Carried Out |
| Sgt J Svoboda | Sgt J Filler | R1777 | KX-M | 1x500lb+3x500lb | Duty Carried Out |
| Sgt A Mžourek | Sgt F Naxera | R1598 | KX-C | 5x500lb+1x250lb | Alternative Target |
| F/O J Stránský | W/O K Weiss | X9742 | KX-F | 5x500lb+1x250lb | Duty Carried Out |
| Sgt O Jambor | Sgt V Pára | R1451 | KX-P | 5x500lb+1x250lb | Duty Carried Out |

Sergeant Josef Tománek would take part in his first operation on this night. He was born on March 12th 1918, in the town of Přerov. Like many of his compatriots, he found his way to England after a perilous journey from his homeland via Slovakia, North Africa and France. He would be operating beside Sergeant M Plecitý, who would be captaining his own crew operationally for the first time. By 17:37hrs, all eight crews had departed into a less than friendly looking

---

[50] *Initially buried Cardigan Cemetery, cremated Pontypridd Crematorium ashes returned to Czechoslovakia post-war.*

overcast. The Group recorded two aircraft failing to take-off and six early returns, none from 311 Squadron. Northern Europe was almost entirely covered by an impregnable layer of cloud, which brought problems for the navigators. Sergeant A Mžourek got as far as Schleswig, a town in the northeastern part of Schleswig-Holstein, Germany, before dropping his bombs on a flak position at 21:20hrs due to engine problems. Five crews were obliged to bomb Kiel on either E.T.A. or the glow of fires or searchlights. Sergeant J Jambor got a glimpse of the southern edge of the harbour. He dropped his bombs from 15,000ft half a mile northeast of the aiming point at 21:40hrs. A small gap in the cloud was enough for Sergeant J Svoboda to see an estuary and what looked like Kiel town before he also dropped at 21:40hrs. The frustrated crews were all safely back at R.A.F East Wretham by 01:06hrs. The raid was a complete failure, 10/10th cloud meant almost all the Group aircraft bombed blindly. Some crews surprisingly reported fires in the Deutsche Werkes, however, due to the clouds an accurate assessment was impossible.

Other than the Training Flight the squadron stood down on the 24th and 25th. The weather was lousy, low clouds, drizzly and the occasional heavy downpour reflected the general mood on the squadron. On the 26th, 3 Group offered five squadrons for an attack on Hamburg. Eight crews from 311 Squadron would be detailed and briefed.

October 26th 1941 : Dace 'B' – Hamburg

| Pilot | 2nd Pilot | Serial | Code | Bomb Load | Result |
|---|---|---|---|---|---|
| Sgt J Svoboda | Sgt J Janek | T2971 | KX-J | 1x500lb+1x250lb+420x4lb | Duty Carried Out |
| Sgt V Ryba | Sgt O Hlobil | Z1167 | KX-A | 1x500lb+1x250lb+420x4lb | Duty Carried Out |
| F/O F Pohlodek | Sgt F Dostál | Z8784 | KX-U | 1x500lb+1x250lb+420x4lb | Alternative Target |
| Sgt A Musálek | Sgt A Šiška | T2553 | KX-B | 1x500lb+1x250lb+420x4lb | Alternative Target |
| Sgt J Fína | Sgt F Bulis | W5711 | KX-H | 1x500lb+1x250lb+420x4lb | Duty Carried Out |
| Sgt O Jambor | Sgt V Pára | X9741 | KX-D | 1x500lb+1x250lb+420x4lb | Duty Carried Out |
| Sgt A Mžourek | Sgt F Naxera | R1777 | KX-M | 5x500lb+1x250lb | Duty Carried Out |
| Sgt V Pánek | Sgt J Lenc | R1161 | KX-X | 5x500lb+1x250lb | Early Return |

The weather for the outward flight to north-western Germany was promising. Three crews would not get the benefit of the clear conditions. Engine problems aboard Wellington R1161 KX-X resulted in the crew of Sgt V Pánek jettisoning his complete load 'live' in the sea off the island of Juist, North Frisians. Another two crews reported engine trouble. Sergeant A Musálek bombed flak concentrations at Westerhever, on the North Sea coast in the Schleswig-Holstein area at 21:10hrs. While Flying Officer F Pohlodek bombed the town of Tönning, located on the northern bank of the Eider River, approximately six miles from its mouth at the North Sea, at 21:17hrs. Neither crew observed any results. What was left of the squadron arrived at the primary target at 21:00hrs to find heavy clouds but bright moonlight. Sergeant J Fína dropped his mixed load in two sticks from 16,500ft south of the aiming point. Opposition was fierce, Hamburg's flak batteries putting up a tremendous barrage while the searchlight batteries were equally active sweeping the sky. Sergeant V Ryba bombed at 21:20hrs, claiming his incendiaries started fires north of the aiming point. Flying with the crew on his first operation was Sergeant Oldřich Hlobil. Five minutes after Sergeant Ryba it was the turn of Sergeant J Svoboda, who dropped his mixed load in one stick from 18,500ft. The crew were unable to observe any results due to the cloud and the glare from the searchlights. Sergeant O Jambor dropped his entire bombload in one stick from 15,000ft at 21:35hrs. They claimed that these started a fire, which was visible for 30 minutes after leaving the target area. On arrival back at their East Anglian bases, the 3 Group crews reported numerous explosions in the target area along with fires that suggested an effective raid, which would be confirmed by local sources with mention of around a dozen serious fires in Hamburg. On the 27th, news arrived on the squadron that a body, believed to be from Wellington T2624 had washed ashore on a beach in Wales.

*One of five crews trained at 311 Squadron OTF from August to October 1941. L-R: Front Gunner Sergeant Jaromír Bajer; Second Pilot Sergeant Karel Hurt, Navigator Sergeants Alois Tolar; Wireless Operator Bedřich Gissübel. Skipper Sergeant Ondřej Špaček and Rear Gunner Sergeant Přibyslav Strachoň. Sergeant Hurt went missing on a cross country flight over the Irish Sea on 23rd October 1941 as a skipper of Wellington Mk.IC T2624. (Pavel Vančata)*

October 28th was Czechoslovak Independence Day, or the birthday of Czechoslovakia and would be suitably honoured. The AOC 3 Group, AVM J.E.A Baldwin CB, OBE, DSO visited R.A.F East Wretham and officiated over a morning parade followed by a religious service given by Squadron Leader Pouchly honouring fallen comrades. Some tour-expired 311 Squadron airmen arrived from various training units dotted around the country to join in the celebrations that involved various sporting activities undertaken in rain and drizzle until retreating into the various Messes for refreshments and entertainment. On the 29th, members of the squadron flew down to Wales to identify the body washed ashore. Sadly, it was confirmed that the naked body found at 09:30hrs on the 27th at New Quay, a seaside town in Ceredigion, Wales, located 19 miles south-west of Aberystwyth on Cardigan Bay, was that of the navigator Pilot Officer František Dittrich. The cause of the crash had not been determined. The cloud base on the day was around 20,000ft above sea level, so there was no need for low flying. Enemy activity in the Irish Sea was discounted, and no 'Friendly fire' incidents were reported. It was a complete mystery. The fact that the body of Pilot Officer Dittrich was naked bears witness to the violence of his death. Was there a catastrophic structural failure? The Wellington had seen operational service with 15 Squadron and then spent considerable time with No.43 Maintenance Group before its arrival on the Czech Training Flight. Here, it appears to have been involved or suffered damage on two occasions. The funeral arrangements for Pilot Officer Dittrich were undertaken by R.A.F Aberporth. Pilot Officer Dittrich was buried on November 4th in the Cardigan Cemetery. On May 8th, 1946, he was exhumed and then cremated, and his English fiancée, Maria Worthington, took his ashes home to his parents for a family funeral. On July 13th, 1946, the remains of František Dittrich were buried in the Dittrich family grave at Hradec Králové-Kukleny.

The squadron was to operate against Brest on the 30th, but for reasons unknown, the operation was cancelled at 13:30hrs. Heavy rain, thundery showers, and hail kept the squadron grounded the following day, bringing the month to a close.

*Tour complete. Pilot Officer Jaroslav Bala and Pilot Officer Leo Anderle. Both would put their skills and vast experience to good use when posted to No.138 (Special Duties) Squadron.*

The squadron operated on eight occasions, flying 46 sorties.[51] One crew was lost on operations, while another from the Training Flight were killed. A total of 140,820 pounds of bombs were dropped throughout the month. Disappointingly, the squadron once again had the lowest serviceability rate in the Group with 61.6%. It was not something the Czechs were happy with and rightly laid the blame on the ageing Wellingtons. Once again, there was reason to celebrate with the award of 13 Czech Medals of Honour, 13 Czech War Crosses, and a further 19 Bars to the Czech Medal of Honour. Both Wing Commander Ken Batchelor DFC and Flying Officer Anthony Foord DFC of the Czech Training Flight were awarded the Czech Flying Badge. The squadron lost a few of its popular members during the month. Two of which were the experienced, Pilot Officer Jaroslav Bala, who had flown his first operation in September 1940, and Pilot Officer Leo Anderle who began his operational career in October the same year. Both would, with two selected crews joined 138 Special Duties Squadron after conversion on Whitley aircraft. Also posted was Pilot Officer Arthur Roman, Bombing Instructor/Armourer Leader. With the Butt Report's damning findings still reverberating around Bomber Command H.Q and questions about the effectiveness of Bomber Command being raised at almost every level of government, the vultures were gathering at the Air Ministry and in Parliament, stirred on by the Senior Service, who were firmly anti-Bomber Command questioning the very existence of a strategic bomber force.

---

[51] *Figure differ depending on what file you use.*

# November 1941 : Under the Spotlight

As welcome as the longer nights were, it came with problems, worsening weather conditions, clouds, and poor visibility, Bomber Command's most significant obstacles to success. If you could not see the ground, you could not accurately navigate. If you could not see the target, it was impossible to bomb precisely. It was that simple. The bravery and tenacity of the bomber crews were not in question. They simply did not have the navigational and bombing aids needed to achieve the desired accuracy. Behind the scenes there were influential individuals eager for Bomber Command to fail. All eyes were now on Bomber Command, as the future of Britain's strategic bomber force was in the balance.

There was a change of Base Commander on the 1st, the respected Group Captain J.A Gray DFC GM was posted to 3 Group H.Q as S.A.S.O. His replacement was Group Captain Montagu Ommanney fresh from a stint at 3 Group H.Q. However within a matter of days he was replaced temporarily by Wing Commander K.M Wasse DFC when the new Base Commander was almost immediately posted on detachment.

The November offensive started on the 1st with a medium sized attack on Kiel, plus two smaller operations on Le Havre and Brest. 311 Squadron would be part of a force of 134 briefed for a return visit to Kiel and Aiming Points 'A'-'B' and the Deutsche Werkes. The squadron was ordered to prepare ten crews and load the majority of the Wellingtons with incendiaries.

### November 1st 1941 : 'Minnow 'A' – Kiel Harbour

| Pilot | 2nd Pilot | Serial | Code | Bomb Load | Result |
|---|---|---|---|---|---|
| Sgt V Ryba | Sgt O Hlobil | Z1167 | KX-A | 1x250lb+530x4lb | Duty Carried Out |
| Sgt J Svoboda | Sgt J Janek | T2971 | KX-J | 1x250lb+530x4lb | Duty Carried Out |
| Sgt A Mžourek | Sgt F Naxera | R1598 | KX-C | 1x250lb+530x4lb | Early Return |
| Sgt O Jambor | Sgt V Pára | X9741 | KX-D | 1x250lb+530x4lb | Duty Carried Out |
| Sgt V Pánek | Sgt J Lenc | R1777 | KX-M | 1x250lb+530x4lb | Duty Carried Out |
| Sgt K Knaifl | Sgt K Danihelka | W5682 | KX-Y | 1x250lb+530x4lb | Early Return |
| F/O F Pohlodek | Sgt F Dostál | Z8966 | KX-E | 1x250lb+530x4lb | Alternative Target |
| Sgt J Fína | Sgt F Bulis | X9742 | KX-F | 1x250lb+530x4lb | Duty Carried Out |
| P/O J Nejezchleba | Sgt J Tománek | W5711 | KX-H | 1x250lb+530x4lb | Duty Carried Out |
| Sgt M Šebela | Sgt B Hradil | W5668 | KX-T | 5x500lb+1x250lb | Duty Carried Out |

Disappointingly the squadron had three crews abort with engine trouble. Sergeant K Knaifl and crew jettisoned their entire bombload at 54:30N – 05:50E, landing back at East Wretham at 21:06hrs. They were followed thirty-three minutes later by Sergeant A Mžourek with a defective port engine. He jettisoned his bombs 'live' almost within sight of the north Frisian Islands at 53:46N – 04:75E. Flying Officer F Pohlodek dropped his bombload on the island of Sylt at 20:45hrs from 15,000ft. The incendiaries were seen to ignite, but no other details were observed. It was not an encouraging start. The first of the squadron crews arrived over a completely cloud-covered Kiel at 21:14hrs. They bombed on E.T.A or on the glow of what appeared to be fires in the general area of Kiel. A few dropped on the flashes of anti-aircraft guns. Sergeant O Jambor was the last of the squadron over the target at 22:05hrs. He also bombed on E.T.A from 13,000ft but reported the glow of a big fire reflected in the clouds. Both 3 Group and Bomber Command H.Qs reported the raid as *'a complete failure'*.

The following day, the crews rested while the squadron's Wellingtons, especially those with reported engine problems were worked on. The squadron was ordered to prepare for an operation to Bremen on the 3rd, but this was cancelled at 13:55hrs. Bombing training over Berners Heath was flown on the 4th in varying weather conditions. Group Captain Montagu Ommanney returned from detachment on the 4th for what would turn out to be a difficult few months. The Ruhr and the Krupps works at Essen was the target for the 5th, but this was cancelled at 16:21hrs due to weather issues.

*An unknown 311 Squadron crew inspecting the 250-pounders that they would be dropping on a target in Germany within a few short hours. (John Costin)*

Finally, on the 6th, the weather improved, and an extensive training programme was to be flown by both the Czech Training Flight and the Operations Flight. The squadron was to re-visit Kiel that night, but Group H.Q cancelled this operation late afternoon. With a break in the weather over England and the continent, Bomber Command H.Q organised a raid on the Nazis capital, Berlin. 3 Group called for a maximum effort from its squadrons which would detail 87 Wellingtons on Berlin and a further 33 crews on Mannheim, Ostend and the Ruhr, bringing the total to a respectable 120 bombers. The Czechs would detail seven crews against Berlin, plus three on Mannheim.

November 7th 1941 : 'Whitebait' – Berlin

| Pilot | 2nd Pilot | Serial | Code | Bomb Load | Result |
|---|---|---|---|---|---|
| Sgt V Ryba | Sgt O Hlobil | Z1167 | KX-A | 2x500lb+360x4lb | Duty Carried Out |
| Sgt M Plecitý | Sgt V Pára | X9741 | KX-D | 2x500lb+360x4lb | Duty Carried Out |
| Sgt J Svoboda | Sgt J Janek | T2971 | KX-J | 2x500lb+360x4lb | Alternative target |
| Sgt A Šiška | Sgt J Filler | T2553 | KX-B | 1x1000lb+3x500lb | Jettisoned |
| P/O J Nejezchleba | Sgt J Tománek | W5711 | KX-H | 1x1000lb+3x500lb | Duty Carried Out |
| F/O F Pohlodek | Sgt F Dostál | Z8966 | KX-E | 5x500lb+1x250lb | Duty Carried Out |
| Sgt J Fína | Sgt F Bulis | R1451 | KX-P | 5x500lb+1x250lb | Duty Carried Out |

Clearly wishing to offset the growing criticism aimed at him personally, and his Command, the AOC Bomber Command Sir Richard Peirse needed to produce an attention grabbing success to rescue both his and Bomber Command's tarnished reputation. There was no better target than the seat of Nazi power, Berlin. Despite warnings of unsuitable weather, Peirse stubbornly refused to change the night's planned raids. The planned operation had already suffered a serious blow when concerns about the weather prompted 5 Group AOC, AVM Slessor, to question the wisdom of going ahead with the operation, and he requested he be allowed to withdraw his force and send it instead to Cologne. Slessor, a highly capable and forceful commander would fall foul of Peirse over his Groups withdrawal.

The Berlin force was the first away and all seven Czech crews had safely departed by 17:54hrs. Stronger-than-forecast winds and icy conditions over the North Sea added to the problems for the Berlin force, which also had to deal with solid cloud for the entire outward flight and over Berlin. Once again, engine problems blighted the squadron's contribution. The first back at 00:01hrs was Sergeant J Svoboda. He dropped his bombs at 20:10hrs on a railway junction at Vienenburg, 24 miles south of Brunswick. The crew of Sergeant A Šiška also experienced engine issues, they dropped their bombload in the 'Ruhr' area at 20:45hrs, explosions were seen below the solid clouds. Berliners would be spared 4,940 lbs of bombs. The remaining Czechs were over the target between 20:45hrs and 21:25hrs. Other than a brief glimpse of gun flashes and explosions seen through the clouds the crews had no option than to bomb on E.T.A or flak concentrations. As expected, Berlin's flak batteries put up a tremendous barrage, with burst exploding at varying heights. A number of burning bombers were seen falling like blazing comets. While on the return route, Sergeant J Fína and crew were attacked by a Bf109 over Alkmaar at 00:15hrs. The fighter inflicted damage to the rear turret hydraulic system and fuselage during its several attacks. It was finally shaken off at 4,000ft.

Sergeant Fína's Wellington sustained extensive damage. Both airscrews were hit, the starboard nacelle and fuel tank punctured, and the engine rendered practically useless. The hydraulics were also hit, resulting in a wheels-up landing back at East Wretham. It had been a masterful piece of flying by the pilots in bringing the crippled Wellington back.

| Date | 07/11/1941 |
|---|---|
| Mark | Mk.IC |
| Serial | R1451 |
| Code | KX-P |
| Taken On Charge | 06/03/1941 |
| Manufacturer | Vickers (Weybridge) |
| Contract | 992424/39 |
| Pilot (s) | Sergeant J Fína / Sergeant F Bulis |
| Flight | Operational |
| Time | 02:03hrs (Not confirmed) |
| Cause | Night Fighter Encounter |

The squadron had been lucky, all the Berlin force had landed safely by 02:47hrs, most already knew the raid was a complete fiasco. Damage to Berlin was minimal and scattered, damage to the reputation of Bomber Command and particular the AOC Sir Richard Peirse was vastly more severe. A total of 21 bombers failed to return due to the German defenders and the terrible weather conditions. A number of crews were obliged to ditch in the North Sea short of fuel.

November 7<sup>th</sup> 1941 : 'Chubb' – Mannheim

| Pilot | 2nd Pilot | Serial | Code | Bomb Load | Result |
|---|---|---|---|---|---|
| Sgt V Pánek | Sgt S Linka | R1777 | KX-M | 3x500lb+410x4lb | Duty Carried Out |
| Sgt M Šebela | Sgt B Hradil | R1458 | KX-Z | 7x500lb | Jettisoned |
| Sgt K Knaifl | Sgt O Jambor | W5682 | KX-Y | 7x500lb | Duty Carried Out |

*Berlin*
*Mannheim*

From :- O.C. 311 Squadron, East Wretham.

To :- Intelligence Officer, Headquarters, No.3 Group.

Date :- 8th. November, 1941.

Ref :- 311/S.2014/Int.

## Combat Reports.

With reference to Form Y 90(BW), dated 8/11/41, the following are supplementary combat reports.

**Para D (1).** Wellington 1C, 311 Squadron, Z 1458 on Mannheim. Roubaix, 12 miles East, 2050 hrs, 10,500 feet, 115 mph I.A.S. Course 166 degrees. Visibility good, moon low on port bow. Scattered searchlights, but not near our aircraft. I.F.F. not used. Three flares shot up from ground, igniting to starboard and slightly higher than our aircraft. No flak. ME 110 first seen 500/600 yards away on starboard beam, going opposite direction, turned and attacked from starboard quarter 500 yards away, slightly above, believed with cannon from nose. Rear gunner replied with short bursts, and Wellington dived and turned to starboard at same time jettisoning bombs. Enemy aircraft not seen again, and our aircraft levelled out at 10,500 feet. No damage suspected to our aircraft, but tyre burst on landing. No casualties. No claim.

        Rear gunner :- Sgt. Plusman.

**Para D (2).**

Wellington 1C, 311 Squadron, P.1451, on Berlin. Alkmaar, 0015 hrs, 18,000 feet, 120 mph I.A.S. Bombs dropped. Course 280 degrees. Visibility good, with moon on port beam, and no cloud. No searchlights operating, and I.F.F. not in use. No flak. An ME 109, without lights was observed on the starboard quarter, below our Wellington 1C. The ME 109 opened fire with cannon and machine guns, hitting our aircraft in the fuselage, just forward of the rear turret, approaching to within 200 yards before breaking away. Our aircraft took evasive action by turns of 30/40 degrees, diving at same time to 14,000 feet. Enemy aircraft then approached again, and made 7 more attacks, always from between 400 and 200 yards astern, while our aircraft snaked down to 4,000 feet. Rear gunner was unable to reply as the hydraulics of his turret had been cut in the first engagement, and he was unable to hand-manipulate quickly enough. At 4,000 feet enemy aircraft was evaded, and not seen again. Rear turret rendered u/s, fuselage hit in several places, and port wing hit. No casualties. No claim.

        Rear gunner :- Sgt. Kamarad.

...........................
Flight Lieutenant, for
Wing Commander, Commanding,
311(Czech) Squadron, Wretham.

*The two encounters with night fighters on November 7th 1941. (Pavel Vančata)*

*Wellington IC R1451/KX-P photographed on 8th November 1941 after belly-landing on return from Berlin. The aircraft has been attacked eight times by enemy night fighter believed to be Messerschmitt Bf 109 over Alkmaar at 00.15. Rear turret has been hit and put out of order during the first attack but the Rear Gunner W/O Jaroslav Kamarád escaped unharmed. (Pavel Vančata)*

The three crews given Mannheim were airborne by 19:45hrs and slowly climbed over the English coast towards northern France. During take-off, the Wellington of Sergeant V Pánek struggled to climb and struck a hedge, and what was also recorded as the *'roof of a dispersal point'*, damaging both tyres. Once airborne, the pilot gingerly retracted the undercarriage, and the crew checked for further damage. With the engines running normally and no issues with the controls, the crew courageously decided to continue the operation. Over the town of Roubai, France, a Me110 night-fighter attacked the Wellington flown by Sergeant M Šebela. During the encounter, the bombload was jettisoned at 20:50hrs. The crew returned to East Wretham, where, on landing at 23:09hrs, a tyre burst, thankfully without causing further damage.

| Date | 08/11/1941 |
|---|---|
| Mark | Mk.IC |
| Serial | W5682 |
| Code | KX-Y |
| Taken On Charge | 10/05/1941 |
| Manufacturer | Vickers (Weybridge) |
| Contract | 71441/40 |
| Pilot (s) | Sergeant Metoděj Šebela / Sergeant Bohuslav Hradil |
| Flight | Operational |
| Time | 03:58hrs |
| Cause | Suspected fighter damage. |

The weather over Mannheim was marginally better than Berlin, with an occasional break in the clouds. Sergeant K Knaifl bombed at 22:30hrs from 15,000ft. They managed to identify the town and observed their bombs bursting near the railway station in the centre of the city, starting some fires. Twenty-seven minutes later, Sergeant V Pánek, having identified the town centre, dropped their bombs from 14,000ft. On return to East Wretham at 03:58hrs, Sergeant V Pánek brought his Wellington into land with both tyres presumably punctured, unsurprisingly the undercarriage collapsed and the Wellington skidded for 1,400 yards before coming to a halt near a dispersal hut.

| Date | 08/11/1941 |
|---|---|
| Mark | Mk.IC |
| Serial | R1777 |
| Code | KX-M |
| Taken On Charge | 18/05/1941 |
| Manufacturer | Vickers Armstrong (Chester) |
| Contract | 992424/39 |
| Pilot (s) | Sergeant V Pánek / Sergeant S Linka |
| Flight | Operational |
| Time | 03:57hrs |
| Cause | Pilot Error |

Returning crews reported a number of fires on leaving the target. Despite their optimism Mannheim reported that no bombs fell on the city! Seven Wellingtons failed to return from this operation. The night would witness the loss of 37 aircraft, more than double the previous highest for a night raid, for little, if any, tangible results. Why the raid went ahead despite the warning of stormy weather falls directly at the feet of Peirse. Known for his arrogance, he would be summoned to a meeting at Chequers with the Prime Minister on the evening of November 8th to explain the loss of 414 night bombers and 112-day bombers over the previous five months and the debacle over Berlin the previous night. Within days of the meeting with the Prime Minister, Bomber Command H.Q found its bombing offensive strictly limited, if only temporary, while the Air Ministry reviewed its current policies. The decision had been made, Bomber Command's operations over Germany were virtually on hold. To the crews of 311 (Czech) Squadron, all they were concerned about was it was a loss free operation. Tactics and politics were not their concern, survival was.

Training, both day and night was flown on the 8th. Bomber Command licked its wounds after the disaster over Berlin and Mannheim. It was a short reprieve. On the 9th, the squadron found itself being briefed for Hamburg, nine crews were selected.

### November 9th 1941 : 'Dace B' – Hamburg – Blohm and Voss Shipyards

| Pilot | 2nd Pilot | Serial | Code | Bomb Load | Result |
|---|---|---|---|---|---|
| Sgt M Plecitý | Sgt F Naxera | X9741 | KX-D | 2x500lb+420x4lb | Duty Carried Out |
| Sgt V Ryba | Sgt O Hlobil | Z1167 | KX-A | 2x500lb+420x4lb | Duty Carried Out |
| Sgt M Šebela | Sgt J Filler | R1598 | KX-C | 1x1000lb+3x500lb+1x250lb | Duty Carried Out |
| F/O J Stránský | W/O K Weiss | Z1111 | KX-N | 1x1000lb+3x500lb+1x250lb | Duty Carried Out |
| Sgt J Fína | Sgt F Bulis | X9742 | KX-F | 6x500lb | Duty Carried Out |
| Sgt S Linka | Sgt B Hradil | W5682 | KX-Y | 6x500lb | Duty Carried Out |
| F/O F Pohlodek | Sgt F Dostál | Z8966 | KX-E | 6x500lb | Alternative Target |
| P/O J Nejezchleba | Sgt J Tománek | W5711 | KX-H | 6x500lb | Alternative Target |

3 Group would contribute 31 Wellingtons and 5 Stirlings to an overall force of 107 bombers briefed for Hamburg. Weather over the East Anglian bases at take-off was good with clear skies, but bitterly cold. Once again, the squadron would frustratingly record the early return of two experienced crews. The first to return was that of Pilot Officer J Nejezchleba. They experienced severe icing and decided to drop their bombs on gun positions on the island of Terschelling at 18:30hrs. Selecting to take a more northerly route to Hamburg, the crew of Flying Officer F Pohlodek got as far as Pellworm, one of the North Frisian Islands on the North Sea coast of Germany, when they too, experienced icing issues. They dropped their bombs 'live' at 19:15hrs. The first of the squadron crews arrived over Hamburg at 19:50hrs and, to their delight, found it virtually cloud free, but with some ground haze. Sergeant V Ryba was able to identify the docks and watched as his mostly incendiary load exploded on the northwest side of the Alster. Sadly icing prevented thirty 4lb incendiary bombs from dropping. Flying a brand new Wellington, Flying Officer J Stránský could clearly see the docks and the Elbe River before he dropped his all-high explosive load 1 mile west of the Alster. The remaining crews all claimed to have bombed the target. Sergeant M Plecitý was at 14,000ft, and he reported on return to base, *'Big red fire started, visible five minutes after leaving'*. By 20:20hrs the Czechs were on their way home. The raid appeared effective, 3 Group H.Q recorded, *'The attack seemed successful, the majority of aircraft dropped bombs in the target area and a satisfactory number of bursts were seen.'* Bomber Command H.Q was slightly more guarded in their appraisal, especially after the recent tendency to exaggerate the success.

There followed five days of poor weather that restricted flying to a minimum. On the 14th, fog settled over the region, it was an ideal opportunity to carry out some blind flying training and cross country flights. Finally on the 15th an improvement in the weather would see 3 Group ordered to attack the Germania Werkes at Kiel. 47 Wellingtons drawn from 7 squadrons would be the only aircraft over the target. A further nineteen 3 Group crews would attack the docks at Emden. The Czechs would provide nine crews on Kiel. Clear weather throughout the day looked promising for that night, but this would change dramatically and create numerous problems for the crews.

### November 15th 1941 : 'Python' – Kiel – Germania Werkes

| Pilot | 2nd Pilot | Serial | Code | Bomb Load | Result |
|---|---|---|---|---|---|
| Sgt M Šebela | Sgt O Hlobil | Z1167 | KX-A | 1x1000lb+3x500lb | Early Return |
| Sgt J Svoboda | Sgt J Janek | T2971 | KX-J | 1x250lb+510x4lb | Alternative Target |
| P/O J Nejezchleba | Sgt J Tománek | W5711 | KX-H | 1x1000lb+3x500lb | Duty Carried Out |
| Sgt S Linka | Sgt J Lenc | Z8966 | KX-E | 1x1000lb+3x500lb | Missing |
| Sgt M Plecitý | Sgt K Danihelka | X9741 | KX-D | 1x250lb+510x4lb | Abandoned |
| Sgt K Knaifl | Sgt V Pánek | X9742 | KX-F | 1x250lb+510x4lb | Alternative Target |
| F/O J Stránský | W/O K Weiss | Z1111 | KX-N | 1x1000lb+3x500lb | Duty Carried Out |
| Sgt A Šiška | Sgt J Filler | T2553 | KX-B | 1x1000lb+3x500lb | Duty Carried Out |
| Sgt A Mžourek | Sgt F Naxera | R1598 | KX-C | 1x1000lb+3x500lb | Duty Carried Out |

By take-off time, the weather had closed in, and frost had begun to form on any exposed surface. The night started on the wrong foot for the Group. Eleven aircraft failed to take-off. Over at R.A.F Honington, five Wellingtons of 9 Squadron were withdrawn due to hoar frost forming on the exposed Wellingtons. Only three managed to get airborne in time, but these attacked Boulogne. The last of the Czech crews departed at 21:33hrs on what would be another challenging night for the squadron and the Group. Within 90 minutes, the crew of Sergeant M Šebela were back at their dispersal, having dropped their bombs 'live' into the North Sea after encountering severe icing. They would not be the only ones, a further six 3 Group crews returned early due to icing. It was soon apparent that the winds and weather conditions given at the briefing were inaccurate. Strong, variable winds blew unwary crews northwards and way off their intended course. Sergeant M Plecitý and crew found themselves over Sweden. They eventually identified the city of Goteborg before jettisoning their bombs into the Kattegat and turning for home. They landed at R.A.F Church Fenton after being airborne for 8 hours 23 minutes.[52] Two other crews were blown northwards. The small town of Schleswig in the northeastern part of Schleswig-Holstein, Germany, was chosen by the crew of Sergeant K Knaifl to receive their bombload at 01:00hrs. Bursts were seen, but no details. Sergeant J Svoboda dropped his bombload on Flensburg at 01:23hrs, creating several fires which were visible 20 minutes after departure. One crew that does not appear to have been troubled by the winds was Pilot Officer J Nejezchleba, who bombed at 00:36hrs. They bombed on ETA and flak flashed from 15,000ft over a completely cloud covered Kiel. Other than faint explosions beneath the clouds nothing was observed. The remaining three crews were over Kiel between 01:00hrs and 01:45hrs, each bombing on ETA or flak concentrations. Flak was heavy over the target, but thankfully the searchlights were absent due to the 10/10th cloud. Having dropped their bombs, the crews now had to get home. Recognising the strong winds, the navigators worked feverously on a route home, none wanting to end up lost over the North Sea. One crew who experienced problems was Sergeant S Linka. They had been blown north of Kiel and bombed a target of opportunity. Relying on Astro navigation, the crew began the long sea crossing. A lack of oxygen however, forced the crew to reduce their altitude, but in doing so, they began to encounter clouds, which restricted the view of the stars. When they entered a clear patch of sky, the Astro hatch was found to be covered in thick ice. Unsure of their exact location, a QDM was requested. This, however, was not immediately responded to, and no acknowledgement was received. When a QDM was finally received, the wireless operator was sceptical that the QDM was intended for them. Passing the QDM to the navigator, he raised his suspicions. Despite of his concerns, the navigator used it as it corresponded with his plotted D/R position. The wireless operator again requested a QDM, but once again nothing was received. Confident of his ability, the navigator, Pilot Officer Jiří Engel, decided to carry on until landfall was made. The Wellington eventually crossed the coast, flying well above the cloud base at 2000ft.

Not sure of their exact whereabouts and conscious of the danger of hills and high ground, and even balloons, the pilot kept at a safe height. Finally, a clear patch of cloud was observed, and cautiously, they descended, but instead of land, they could only see water. The navigator thought it was improbable given the time they were still over the North Sea and correctly guessed that they were over the Irish Sea. Realising this, he instructed the pilot to fly 090 degrees to pick up the coastline. On nearing the coast, searchlights appeared, which caused some doubt and anxiety amongst the crew. The pilot believed these were from the Dutch coast, and a disagreement between the crew followed. Perhaps using his rank, the navigator instructed the pilot to keep on the same heading. The searchlight battery, realising the Wellington was lost, directed the beams to the first suitable airfield. This simple act confirmed the navigator's suspicion that this was the west coast of England. On arriving at the landing ground, the pilot refused to take responsibility, believing it was too small to take a Wellington safely. He suggested the crew either bail out or prepare for a ditching off the coast. The crew decided to stay with the Wellington and ditch. It was a tragic error of judgment. At 05:00hrs, the Wellington came down in the sea 20 miles southwest of the St Bees Lighthouse, Cumbria. Sadly, only two of the crew were rescued by a Danish fishing boat, '*Lydia II*'. Both were slightly injured and suffering from shock and exposure. They were both taken to the West Cumberland Hospital, Whitehaven. Pilot Officer Jiří Engel was discharged on 20th and sent to R.A.F Silloth while 2nd pilot, Sergeant Jan Lenc recovery would take longer.[53]

---

[52] *Conflicting details. One report records crew reached and bombed Kiel.*

[53] *Jan Lenc was shot down by German fighter over the Bay of Biscay when he captained Wellington DV665 B on August 18th, 1942. Jiří Engel has been later retrained to a pilot and survived the war only to die after a routine appendix operation in a Prague hospital on October 26th, 1945.*

## Vickers Wellington Mk.IC Z8966 KX-E

| Manufacturer | Vicker Armstrong (Weybridge) | |
|---|---|---|
| Contract | B.71441/40 | |
| Taken on Charge | 25/10/1941 via No.8 MU. | |
| Cat E Missing | 16/11/1941 | |
| Struck Of Charge | 17/11/1941 | |
| Total Flying Hours | N/K | |
| Take-Off Time | 21:29hrs | |
| Bomb Load | 1x1000lb+3x500lb | |
| | **CREW** | |
| Captain | Sergeant Stanislav **Linka** 787395 RAFVR. Age 26. | Runneymede Panel 47 |
| Second Pilot | Sergeant Jan **Lenc** 787171 RAFVR | Survived |
| Navigator | Pilot Officer Jiří **Engel** 825971 RAFVR | Survived |
| Wireless Operator | Flying Officer Jan František **Parolek** 82626 RAFVR. Age 24. | Runneymede Panel 30 |
| Front Gunner | Sergeant Pavel **Skutek** 787888 RAFVR. Age 20. | Runneymede Panel 52 |
| Rear Gunner | Sergeant Arnošt **Václavek** 787479 RAFVR. Age 24. | Runneymede Panel 53 |
| Posting History | Sgt S Linka via Wilmslow Depot, 28/02/1941<br>Sgt J Lenc via RAF Church Fenton, 05/09/1941 | |
| Operations Flown | Sgt S Linka : 25 / Sgt J Lenc : 7 | |
| Remembered | RUNNYMEDE MEMORIAL | |

*The skipper of Wellington Mk.IC Z8966 KX-E, Sergeant Stanislav Linka and his co-pilot Sergeant Jan Lenc. Indecision and a lack of firm leadership resulted in the needless loss of the crew.*

The raid was a disaster. 3 Group H.Q reported that just eight crews actually bombed the Kiel area, most attacking targets of opportunity or jettisoned over the sea. The report finished by stating, *'Altogether a very unsuccessful night'*. Six crews were missing, plus two ditching and 3 aircraft crashed on take-off or return. The results heaped more pressure on the C-in-C Sir Richard Peirse, whose leadership qualities and decision-making abilities were now closely scrutinised by the Prime Minister and the Air Ministry. Sadly for 311 Squadron, the operation resulted in the loss of an experienced crew, which should have been avoided.

The Training Flight reported another incident during the early hours of the 16th. Sergeant Maxmilián Politzer was preparing to take-off in Wellington N2772 on a dual instructional flight at 03:30hrs. Opening up both engines to take-off, the Wellington swung off the runway and hit a stationary M.T vehicle, bringing the flight and the Wellington to an abrupt halt.

| Date | 16/11/1941 |
|---|---|
| Mark | Mk.IC |
| Serial | N2772 |
| Code | KX-K |
| Taken On Charge | 28/09/1941 via No.18 MU. |
| Manufacturer | Vickers Armstrong (Chester) |
| Contract | B124362/40 |
| Pilot (s) | Sergeant Maxmilián Politzer / ? |
| Flight | Training (Czech Training Flight) |
| Time | 03:30hrs |
| Cause | Swung on take-off. |

There was some criticism aimed at the unnamed instructor who it is believed should have regained control sooner before the swing. A day of clear skies and unlimited visibility followed on the 18th allowing both flights to carry out cross-country and night flying training. It was followed by three days of filthy weather, fog followed by rain prevented almost all flying activities. It was not until the 22nd that training recommenced and orders were received at R.A.F East Wretham that operations were on that night. This was later cancelled. On November 23rd two 'Freshmen' crews finally lifted off from East Wretham grass runway and headed towards Dunkirk.

### November 23rd 1941 : CC25A – Dunkirk Docks

| Pilot | 2nd Pilot | Serial | Code | Bomb Load | Result |
|---|---|---|---|---|---|
| Sgt F Naxera | Sgt V Žežulka | R1598 | KX-C | 7x500lb+120x4lb | Duty Carried Out |
| Sgt J Fína | Sgt K Mazurek | W5682 | KX-Y | 7x500lb+120x4lb | Duty Carried Out |

Sergeant František Naxera would carry out his first operation in the left-hand seat, he was joined by 24-year-old Prague-born Sergeant Vratislav Žežulka. It would be the first pairing of Sergeant Jiří Fína and Sergeant Karel Mazurek. Group H.Q requested that its 11 Wellington squadrons prepare their 'Freshmen' crews for a raid directed at the Docks of Dunkirk. A total of 36 Wellingtons and a solitary Stirling from 15 Squadron were made ready. The Group's two operational Stirling squadrons would tackle Brest. Sergeant F Naxera departed at 17:15hrs, followed ten minutes later by Sergeant J Fína. They and the rest of the force encountered ten-tenths cloud, which continued on the entire route to Dunkirk. Sergeant J Fína identified the French coast through the haze. They dropped their bombload at 19:13hrs from just 10,000ft, their 500-pounders being observed to explode near a large fire. Not so lucky was the crew of Sergeant F Naxera. They were over the target at 19:43hrs, unable to identify any landmarks, they bombed in one stick on ETA. No results were seen. Most of the participating crews either jettisoned their bombs or returned to their bases with them. The raid was another disaster, made worse when a 15 Squadron Stirling N3671 LS-H flown by Pilot Officer Fink dropped their bombs near Canterbury, Kent, thinking it was Dunkirk! A navigation error made the crew believe they were over the target! The damage was thankfully minor. Two buildings were damaged, two civilians were injured, and four horses were killed. Once again the weather intervened, the squadron remained grounded on the 24th and a planned raid on the 25th against Emden was cancelled at 13:35hrs.

The squadron would provide five crews on the 26th. Two 'Freshmen' crews would bomb the docks at Ostend while three had been selected to attack the docks of Emden. Sergeant Bulis preparing for his debut as captain was in for a disappointment when the take-off of Z1105 KX-R was cancelled for unknown reason.

### November 26th 1941 : CC13(A) – Ostend

| Pilot | 2nd Pilot | Serial | Code | Bomb Load | Result |
|---|---|---|---|---|---|
| Sgt F Naxera | Sgt V Žežulka | R1598 | KX-C | 7x500lb+120x4lb | Duty Not Carried Out |
| Sgt J Fína | Sgt K Mazurek | W5682 | KX-Y | 7x500lb+120x4lb | Duty Not Carried Out |

Both had departed by 17:18hrs and once airborne, they would join 13 other crews drawn from 3 Group who made up the majority of the small force of just 18 bombers given the docks as a target. On arriving at Ostend, the crews found it completely covered by an impregnable layer of cloud. Unable to see any ground detail, both crews jettisoned 'live' into the sea. While circling the target, Sergeant F Naxera's Wellington was hit by flak.

### November 26th 1941 : 'Herring' – Emden Docks

| Pilot | 2nd Pilot | Serial | Code | Bomb Load | Result |
|---|---|---|---|---|---|
| Sgt K Knaifl | Sgt J Tománek | R1161 | KX-X | 6x500lb+1x250lb+120x4lb | Early Return |
| Sgt O Jambor | Sgt V Pára | Z1098 | KX-U | 6x500lb+1x250lb+120x4lb | Duty Carried Out |

Bomber Command assembled a group of 100 bombers to attack Emden, with 56 being provided by 3 Group. The crew of Sergeant K Knaifl got as far as Norden, a town in Lower Saxony, Germany, located on the North Sea coast, before they aborted with engine trouble. The crew selected a flak and searchlight position near Norden as a target, dropping their bombs at 19:17hrs. Sergeant O Jambor, unable to locate the cloud-covered target, did manage to pinpoint the town of Dollart, located south of Emden. From here, they bombed on ETA at 19:38hrs from 12,000ft. No results were observed. Both operations on this night achieved little if any success. 3 Group summed up the Emden operation as a *'complete failure'*. The restriction on attacking targets in Germany was again put to one side. Bomber Command turned its attention towards Düsseldorf, perhaps to save its battered reputation and achieve some tangible success on what had been, to date, a disastrous November. Düsseldorf was a vital cog in the German war machine, with many factories and workshops supplying essential equipment. It was also the home of the Rhenania Ossag refinery. Positioned on the River Rhine in the very heart of the industrial Ruhr, it was surrounded by Duisburg to its north, Cologne to the south and flanked on either side by Mönchengladbach and Wuppertal. The Group provided 49 crews from its Wellington force while a further seven would re-visit Ostend. 311 Squadron would offer eight crews on this night.

### November 27th 1941 : 'Perch 'B' – Düsseldorf

| Pilot | 2nd Pilot | Serial | Code | Bomb Load | Result |
|---|---|---|---|---|---|
| Sgt J Svoboda | Sgt Z Sichrovský | T2971 | KX-J | 3x500lb+540x4lb | Early Return |
| Sgt K Knaifl | Sgt J Tománek | R1161 | KX-X | 3x500lb+540x4lb | Alternative Target |
| Sgt O Jambor | Sgt V Pára | Z1098 | KX-U | 3x500lb+540x4lb | Duty Carried Out |
| F/O J Stránský | W/O K Weiss | Z1111 | KX-N | 1x1000lb+5x500lb+1x250lb | Duty Carried Out |
| Sgt V Ryba | Sgt O Hlobil | Z1167 | KX-A | 3x500lb+540x4lb | Alternative Target |
| Sgt M Plecitý | Sgt K Danihelka | W5682 | KX-Y | 3x500lb+540x4lb | Duty Carried Out |
| Sgt A Šiška | Sgt J Filler | T2553 | KX-B | 1x1000lb+5x500lb+1x250lb | Duty Carried Out |
| P/O J Nejezchleba | F/Sgt J Doktor | W5711 | KX-H | 1x1000lb+5x500lb+1x250lb | Duty Carried Out |

This night saw the return of the recently promoted Flight Sergeant J Doktor, who had disappeared from the squadron roster in June 1941. He suffered facial lacerations, a broken nose and a suspected concussion in a car accident in Croxton on June 24th. Considered out of danger, he was discharged from hospital on July 15th and immediately sent on sick leave. This was followed by a stay at RAF Hospital Halton between September 12th and October 8th, followed by another week stay starting November 7th. It was not until November 21st that Flight Sergeant Doktor flew again. Due to his injuries, he was required to be medically checked before resuming flying. This was undertaken during a high-altitude training flight with Squadron Leader J Šejbl, closely checked by Squadron Medical Officer F/Lt MUDr. Karel Plaňanský. Four days later, Flight Sergeant Doktor completed another high-altitude training flight with the squadron CO Wing Commander J Ocelka DFC, who found him fit for operational flying.

*Poor quality but an excellent photograph which shows how narrow the cockpit of the Wellington was. Here Flying Officer Josef Stránský and Sergeant František Bulis are seen at the controls of a squadron Wellington. (John Costin)*

All the crews had safely departed by 17:28hrs and headed out over the North Sea to make landfall in the region of the Scheldt estuary. Of the 56 crews detailed and briefed, just two failed to take-off, while just four returned early. Unfortunately, one of those was again a 311 Squadron crew skippered by Sergeant J Svoboda with his second pilot, Sergeant Zdeněk Josef Sichrovský, who was on his first operation. They encountered severe icing while en route and unwilling to continue, dropped their bombs at 18:30hrs on a canal junction at Heist, Zeebrugge. The bombs were seen to burst, followed by an explosion some minutes later. The crew had a brief encounter with a Bf109 fighter at 18:40hrs.

Sergeant K Knaifl skippered another crew unable to reach Dusseldorf. They reported that they were unable to climb to a safe bombing height and selected to attack the town of Erkelenz, located 12 miles southwest of Mönchengladbach. They deposited their bombs from 10,000ft at 19:10hrs creating a series of fires from their incendiaries. One crew who reached Düsseldorf and managed to pinpoint their position was captained by Sergeant O Jambor. The crew was in the

*Sergeant J Svoboda combat report. Note the reference to the HE113. Even today, it is still being determined precisely whom this effort was intended to impress, foreign air forces or Germany's public. Whoever it was, it was a successful deception. (Pavel Vančata)*

minority. They bombed from 12,000ft at 19:15hrs, reporting that their bombs landed close to the Aiming Point. Sergeants M Plecitý, A Šiška, Flying Officer J Stránský and Pilot Officer J Nejezchleba were over the target between 19:05hrs and 19:55hrs. Sergeant M Plecitý reported his bombs exploding near Aiming Point 'A'. Flying Officer Stránský's rear gunner, Sergeant Vilém Orlík reported their bombs created a large fire on departure. Unable to locate the primary due to cloud, Sergeant V Ryba and crew chose to share their bombload over the general area of Düsseldorf, Cologne and Remscheid in equal measures. All the Czechs returned safely to East Wretham apart from Sergeant Jambor, who landed at the Fighter Command airfield R.A.F Kenley with wireless failure. He left Kenley at 17:07hrs the following afternoon and flew back to East Wretham.

All initial indications seemed to confirm that the raid achieved some success, prompted by favourable reports from some of 3 Group's more experienced crews. The Group and Command H.Q had been in this position before, trusting the crews often inflated claims. However, their hopes were dented when other reports began to filter in that crews also reported heavy bombing west and south of Düsseldorf. Despite this, Bomber Command H.Q reported that 52 crews successfully bombed the city and its Marshalling Yards. This is at odds with the two HE bombs and two hundred incendiaries reported by local sources in Düsseldorf. Cologne, the much-bombed cathedral city, recorded damage to 119 properties.

Weather prevented the majority of flying practice on the 28th, all apart from Wellington R1228 flown by Flight Lieutenant Josef Štrégl. He was airborne on an instrument training flight and while landing at R.A.F Brize Norton at 11:50hrs made a hash of his approach and landing. With insufficient speed, the Wellington hit the runway too hard damaging the tailwheel. The aircraft was quickly repaired on site returning to the Training Flight on December 13th.

| Date | 28/11/1941 |
| --- | --- |
| Mark | Mk.IC |
| Serial | R1228 |
| Code | KX-? |
| Taken On Charge | 03/08/1941 via 305 (Polish) Squadron. |
| Manufacturer | Vicker Armstrong (Chester) |
| Contract | B.992424/39 |
| Pilot (s) | Flight Lieutenant Josef Štrégl |
| Flight | Training (Czech Training Flight) |
| Time | 11:50hrs |
| Cause | Heavy Landing |

Flight Lieutenant Josef Štrégl joined the R.A.F at Cosford in August 1940. Prior to his arrival on 311 Squadron he had served in London with the C.I.G. On September 29th 1941 he arrived on 311 Training Flight for operational training.

On the 29th, flying restrictions continued. It would not be until the 30th that two 'Freshmen' crews found themselves being briefed for operations. The last night of the month brought a major operation against Hamburg involving a force of 180 aircraft, the crews of which were briefed to attack a number of aiming points including shipyards and railway installations. 3 Group detailed 49 Wellingtons and two Stirlings for the main event and a further 16 Wellingtons and three Stirlings for a return to Emden. The two Czech crews would visit Emden.

November 30th 1941 : 'Herring 'A' – Emden

| Pilot | 2nd Pilot | Serial | Code | Bomb Load | Result |
| --- | --- | --- | --- | --- | --- |
| Sgt J Fína | Sgt K Mazurek | W5682 | KX-Y | 6x500lb+1x250lb+120x4lb | Duty Carried Out |
| Sgt F Naxera | Sgt V Žežulka | Z1111 | KX-N | 6x500lb+1x250lb+120x4lb | Duty Carried Out |

Both crews were airborne by 17:09hrs, and for once, conditions were excellent. A clear sky with no clouds and a bright moon was encountered along the entire route and over the target areas. The first bombers arrived over Emden around

19:10hrs and apart from some haze, managed to identify the dock area without trouble. Sergeant J Fína was the first to bomb at 19:27hrs. He reported, *'Bombs dropped in the Falderndelft area. Ground detail clearly visible and recognised. Fires seen by rear gunner on leaving, believed to be the result of own bombs'.* Sixteen minutes later, Sergeant F Naxera was over the docks at 15,000ft. The explosions from their bombs were clearly observed, but haze prevented observation of the incendiaries. By 22:20hrs, the raid was over, the crews left behind a significant fire.

Thus ended November 1941, bad weather had meant that the Group operated on just 17 of the 30 nights reducing its overall contribution. There were some positives for 311 Squadron. The loss of one crew was regrettable, however, considering the severe weather conditions throughout the month and the targets attacked it was an acceptable return. The squadron had operated on eight nights flying 53 sorties and dropping in the process 137,760lb of bombs. Serviceability had risen slightly to 66.4% pipping RAF Marham's 115 and 218 Squadrons and the Stradishall based 214. This was particularly pleasing for the hardworking groundcrews, who toiled out at the windswept and freezing dispersals. The award of the DFC to the recently posted Flight Lieutenant John Gellner was gazetted on the 15th, and if ever an award was justified it was to this unassuming and courageous airman. He had flown a total of 194 hours 45 minutes since the outbreak of war and completed an impressive 37 operations.

> *On the night of 1-2 July 1941, he was Navigator of Wellington R1015, captained by Pilot Officer V. Korda, detailed to attack the German cruiser Prinz Eugen at Brest. Though conditions were good and the docks visible, the cruiser could not be seen due to darkness. It was then decided to drop stick across its position, endeavouring to get the first bomb on the jetty. Despite intense flak, four runs across the target were carried out and a determined attack made on the last run. The heaviest bomb and another hit the jetty, lighting up the area, and it was estimated the remainder of the stick position of the cruiser straddled the position of the cruiser. Intelligence reports later stated that two direct hits were obtained on that night, severely damaging the cruiser and killing many of the crew. These hits were credited to either 9 or 311 Squadrons, and it is highly probable that one of their hits was obtained by Pilot Officer Gellner. Since January 1941, this officer has taken part in thirty-seven major operations, and his skill as a Navigator has been exceptional throughout. His precision and accuracy, particularly in astro navigation, would be difficult to exceed. He has had a log published in the August Bomber Command Monthly Navigational Summary under the heading of 'Meritorious Flights'. He has shown conspicuous courage and devotion to duty in his determination to hit his target. He frequently carried out three or four runs to satisfy himself as to his target, and often dropped at least two sticks of bombs. His splendid example and instruction have been of the greatest asset in raising the standard of navigation and bombing in 311 Squadron.*
>
> *Jos Ocelka*
> *Wing Commander, Commanding No. 311 (Czech) Squadron RAF*
> *Date: 17 September 1941*

Eighteen airmen were awarded the Czech Gallantry Award, eleven Czech War Crosses were won, plus four Bar to the Czech War Cross and five 2nd Bars. The Czech Training Flight continued to produce well-trained crews throughout the month despite the weather conditions.

# December 1941 : A Not So Merry Christmas

Widespread fog settled over East Anglia on December 1st, effectively bringing the squadron to a standstill for the next five days. An operation planned for the 1st was cancelled, as was one arranged for the 4th. It would not be until the 6th that the Czechs managed to get back into the air with several cross-country flights.

After an enforced lay-off due to the weather, Bomber Command was out in force of the 7th with attacks planned against, Brest, Aachen and Boulogne. 3 Group would detail and brief nine Stirlings and 24 Wellingtons against *Scharnhorst* and *Gneisenau* at Brest. A further 28 Wellingtons would bomb the railway bridge at Aachen while eleven 'Freshmen' crews would attack Boulogne. For the Czechs, they had been chosen to attack Aachen with ten crews.

### December 7th 1941 : GH733 – Aachen Railway Bridge

| Pilot | 2nd Pilot | Serial | Code | Bomb Load | Result |
|---|---|---|---|---|---|
| Sgt V Pánek | Sgt Z Sichrovský | T2971 | KX-J | 3x500lb+1x250lb+420x4lb | Duty Carried Out |
| F/O J Stránský | F/Sgt J Doktor | Z1111 | KX-N | 3x500lb+1x250lb+420x4lb | Duty Carried Out |
| Sgt F Bulis | Sgt V Šponar | Z1155 | KX-F | 3x500lb+1x250lb+420x4lb | Duty Carried Out |
| Sgt O Jambor | F/O O Hořejší | Z1098 | KX-U | 3x500lb+1x250lb+420x4lb | Duty Carried Out |
| Sgt B Hradil | Sgt M Politzer | X9741 | KX-D | 3x500lb+1x250lb+420x4lb | Duty Carried Out |
| Sgt J Janek | Sgt K Pospíchal | T2553 | KX-B | 3x500lb+1x250lb+420x4lb | Alternative Target |
| Sgt M Šebela | F/Sgt J Kalenský | Z1105 | KX-R | 7x500lb+1x250lb | Duty Carried Out |
| Sgt J Fína | Sgt K Mazurek | W5682 | KX-Y | 7x500lb+1x250lb | Duty Carried Out |
| Sgt F Naxera | Sgt V Žežulka | R1598 | KX-C | 7x500lb+1x250lb | Duty Carried Out |
| W/Cdr Ocelka DFC | Sgt O Hlobil | Z1167 | KX-A | 7x500lb+1x250lb | Duty Carried Out |

There were some unfamiliar faces at the briefing, with new arrivals from the Training Flight. Sergeants Vladimír Šponar, Karel Alexander Pospíchal, Flight Sergeant Josef Kalenský and Flying Officer Oldřich Josef Hořejší would be carrying out their first operational trip with the squadron. Two pilots, Sergeant F Bulis and Sergeant B Hradil, would be captaining a crew for the first time. Both would be flying relatively new Wellingtons on the squadron. A welcomed name was that of the squadron commander, Wing Commander J Ocelka DFC, who would be flying his first operation since the October 14th raid against Nuremberg. The squadron took off from R.A.F East Wretham between 01:51hrs and 02:01hrs into a distinctly unfriendly looking night sky. The Albert Canal and its snow-covered paths were briefly glimpsed by Wing Commander J Ocelka DFC and Sergeant V Pánek while en route, they were in the minority seeing any ground detail. The squadron arrived over Aachen around 03:00hrs and found the target area concealed beneath cloud and low level industrial haze. All apart from Sergeant J Janek bombed on ETA or flak positions, bombing heights varied from 11,000ft to 16,000ft. By 04:05hrs, the squadron was on its way home having dropped 30,900lb of bombs and incendiaries. Sergeant J Janek was unable to find, let alone bomb the primary, so he set about visiting and then bombing Mönchengladbach and Dusseldorf. Ominously, he reported he dropped his bombs on existing fires, neither city was on the list of targets to be attacked that night!

The weather had once again got the better of the crews. Bomber Command H.Q reported that 64 crews claimed to have bombed the target. Aachen reported an attack by just 16 aircraft and a further report claimed only five high-explosive bombs and nine incendiaries fell in the city. The failure of the raid, would not get a mention in any British papers or radio broadcast because news was filtering in from the Pacific, the Japanese had attacked the American Naval Base at Pearl Harbour in Hawaii. The Japanese attack was the topic of conversation in each Mess as graphic updates were broadcast on the wireless. A planned raid on Boulogne was cancelled on the 8th and harsh weather kept the squadron grounded the following day. On the 11th, Op's were back on. Two 'Freshmen' crews were chosen to bomb the docks at Le Havre.

*Squadron pilots Flight Sergeant Jaroslav Doktor a Sergeant Hugo Dostál standing by Ford V8 Model 62 at East Wretham. (Jaroslava Rozumová)*

December 11th 1941 : CC24 – Le Havre

| Pilot | 2nd Pilot | Serial | Code | Bomb Load | Result |
|---|---|---|---|---|---|
| Sgt F Dostál | F/O O Hořejší | Z1167 | KX-A | 13x250lb+120x4lb | Duty Carried Out |
| Sgt V Pára | Sgt O Špaček | Z1111 | KX-N | 13x250lb+120x4lb | Duty Carried Out |

Group allocated 17 Wellingtons and two Stirlings, both provided by 149 Squadron. They were in the process of converting over from the Wellington, a bomber they had been operating since January 1939. Bombs began dropping on a cloud-covered target around 18:15hrs, but it was not until 18:50hrs that Sergeant V Pára arrived over the target. His bombs were dropped on the estimated area of the docks from 11,000ft. Joining Vladimír Pára, who was captaining a crew for the first time, was Sergeant Ondřej Špaček, Le Havre would be his first experience of a bombing raid. Probably the last bomber over the docks at 19:02hrs was Sergeant H Dostál. He, like Pára, was experiencing the role of skipper for the first time. The crew identified and bombed the target using explosions and flak flashes. The 'Freshmen' crews were given another outing the following night when two were chosen to visit the docks at Dunkirk.

December 12th 1941 : CC25A – Dunkirk

| Pilot | 2nd Pilot | Serial | Code | Bomb Load | Result |
|---|---|---|---|---|---|
| Sgt F Dostál | F/O O Hořejší | Z1167 | KX-A | 14x250lb+120x4lb | Duty Carried Out |
| Sgt V Pára | Sgt O Špaček | Z1111 | KX-N | 14x250lb+120x4lb | Duty Carried Out |

3 Group were given the sole responsibility of attacking Dunkirk, with just nine crews. It was not until 04:45hrs that the first Wellington rumbled down East Wretham's runway. Unlike the previous night, the sky was clear and promised to remain so to the target. Sergeant V Pára was the first over Dunkirk at 12,000ft. The docks were clearly seen. The crew

dropped their mixed load in the target area at 06:16hrs, starting two small fires. Half an hour later, it was the turn of Sergeant F Dostál. Dunkirk's defences were putting up a tremendous amount of flak, which, coupled with the searchlights, made the bomb run dangerous. The crew dropped their bombs in one stick from 13,000ft over the southwest area of the docks. Observation of the results was impossible as Sergeant F Dostál took violent, evasive action. It was daylight when the crews landed at a gusty and wet East Wretham. For the next two nights the squadron prepared for operations. On the 13th, Wilhelmshaven was the intended target, but this however was cancelled at 16:0hrs. The following night Cherbourg docks were scrubbed at 21:45hrs. The weather had cleared sufficiently on the 16th for Bomber Command H.Q to send a sizable force to the docks at Wilhelmshaven while the Group's 'Freshmen' crews were given another opportunity to re-visit Oostende.

December 16th 1941 : CC13A – Oostende Docks

| Pilot | 2nd Pilot | Serial | Code | Bomb Load | Result |
|---|---|---|---|---|---|
| Sgt F Dostál | F/O O Hořejší | Z1167 | KX-A | 14x250lb+120x4lb | Duty Carried Out |
| Sgt V Pára | Sgt O Špaček | Z1111 | KX-N | 14x250lb+120x4lb | Duty Carried Out |

All nine crews had departed by 17:31hrs, with the Oostende 'Freshmen' pair crossing the English coast over Southwold. Only seven other 3 Group crews joined them. Accompanying the small force would be 23 Wellingtons drawn from 1 Group. Excellent conditions were encountered en route, but clouds were found over the target. Upon arrival over the docks, a large red fire was easily visible courtesy of 1 Group, which opened the attack at 18:00hrs. Moderate flak was observed from the docks and from offshore. Sergeant F Dostál dropped his bombs in two sticks from 13,000ft at 18:28hrs. These landed in the dock area, the incendiaries being seen to ignite. Two minutes later, and also at 13,000ft, Sergeant V Pára's bomb aimer let their bombs go in two sticks in the general area of the docks, but the exact position was unobserved due to clouds. Bursts and incendiaries were seen, but no other results.

December 16th 1941 : 'Kipper' – Wilhelmshaven Docks

| Pilot | 2nd Pilot | Serial | Code | Bomb Load | Result |
|---|---|---|---|---|---|
| Sgt J Svoboda | Sgt Z Sichrovský | T2971 | KX-J | 2x500lb+420x4lb | Duty Carried Out |
| Sgt V Pánek | Sgt K Knaifl | R1161 | KX-X | 2x500lb+420x4lb | Alternative Target |
| Sgt F Bulis | Sgt V Šponar | Z1155 | KX-F | 2x500lb+420x4lb | Duty Carried Out |
| Sgt A Šiška | Sgt J Tománek | T2553 | KX-B | 2x500lb+420x4lb | Duty Carried Out |
| P/O J Nejezchleba | F/Sgt J Doktor | W5711 | KX-H | 6x500lb | Duty Carried Out |
| Sgt M Šebela | F/Sgt J Kalenský | Z1105 | KX-R | 6x500lb | Duty Carried Out |
| Sgt J Janek | Sgt K Pospíchal | X9733 | KX-L | 6x500lb | Duty Carried Out |

Bernát
The squadron was accompanied by a relatively small force of just ten Wellingtons from 3 Group. Once again, there were no early returns as the squadron flew across the North Sea to a point 40 miles north of the island of Wangerooge before turning towards Wilhelmshaven. The Wellingtons of 1 Group were briefed first over the target at 19:30hrs. They had the benefit of reasonable weather conditions on arrival. The Czechs were close behind. Sergeant F Bulis reported no activity on the ground but plenty of heavy flak working in conjunction with searchlights and murderous light flak up to 15,000ft. The western suburbs attracted most of the squadron crews. Sergeant J Svoboda and Pilot Officer J Nejezchleba both claimed to have bombed this area. Sergeant F Bulis frustratingly had all his bombs hang-up, these were later jettisoned west of the town centre. The crew of Sergeant A Šiška dropped their incendiaries from 16,000ft, starting fires in the town centre. The 500-pounders of Sergeant M Šebela were seen to explode across the target area. There is some confusion regarding Sergeant V Pánek. The ORB states he *'attacked target'* however, two other documents report that the crew bombed either the port of Delfzijl, Northern Holland or Emden at 17:22hrs. The last over the target were the lumbering Whitleys of 4 Group. They reported good fires in the Rustringen area and the western suburbs of Wilhelmshaven. Despite this, the raid achieved little worthwhile damage. Bomber Command was still unable to shrug off the condemning Butt Report, the pressure was growing. Wintery weather ensured that the operational contribution of 311 Squadron leading up to Christmas was minimal. A planned operation to Brest on the 18th was

cancelled at 15:30hrs due to weather. Patchy fog made an appearance at East Wretham on the 19th and would stay for the next three days. Regardless of the weather, the Training Flight continued flying, it was valuable experience, if but dicey. One of the crews flying on a cross-country flight on the 21st was 22-year-old Sergeant Vladimír Hanzl. Attempting to land downwind, the inexperienced pilot overshot his landing at R.A.F Honington. Trying to prevent the Wellington from hitting the boundary fence and hedge he swung the Wellington, it was just too violent for the undercarriage, which buckled and collapsed.

| Date | 21/12/1941 |
| --- | --- |
| Mark | Mk.IC |
| Serial | X9806 |
| Code | KX-? |
| Taken On Charge | 28/09/1941 |
| Manufacturer | Vickers Armstrong (Chester) |
| Contract | 124362/40 |
| Pilot (s) | Sergeant Vladimír Hanzl |
| Flight | Training (Czech Training Flight) |
| Time | 15:30hrs |
| Cause | Overshot landing. |

The ex 75(NZ) Wellington was Cat AC/FA. It would be repaired on site and eventually return to service on December 31st with the newly named and formed 1429 Czech Operational Training Flight (COTF), which had been created on December 16th. On the 22nd, the squadron was ordered to prepare for operations that night. This was cancelled early afternoon, giving the ground crews an early reprieve. A cross-country flight was flown on the 23rd, which coincided with a visit by Air Vice Marshal Karel Janoušek. The following day, 3 Group's AOC Air Vice Marshal J.E.A Baldwin CB, OBE, DSO visited the squadron. He was there to present Wing Commander Ocelka DFC the official squadron crest. In freezing weather conditions, the squadron was paraded, and a copy of the crest was handed over. Also in attendance was R.A.F Honington's Base Commander, Group Captain Ommaney. On completion of the ceremony all quickly retreated to the warmth of the Airmen's Mess where they were served dinner. This was followed by a party in the Officer's Mess. Throughout the day, the usual morale-boosting messages were received. Little if any notice was taken of them, there were far more important festive matters to attend to. The squadron padre, Squadron Leader František Pouchlý, held a High Mass at midnight, which was as expected well attended by the religious Czechs .

*Christmas messages 1941.*

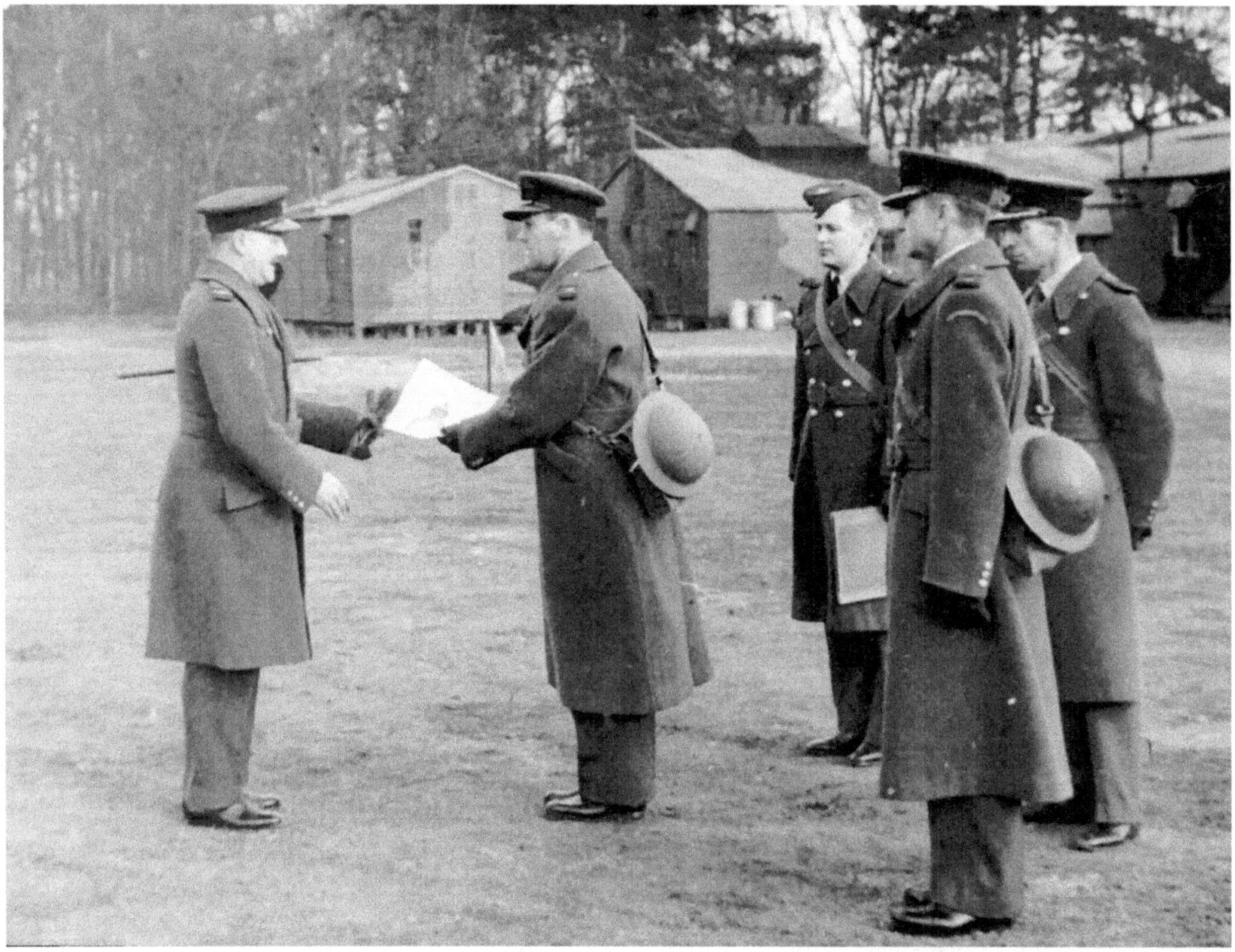

*Handover of the 311 Squadron badge at East Wretham on 24th December 1941. L-R: 3 Group Commanding Officer AVM J. E. A. Baldwin CB, DSO, OBE, A/W/Cdr Ocelka DFC, A/W/Cdr Kenneth Stewart Batchelor DFC, A/G/Cpt Alois Kubita and A/G/Cpt Montagu Douglas Ommanney, RAF Station Honington's new Commanding Officer since 1st November 1941. (Zdeněk Hurt)*

355480/31/O.4. 6 December, 1941.

Presentation of Unit Badges.

Sir,

I am directed to forward for presentation to No.311 (Czechoslovak) Squadron the original painting of the badge and motto as approved by His Majesty the King.

2. The design and motto have been registered with the Inspector of Royal Air Force Badges of the College of Arms and the original is to be preserved by the Unit.

I am, Sir,
Your obedient Servant,

(Sgd) W. R. GIBBS, G/Capt
for Air Vice-Marshal,
Director-General of Organisation.

The Air Officer Commanding-in-Chief,
Headquarters, Bomber Command,
Royal Air Force,
c/o G.P.O.,
High Wycombe, Bucks.

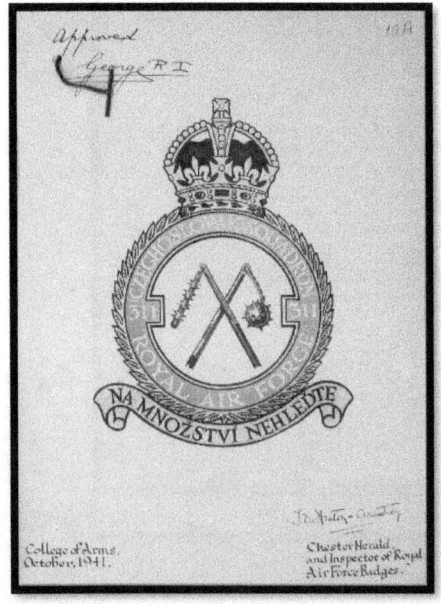

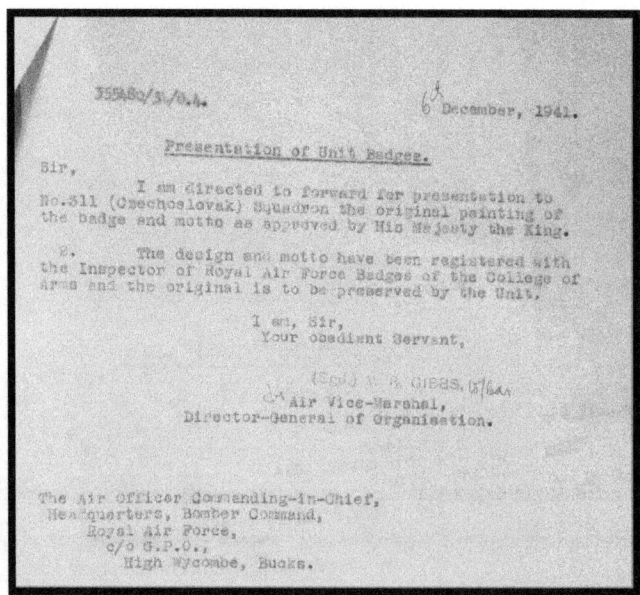

CREST FOR NO. 311 (CZECH) SQUADRON.
------------------------------------

The Husittian Period meant for the Czech Nation the culminating point in their history. Enlightened with new spiritual ideas, the Czech Nation took the lead in the great religious reformation which influenced all the other advanced European nations so much.

Through the leader in this great spiritual fight, JAN HUS, and his ideas, which had so much in common with John Wycliffe's were condemned by the Council of Constanz (1414-1416) the Czech Nation decided to follow him and thus had to stand up against nearly the whole of Europe for ideals which in the later centuries were generally accepted and recognised by the civilised nations of Europe.

On account of these spiritual arguments a real war broke out which proved to be one of the most cruel in history. Here we meet a new figure in Czech history, JAN ZIZKA, who soon became the military head of the Czech nation.

Out numbered by more than ten to one by stronger armies, crusaders whom King SIGISMUND and his mighty allies had sent against him, he defeated his enemies in 12 battles and drove them deeply into Germany and Hungary without a single defeat for himself and succeeded even in reaching the Baltic.

All this was achieved by his great genius in inventing new methods of warfare and the introduction of new weapons and - last but not least - by the invincible spirit and Patriotism of his followers.

He introduced the first tank into the history of warfare, the armed waggon, and built the first mobile field-fortifications, the "waggon forts". He invented new and terrifying arms of which the most famous was the cudgel on which a ball with iron spikes were fastened and which was called "Morning Star". When approaching the enemy, slogans like "Beat them, kill them and never Pardon them" or "Never regard their numbers" were heard and a battle cry "Hrrrr - advance" usually terrified the enemy so much that he fled from the battlefield without a fight.

In order to uphold this heroic tradition in our present Army and Air Force, the Czech Bomber Squadron wishes to incorporate into its crest the Husittian Morning Star which made the enemy flee in battles of the past.

The Morning Star and the battle cry "Hrrrr, na ne!" has for us the significance of strength and determination and will be our symbol for victory.

*How it all began.*

Christmas Day 1941 was celebrated in the traditional way, with food, drink, songs and laughter, anything not to dwell on the war and the weather. An official party was held in the Officer's Mess, where a riotous time was had. On Boxing Day, the unwelcome news that the squadron would be needed that night was received, but to the relief of all, this was cancelled at 16:00hrs. The following day, the squadron was ordered to prepare and make ready for Operation 'Fuller', a plan for combined operations against the German Capital ships birthed in Brest should they try to break out. This order was subsequently cancelled at 13:25hrs. Nonetheless, all aircraft and crews were to be ready to operate by 08:00hrs the next morning. Over at the Training Flight flying practice continued. Sergeant Jan Bláha was scheduled for night Circuit and Landing training at the controls of Wellington P9280. While taxiing at 21:00hrs, the undercarriage suddenly collapsed without warning. No blame was directed at the pilot, and it was considered that the pounding the undercarriage had received with all the heavy landings had weakened it to the extent that it collapsed due to extensive wear.

| Date | 27/12/1941 |
|---|---|
| Mark | Mk.Ic |
| Serial | P9280 |
| Code | KX-? |
| Taken On Charge | 05/12/1941 |
| Manufacturer | Vickers Armstrong |
| Contract | 549268/36 |
| Pilot (s) | Sergeant Jan Bláha |
| Flight | Training |
| Time | 21:00hrs |
| Cause | Undercarriage collapsed (1429 COTF) |

The crews were duly available the following morning, but no follow-up orders were received. Weather conditions on the morning of the 28th were abysmal, low clouds and rain had replaced the fog. It came as some surprise given the conditions that the squadron was informed it would be operating that night. A return trip to Wilhelmshaven was planned. Four 3 Group squadrons would be involved. Honington's 9 and East Wretham's 311 Squadron, and Marham's pairing of 115 and 218 Squadron. It would be a record breaking night for the Czechs, they would offer 13 Wellingtons, its largest contribution so far. The recent lay-off and the preparations the previous day found the squadron at its operational peak.

December 28<sup>th</sup> 1941 : 'Kipper' – Wilhelmshaven Docks

| Pilot | 2nd Pilot | Serial | Code | Bomb Load | Result |
|---|---|---|---|---|---|
| Sgt M Šebela | F/Sgt J Kalenský | Z1105 | KX-R | 1x1000lb+2x500lb | Duty Carried Out |
| Sgt O Jambor | F/Sgt A Jedounek | Z1111 | KX-N | 4x500lb+1x250lb | Early Return |
| Sgt F Naxera | Sgt O Špaček | X9877 | KX-V | 4x500lb+1x250lb | Duty Carried Out |
| Sgt J Fína | Sgt K Mazurek | W5682 | KX-Y | 4x500lb+1x250lb | Duty Carried Out |
| Sgt F Dostál | F/O O Hořejší | R1082 | KX-I | 4x500lb+1x250lb | Duty Carried Out |
| Sgt J Svoboda | Sgt Z Sichrovský | T2971 | KX-J | 1x250lb+420x4lb | Duty Carried Out |
| W/Cdr Ocelka DFC | Sgt O Hlobil | Z1167 | KX-A | 1x250lb+420x4lb | Duty Carried Out |
| Sgt M Plecitý | Sgt K Pospíchal | Z8838 | KX-Z | 1x250lb+420x4lb | Duty Carried Out |
| Sgt A Šiška | Sgt J Tománek | T2553 | KX-B | 1x250lb+420x4lb | MISSING |
| Sgt V Pánek | W/O K Weiss | R1161 | KX-X | 1x250lb+420x4lb | Duty Carried Out |
| Sgt J Janek | Sgt M Politzer | X9733 | KX-L | 1x1000lb+2x500lb | Duty Carried Out |
| Sgt B Hradil | Sgt A Nožička | X9741 | KX-D | 1x1000lb+2x500lb | Duty Carried Out |
| P/O J Nejezchleba | F/Sgt J Doktor | W5711 | KX-H | 1x1000lb+2x500lb | Duty Carried Out |

The first Wellington rumbled down the runway at 17:10hrs and slowly climbed into the overcast sky. The 37 Wellingtons provided by 3 Group would be the first over the target at 19:30hrs closely followed by the Wellingtons of 1 Group. Once over the East Coast, the weather began to improve and by the time the bombers flew past the snow-covered Frisian islands the weather was almost perfect. There was just one early return, Sergeant O Jambor suffering electrical and wireless failure aboard Wellington Z1111 KX-N. The bombs were dropped 'Safe' into the sea, they were back at their dispersal within 90 minutes.

The early arrivals found Wilhelmshaven eerily quiet and clear of cloud. The unwelcome sight of balloons at heights up to 10,000ft would add to the stress. One of the first over the target was Wing Commander Ocelka DFC. He dropped his entire bombload in one stick over the dock area from just 12,000ft. Within seconds, the flash of exploding bombs and twinkling incendiaries began to appear below. Such were the conditions Sergeant M Plecitý reported that his bombload landed just 500 yards from the aiming point. Flying at 16,500ft was the crew of Sergeant F Naxera. Upon their arrival, the crew reported large fires in the town centre. They bombed in two sticks, with the bombs bursting across the Marshalling Yards. Both heavy and light flak were intense and accurate, but strangely, given the conditions, Wilhelmshaven's numerous searchlights were ineffective. Sergeant J Janek and crew were approaching the target area at 16,500ft when at 19:52hrs they had a brief encounter with what was recorded as a Ju88. Sergeants F Dostál and J Fína both noted healthy fires on arrival. Sergeant Fína reported a large fire north of the Bauhafen, which was the construction harbour and one of the oldest parts of the Wilhelmshaven port facilities. It was here that he decided to drop his bombs, which exploded just south of the basin. The Dostál crew dropped in two sticks. The first was seen to burst on the quay at Tirpitzhafen, while the second stick landed in the town centre.

*Sergeant Alois Šiška.*

The crew of Sergeant A Šiška had successfully dropped their bombs when they were hit by flak in the port engine. Soon after leaving the target area, the engine began to emit sparks and then flames. The engine was shut down and the extinguisher button was pressed. On the remaining engine, the crew headed home and the long flight over the North Sea. Although feathered, the port propeller continued to spin, resulting in an ominous grinding sound. A fix was requested and received from Hull at 21:06hrs. The noise from the port engine got progressively worse until, without warning, the propeller disintegrated, severing control links. It was obvious to Sergeant A Šiška that England would not be reached, and a ditching was unavoidable. Showing remarkable calm and flying skill, the pilots managed to ditch the Wellington in the North Sea. All but the rear gunner reached the safety of the large circular dinghy. Sergeant Rudolf Skalický sadly went down with the Wellington. For the crew huddled in the dinghy, it was the start of a battle of survival in the harshest of all environments, the North Sea in winter. The following day, the crew looked expectantly up towards the grey sky, hoping for the sound of a friendly aircraft. On the second day, a Lockheed Hudson (AM598 RR-P) of 407 RCAF Squadron, one of three sent out to search for the crew passed over the dinghy. The freezing crew hurriedly fired a flare, which thankfully was seen and acknowledged. The Hudson circled and the pilot Squadron Leader Anderson ordered the dropping of a Thornaby Bag, but the crew appeared too exhausted to paddle towards it. The position of the dinghy was recorded at 52:55N – 03:27E. New Year's Eve found the crew in serious trouble, made worse by the fact that their rations had almost gone. New Year's Day was even worse, the weather had turned ugly and it was bitterly cold. January 2nd, 1942, the young second pilot, Sergeant Alois Tománek succumbed to exhaustion and exposure. His body was carefully given up to the sea. The following day, Flying Officer Josef Mohr died in the arms of Flying Officer Josef Šcerba, the crew's Wireless Operator.[54] On January 3rd, the dinghy was spotted

---

[54] *The dates of death for the two crew members is uncertain, it is considered that the dates were the night of January 2nd to 3rd.*

by two Dutch schoolboys, 11-year-old Thomas Zuiderland and his 9-year- old younger brother. Showing maturity beyond their tender age, they quickly raised the alarm.

> *"It was about 3.30 PM. We saw a yellow object drifting in the sea which came closer. It turned out to be a big circular inflatable dinghy. Three men were in the dinghy, dressed in flying clothing. At the moment the dinghy reached the lower part of the dyke, some people from Petten had arrived together with some Germans from Camperduin. Three men from Petten acted immediately. They pulled the dinghy up the dyke. The three men in the dinghy were completely exhausted and couldn't do anything. Then I saw there was another fourth man at the bottom of the dinghy. He was dead. Two of the men were carried away by citizens from Petten. The third, with terrible looking feet, was pulled on the horizontal tube of a bike so he could hold himself to the steer. All three were brought to the orphanage house called 'Trein 8.28', at that time a residence of the German coastal guards. The German soldiers stayed behind at the dinghy and the dead air man. I learnt that it concerned Czechs in RAF service. Such an adventure which happened to me as a child has always been on my mind."*[55]

After receiving first aid from a Czech born Doctor at the orphanage house, the surviving crew members were taken to the Marinelazerett (Navy Hospital) at Heiloo (often referred to as Alkmaar)[56] Alois Šiška was suffering from severe frostbite to his feet and legs, which were turning gangrenous. Thankfully, excellent medical care by the Hospital nurses and doctors, plus a large dollop of luck, meant that the planned operation to amputate both his lower legs was not necessary as the care provided prevented the gangrene from further infection. In time, the infection abated, but damage to his legs was irreversible. It was not until post war that treatment by the pioneering plastic surgeon, Sir Archibald McIndoe brought some relief and improvement.

<u>Vickers Wellington Mk.IC T2553 KX-B</u>

| Manufacturer | Vickers Armstrong (Weybridge) | |
|---|---|---|
| Contract | B39600/39 | |
| Taken on Charge | 15/10/1940 via No.24 MU. | |
| Cat E Missing | 29/12/1941 | |
| Struck Of Charge | - | |
| Total Flying Hours | - | |
| Take-Off Time | 17:16hrs | |
| Bomb Load | 1x250lb+420x4lb | |
| | | |
| | **CREW** | |
| Captain | Sergeant Alois **Šiška** 787493 RAFVR | PoW |
| Second Pilot | Sergeant Josef **Tománek** 787501 RAFVR. Age 23. | Runnymede Panel 53 |
| Navigator | Flying Officer Josef **Mohr** 82622 RAFVR. Age 29. | Plot 1. Row D. Grave 4[57] |
| Wireless Operator | Pilot Officer Josef **Šcerba** 82633 RAFVR | PoW |
| Front Gunner[58] | Sergeant Rudolf **Skalický** 787283 RAFVR. Age 23. | Runnymede Panel 52 |
| Rear Gunner | Sergeant Pavel **Svoboda** 787399 RAFVR | PoW |
| | | |
| Posting History | Sgt A Šiška posted via Cosford Depot, 11/04/1941. Sgt J Tománek posted via Wilmslow Depot, 22/05/1941. | |
| Operations Flown | Sgt A Šiška : 18 / Sgt J Tománek : 8 | |
| Remembered | RUNNYMEDE MEMORIAL | |
| Buried | BERGEN GENERAL CEMETERY | |

---

[55] *I am indebted to historian Hans Nauta for allowing me to use this information.*
[56] *Sent by Pavel Vancata, via Hans Nauta.*
[57] *Initially buried Grave 32, Row 2.*
[58] *Compared to the ORB entry, the Air Gunners switched places in fact. Sgt Svoboda was manning the front turret while Sgt Skalický sat in the rear turret.*

Theo Boiten in the groundbreaking series of *Nachtjagd Combat Archives* credits the following as possible candidates who inflicted the damage to Wellington T2553 KX-B. The Marine Flak Brigade at Wilhelmshaven. Flakgrudo II. Marine Flak Brigade in Innenjade at 21:19hrs, or 'Heavy' Flak Battery III Einfahrt. Pilot Officer Šcerba was repatriated on September 16th 1944.

The squadron began landing back at R.A.F East Wretham around 22:14hrs. However, it was not until 00:07hrs that the last crew touched down. The crews were excited about the results. After the previous cloud-covered raids, the opportunity to bomb a cloud-free target was welcomed, and the squadron made it count. When the last crew cleared the target at 20:15hrs, the aiming point, plus the docks, and Bauhafen were in flames. The success sadly came with the loss of the popular Sergeant A Šiška and crew. The return of widespread fog on the 29th kept the squadron grounded for the rest of the month. On the 31st, a New Year's Party in the Officers Mess. It had been a frustrating month for the squadron due primarily to the weather. The squadron had flown on just five nights and despatched 36 sorties for the loss of one crew.

1941 ended with Bomber Command's future in the balance. Despite the Ministry of Information's endeavours to convince the British public that Bomber Command was inflicting grievous damage on Germany, over and above that inflicted on London, Coventry, Birmingham, Bristol and Portsmouth, the reality was very different. The Butt report exposed Bomber Command's many failings. A certain amount of complacency had set in at senior level. Their often blinkered faith in the reports from the returning crews had created overconfidence, which at times bordered on arrogance. Bomber Command H.Q did its utmost to convince itself and the Air Ministry that its raids were inflicting severe damage on German industry. Some of the blame can be attributed directly to Air Marshal Sir Richard Peirse's C-in-C Bomber Command and his subordinates, who seemed out of touch or unwilling to accept intelligence, which was often looked at with some suspicion that all was not well with the bombing campaign. Even when the conditions were favourable, the decision to commit small numbers of bombers meant worthwhile damage was often missed. Too frequently, small numbers of bombers had been despatched to various targets in marginal weather for little or no tangible results. These raids had little or no impact on the enemy's capacity to wage war. There was no blame attached to the crews who, night after night, climbed aboard their bombers and headed towards Germany in sometimes atrocious weather conditions. Not all the blame can be aimed at Peirse. He was in an untenable position. Equally to blame were those in the Air Ministry who, like the weather, changed almost daily. Unrealistic demands were placed on Bomber Command. Ever-changing priorities, in their hundreds by year-end, and an ever-growing list of targets had forced Peirse to spread his resources too thinly. A more forceful leader could have challenged his superiors or argued against such futile and often weekly policy changes.

For the Czechs of 311 Squadron, what happened at Bomber Command H.Q and the Air Ministry were of little interest. Their concerns revolved around survival and the squadron, which they could at least influence. It had been a tough year for the squadron. Frustration, anger, and unrest had blighted the early months. Stern action was needed, individuals were removed, and changes were made. However, it was not until the appointment of Wing Commander J Ocelka DFC that true harmony seemed to settle at East Wretham. His appointment was a stroke of genius. Highly respected, honest and brave, his influence on the squadron's success and the wellbeing of individual airman could not be overestimated. He was not alone. Equally influential, but in their own way were Josef Schejbal, Jindřich Breitcetl and Josef Šejbl. By the end of 1941, 311 (Czech) Squadron had forged and earned a reputation as a first-rate squadron and would start 1942 in a confident and buoyant mood. The new replacement crews, fresh from the Training Flight, were keen, courageous, and ready for the battle ahead and quickly settled into squadron life.

# 1942

## January 1942 : Those Bloody Toads Again

Bad weather prevented the squadron from participating in any offensive operations until the 5th. The Naval obsession with the 'Toads' in Brest and the northern ports would dominate the coming month's activities. Not that the crews were complaining. Given the filthy weather, a short trip to Brest or the channel ports was considerably more preferable to a trip over Germany.

The squadron would provide ten crews for the night's activities, 3 Group would provide a further 38 crews briefed to attack the Arsenal Power Station and docks at Brest.

### January 5th 1942 : Brest

| Pilot | 2nd Pilot | Serial | Code | Bomb Load | Result |
|---|---|---|---|---|---|
| Sgt O Jambor | F/Sgt A Jedounek | Z1098 | KX-U | 5x500lb+420x4lb | Duty Carried Out |
| Sgt K Danihelka | Sgt O Hlobil | Z1167 | KX-A | 5x500lb+420x4lb | Duty Carried Out |
| Sgt V Pánek | W/O K Weiss | T2971 | KX-J | 5x500lb+420x4lb | Duty Carried Out |
| Sgt M Plecitý | Sgt K Pospíchal | Z8838 | KX-Z | 5x500lb+420x4lb | Duty Carried Out |
| Sgt B Hradil | Sgt A Nožička | X9741 | KX-D | 5x500lb+420x4lb | Duty Carried Out |
| P/O J Nejezchleba | F/Sgt J Kalenský | W5711 | KX-H | 7x500lb | Duty Carried Out |
| Sgt F Naxera | Sgt V Žežulka | R1598 | KX-C | 5x500lb+420x4lb | Duty Carried Out |
| Sgt F Dostál | F/O O Hořejší | R1802 | KX-P | 7x500lb | Duty Carried Out |
| Sgt F Bulis | Sgt V Šponar | Z1155 | KX-F | 7x500lb | Duty Carried Out |
| Sgt V Pára | Sgt O Špaček | X9877 | KX-V | 7x500lb | Duty Carried Out |

Despite the heavy showers throughout take-off, the last crew managed to get away from R.A.F East Wretham at 03:32 hours. It was a later than normal departure. Once airborne, the weather conditions en route, although freezing, were generally free of icing and clouds. The first three crews arrived over the partially obscured target a few minutes after 05:00hrs. Sergeants K Danihelka, F Naxera and V Pára reported no flares or fires on arrival, and frustratingly the aiming point was not visible. The crew of V Pára dropped their bombs on flak concentrations. At the same time, Sergeant K Danihelka caught a fleeting glance of the docks. He dropped in one stick from 15,000ft between Port Militaire and Basin No.1. All his bombs were seen to burst across the docks and nearby warehouses. The later arrivals benefitted from the fires, which aided their bomb runs. The mixed HE and incendiary load dropped by Sergeant O Jambor straddled dry dock No.1, starting three fires just north of the docks. This dock also received the bombload of Sergeant B Hradil, although the ORB reports he dropped on flak positions. Pilot Officer J Nejezchleba dropped his bombs east of the river on the Arsenal at Port Militaire bursts were seen but no other results. The crew however reported, *'Large glare seen over the target on leaving'*. One of the last crews over the target at 06:15hrs was Sergeant F Bulis. He dropped his bombload in one stick from 15,000ft, which was seen to burst on Dock No.6 close to the Pont de l'Harteloire bridge. It was daylight by the time the last crews touched down at various bases. Sergeant F Dostál landed at Castle Camps while Sergeant O Jambor landed at Newmarket. It had been an excellent night's work by the squadron, a clean slate, no early returns and 39,760lb of bombs dropped. It was Brest again on the 7th, 3 Group would detail and brief 35 crews, 11 of which would be provided by 311 Squadron. There were two Aiming Points. The Arsenal Power Station would be the primary target, while just seven Wellingtons of 99 Squadron were allocated the cruisers. It would be another late departure.

## January 7th 1942 : Brest

| Pilot | 2nd Pilot | Serial | Code | Bomb Load | Result |
|---|---|---|---|---|---|
| Sgt O Jambor | F/Sgt A Jedounek | Z1098 | KX-U | 4x500lb+1x250lb | Early Return |
| Sgt B Hradil | Sgt A Nožička | Z1111 | KX-N | 4x500lb+1x250lb | Duty Carried Out |
| P/O J Nejezchleba | F/Sgt J Kalenský | W5711 | KX-H | 4x500lb+1x250lb | Duty Carried Out |
| Sgt J Fína | Sgt M Politzer | W5682 | KX-Y | 4x500lb+1x250lb | Duty Carried Out |
| Sgt F Dostál | F/O O Hořejší | R1802 | KX-P | 4x500lb+1x250lb | Duty Carried Out |
| Sgt F Bulis | Sgt V Šponar | Z1155 | KX-F | 4x500lb+1x250lb | Duty Carried Out |
| Sgt M Plecitý | Sgt K Pospíchal | Z8838 | KX-Z | 4x500lb+1x250lb | Duty Carried Out |
| Sgt F Naxera | Sgt V Žežulka | R1598 | KX-C | 4x500lb+1x250lb | Duty Carried Out |
| Sgt V Pára | Sgt O Špaček | X9877 | KX-V | 4x500lb+1x250lb | Duty Carried Out |
| Sgt V Pánek | W/O K Weiss | T2971 | KX-J | 4x500lb+1x250lb | Early Return |
| Sgt K Danihelka | Sgt O Hlobil | Z1105 | KX-R | 4x500lb+1x250lb | Duty Carried Out |

Unlike the previous trip, the squadron reported two early returns. The first to return in daylight at 07:16hrs was the crew of Sergeant V Pánek, who reported low oil pressure in the port engine. He was followed at 09:20hrs by Sergeant O Jambor with a misfiring starboard engine. Both crews jettisoned their bombs 'safe' into the sea. With an almost solid cover of cloud over Brest, no crew could identify the port, let alone positively identify the aiming point. Between 05:53hrs and 06:25hrs, the crews bombed on either E.T.A, the glow of fires beneath the clouds or flak concentrations. Sergeant F Dostál reported that on leaving the target, three fires were observed approximately a mile apart, these he bombed from 15,600ft. The raid was a failure, cloud getting the better of the crews on this occasion. The following day, Bomber Command parted ways with Air Marshal Sir Richard Peirse, he had lost the confidence of the Prime Minister Winston Churchill, Sir Charles Portal and the Air Ministry. His temporary replacement was 3 Group's AOC, AVM J.E.A Baldwin CB, OBE, DSO.

Dissatisfied with the results on the 7th, Bomber Command assembled another small force on the 9th for a re-visit. 3 Group would offer 311 Squadron, which would provide 11 aircraft and RAF Marham's 218 Squadron who provided three Wellingtons plus four Stirlings of 149 Squadron.

## January 9th 1942 : Brest

| Pilot | 2nd Pilot | Serial | Code | Bomb Load | Result |
|---|---|---|---|---|---|
| Sgt B Hradil | Sgt A Nožička | Z1111 | KX-N | 6x500lb | Duty Carried Out |
| Sgt V Pánek | W/O K Weiss | R1161 | KX-X | 6x500lb | Early Return |
| Sgt F Dostál | F/O O Hořejší | W5711 | KX-H | 6x500lb | Duty Carried Out |
| Sgt F Bulis | Sgt V Šponar | R1021 | KX-W | 6x500lb | Duty Carried Out |
| Sgt O Jambor | F/Sgt A Jedounek | Z1098 | KX-U | 6x500lb | Duty Carried Out |
| Sgt F Naxera | Sgt V Žežulka | T2971 | KX-J | 6x500lb | Duty Carried Out |
| Sgt J Fína | Sgt K Mazurek | W5682 | KX-Y | 6x500lb | Early Return |
| Sgt M Plecitý | Sgt K Pospíchal | Z8838 | KX-Z | 6x500lb | Duty Carried Out |
| Sgt J Janek | Sgt M Politzer | R1598 | KX-C | 6x500lb | Duty Carried Out |
| Sgt V Pára | Sgt O Špaček | X9877 | KX-V | 6x500lb | Duty Carried Out |
| W/Cdr Ocelka DFC | Sgt K Danihelka | Z1167 | KX-A | 6x500lb | Duty Carried Out |

Wing Commander Ocelka DFC decided to operate on this night, taking with him the relatively inexperienced crew of Sergeant Karel Danihelka. It was another very late take-off for the squadron, something that was not particularly liked by the crews. Soon after departure, the crew of Sergeant J Fína reported back to base they were returning with a rapidly failing port engine.[59] Having turned for home, the bombload was jettisoned. Unsure of their exact location, the crew,

---

[59] *Reports vary on the cause. A fire is mentioned in a crew report, but the Air81 records vibration.*

believing that engine vibration had affected the compass, headed for the south coast and the first available airfield. Thankfully, searchlights were observed, and making for them, the crew crossed the south Devon coast. At 06:15hrs, the Wellington force landed in lousy visibility at Nettley Farm, hitting a haystack in the process which violently swung the Wellington 180 degrees. They had landed in a field close to the main road to Princetown, 3 miles from Tavistock. All the crew, Sergeants Jiří Fína, Karel Mazurek, Josef Svoboda, František Šipula and František Raiskup, plus Flying Officer Karel Sláma were safe but bruised, but otherwise alive and well. Both engines were damaged, and the Wellington's wing and nose were badly smashed.

| Date | 09/01/1942 |
|---|---|
| Mark | Mk.Ic |
| Serial | W5682 |
| Code | KX-Y |
| Taken On Charge | 10/05/1941 via No.33 MU. |
| Manufacturer | Vickers Armstrong (Weybridge) |
| Contract | B7144/40 |
| Pilot (s) | Sergeant Jiří Fína / Sergeant Karel Mazurek |
| Flight | Operations |
| Time | 06:15hrs |
| Cause | Port engine failure |

Such was the damage to W5682 that this veteran Wellington, with 36 raids to her credit, was Struck off Charge on January 24th. Sergeant V Pánek's crew was another that did not reach Brest. They suffered starboard engine trouble and jettisoned the bombs into the sea.

The remaining crews began to arrive over Brest at 05:50hrs. The Wellingtons of 1 Group had opened the attack and stirred up the defences, with the flak over Brest in full swing and very accurate. The only crew to observe any ground detail was Sergeant B Hradil. Flying at 14,500ft, he observed a ship in the entrance of Port Militaire, which, when reported back at base, had the Admiralty at fever pitch. He then dropped his bombs across the town and turned for home. Most of the crews however, had no such luck. They either bombed on ETA or flak concentrations. Sergeant J Janek was in a particularly rebellious mood. Unable to locate the Aiming Point, he decided to take out his frustration on the flak defences by dropping his 500-pounders singly on flak positions. Two Wellingtons reported flak damage, one was serious. Wellington Z8838 KX-Z was hit over the target, and the damage was such that the Wellington force landed at Thetford on return. The crew were uninjured. The other Wellington hit was flown by Wing Commander Ocelka DFC. A burst of flak exploded under the wing, damaging the fabric skin and geodetics. Once again the raid appeared to be unsuccessful, the 3 Group ORB reporting, *'The operation was not very successful, flak was fairly plentiful and a number of aircraft bombed flak concentrations'*. On the 10th, the newly re-titled 1429 Czech Operational Training Flight (formally 311 Czech Operational Training Flight) recorded its first prang for 1942. Sergeant Jan Bláha was aloft on a night circuit and bumps training flight, joined by fellow pupil pilot, Sergeant Vladimír Hanzl. Two successful take-offs and landings had been completed. However, trouble occurred on the third. The Wellington had just cleared the end of the runway but for some reason was slow in climbing away when it hit the uppermost branches of a tree. Full throttle was immediately applied, and the aircraft juddered on, the pilot's quick reaction saving them and the Wellington. An entire circuit was flown to assess the damage, once satisfied, the crew swiftly made their approach landing at 21:45hrs. The crew had been lucky, a few feet lower it could have been far more serious. Damage to the Wellington, which had been on the flight for less than a week, was extensive. The starboard engine, tailplane fabric and geodetics, port tailplane, front turret and bomb doors were badly damaged. Sergeant Vladimír Hanzl had lacerations to his face and glass in an eye. He was taken to West Suffolk General Hospital, the remaining crewmembers were uninjured.

| Date | 10/01/1942 |
|---|---|
| Mark | Mk.Ic |
| Serial | R1269 |
| Code | KX-G |
| Taken On Charge | 31/12/1941 (1429 COTF) via 115 Squadron after being repaired by No.43 Group |
| Manufacturer | Vickers Armstrong (Chester) |
| Contract | 992424/39 |
| Pilot (s) | Sergeant Vladimír Hanzl / Sergeant Jan Bláha |
| Flight | Training Flight |
| Time | 21:45hrs |
| Cause | Insufficient height on take-off (1429 COTF) |

The Wellington was Repaired On Site (RoS) and returned to the strength of 1429 COTF on February 14th. The Wellington would have a lengthy career with various O.T.Us, but it was finally Struck off Charge in March 1944. Away from R.A.F East Wretham, Group Captain John Gray MC, DFC MiD S.A.S.O at 3 Group H.Q took over Command of 3 Group while the A.O.C assumed the duty of C-in-C Bomber Command on the departure of Air Marshal Sir Richard Peirse. Brest was once again the target on the night of the 11th, the novelty was beginning to wearing off! 3 Group would offer five squadrons including 419 RCAF Squadron flying its first raid since its formation in December 1941. The raid would be an all 3 Group effort with the Czechs detailing and briefing seven crews.

### January 11th 1942 : Brest Harbour

| Pilot | 2nd Pilot | Serial | Code | Bomb Load | Result |
|---|---|---|---|---|---|
| Sgt J Janek | Sgt M Politzer | X9733 | KX-L | 6x500lb | Duty Carried Out |
| Sgt J Svoboda | Sgt K Pospíchal | T2971 | KX-J | 6x500lb | Duty Carried Out |
| Sgt K Danihelka | Sgt O Hlobil | Z1167 | KX-A | 6x500lb | Duty Carried Out |
| Sgt V Pára | Sgt A Nožička | X9877 | KX-V | 6x500lb | Duty Carried Out |
| Sgt O Jambor | F/Sgt A Jedounek | R1802 | KX-P | 6x500lb | Duty Carried Out |
| Sgt M Šebela | F/Sgt J Doktor | Z1111 | KX-N | 6x500lb | Early Return |
| Sgt F Bulis | Sgt V Šponar | Z1155 | KX-F | 6x500lb | Duty Carried Out |

Unlike the previous raids, take-off was early with the squadron away by 16:36hrs. Once airborne, the force of just 26 bombers headed towards either the seaside towns of Lyme Regis or Bridport on the Dorset Coast before the long sea crossing. There was one abort. Sergeant M Šebela and crew got as far as Cherbourg's coast before engine problems forced the crew to jettison the bombs into the sea at 19:03hrs. Despite a smoke screen, the crews arriving at 19:00hrs found the docks and the town almost clear. Considerable light flak was experienced emanating from along the estuary and a number of ships offshore. Over the harbour, the anti-aircraft fire was particularly vicious with hundreds of flak bursts ranging in height from 12,000ft to 17,000ft. Sergeant J Svoboda was obliged to bomb in a steep dive. Surrounded by flak, he dropped his bombs in the port area at 19:08hrs. The crew of Sergeant O Jambor was over the aiming point at 19:13hrs. They reported seeing their bombs burst on L'Entrée du Port Militaire and the nearby power station, and close to Basin No.8 and 9. Sergeant V Pára and crew came close to hitting the cruisers. He reported, *'Bombed along the coast in the vicinity of the docks containing two cruisers. Bursts were seen through an effective smoke screen'*. While on their bomb run, the Wellington was bracketed by flak and the crew were lucky to escape unscathed.

Some tangible damage was inflicted on Brest, but the cruisers had again escaped. While the squadrons were over Brest, the weather over the bases had deteriorated, and the crews were diverted on return. Five of 311 Squadron's Wellingtons landed at R.A.F Boscombe Down, while two landed at R.A.F Honington. Sergeant V Pára, who reported flak damage, had the undercarriage of his Wellington collapse on landing. The investigation into the incident does not mention flak damage but reports that the landing was made too far from the flare path. It also states that the pilot opened up again on the first landing, which was too heavy, fracturing the undercarriage. Two of the crew were slightly injured, the second pilot receiving a cut above his left eye, while the Wireless Operator, Sergeant Antonín Bunzl bruised his right leg.

| Date | 11/01/1942 |
|---|---|
| Mark | Mk.Ic |
| Serial | X9877 |
| Code | KX-V |
| Taken On Charge | 06/12/1941 |
| Manufacturer | Vickers Armstrong (Chester) |
| Contract | B929439/40/C.4(c) |
| Pilot (s) | Sergeant Vladimír Pára / Sergeant Antonín Nožička |
| Flight | Operational |
| Time | 22:30hrs |
| Cause | Heavy Landing |

The damage to the Wellington did not initially look too extensive. However, on closer inspection, the aircraft was recategorised, and repairs would have to be undertaken by Vickers. The aircraft never returned to the squadron. The following night, 1429 COTF reported another accident, which was considered avoidable. Pilot Officer Alois Šedivý, who was now instructing, was taking off when the Wellington's tail wheel burst. Instead of aborting the take-off, he continued on. On landing at 19.00hrs, the entire tailwheel assembly collapsed and snapped off.

| Date | 12/01/1942 |
|---|---|
| Mark | Mk.Ic |
| Serial | T2468 |
| Code | KX-K |
| Taken On Charge | 03/06/1941 via 40 Squadron |
| Manufacturer | Vickers Armstrong (Weybridge) |
| Contract | B39600/39 |
| Pilot (s) | Pilot Officer Alois Šedivý |
| Flight | Training |
| Time | 19:00hrs |
| Cause | Tail-wheel burst. (1429 COTF) |

The damage would be repaired on-site, and the Wellington would return to the Training Flight by the end of the month. On the afternoon of the 13th, a message that had been expected arrived at R.A.F East Wretham. Operation 'Fuller' was to be put into operation. All available Wellingtons were hurriedly made ready as snow started to fall and settle across the airfield. Towards evening, the squadron was ordered to stand down. The cancellation was welcomed. An ENSA show titled 'Big Ben' was due to start in the N.A.A.F.I. On the 13th, 3 Group ORB records the formation of 1429 Czech Operational Training Flight with an establishment War/BC/172 (Honington) of 10 + 2 Wellingtons and 2 +1 Oxfords.[60] On the morning of the 14th, the squadron was informed it would be required for operations that night to Hamburg, it was however to remain on standby. The departure from Brest of the 'Toads' was expected at any time. The Hamburg operation was cancelled at 13:25hrs, followed by a 'Fuller' stand-down. On the 15th, Hamburg was again on the agenda, the squadron would provide six crews. Group would provide just 22 crews on Hamburg on this night while three would attack Emden and a further four would harass Soesterberg aerodrome.

---

[60] *It is generally accepted that the flight was formed in December 1941.*

## January 15th 1942 : 'Dace B' – Hamburg

| Pilot | 2nd Pilot | Serial | Code | Bomb Load | Result |
|---|---|---|---|---|---|
| Sgt J Janek | Sgt M Politzer | W5711 | KX-H | 1x500lb+1x250lb+360x4lb | Duty Carried Out |
| Sgt F Naxera | Sgt V Žežulka | R1598 | KX-C | 1x500lb+1x250lb+360x4lb | Duty Carried Out |
| Sgt O Jambor | F/Sgt A Jedounek | Z1098 | KX-U | 1x500lb+1x250lb+360x4lb | Duty Carried Out |
| Sgt F Bulis | Sgt V Šponar | Z1155 | KX-F | 1x1000lb +2x500lb+1x250lb | Duty Carried Out |
| Sgt M Šebela | F/Sgt J Doktor | Z1105 | KX-R | 1x1000lb +2x500lb+1x250lb | Duty Carried Out |
| P/O F Taiber | Sgt K Pospíchal | Z1167 | KX-A | 1x1000lb +2x500lb+1x250lb | Early Return |

The crews were away between 17:30hrs and 17:36hrs. Once airborne, they headed for the departure point over the town of Happisburgh on the Norfolk coast. Varying cloud layers over the North Sea did not bode well for a clear target. While over the North Sea, a SYKO message[61] was sent to all 3 Group crews. Somehow, wrong wind speeds and directions were given out at the briefing. The message updated the crews with the new data. It was not a promising start. Already promoted Pilot Officer František Taiber who had returned to operational flying after spending the last year as an instructor with Czech Training Flight was forced to abandon the operation. There is some confusion surrounding this crew's return. The ORB reports that the crew encountered severe icing and bombed a radio beacon on an island off the Dutch coast. Two other reports state that the crew experienced engine trouble and jettisoned the bombs 'live' at position 53-47N / 06-07E. The Group's contribution was further depleted when a further five crews aborted. The Squadron arrived over Hamburg exactly as briefed at 20:35hrs. They were the first over the target and unsurprisingly reported little bombing activity below. It may have been quiet on the ground, but above Hamburg the flak was intense. Both heavy and light flak was encountered, the light flak was described as murderous. There were also reports of Flak ships positioned along the Elbe estuary. The early arrivals, Sergeants Naxera, Jambor, Bulis and Šebela, all reported *'Nothing seen on arrival'*. They bombed what they believed to be Hamburg or flak concentrations. Turning for home, Sergeant O Jambor, flying at 15,000ft, glimpsed what he later described as *'burning streets'*. Probably the last squadron crew over the target was Sergeant J Janek at 21:10hrs. He reported a strong glow below the clouds on arrival but opted to bomb flak concentrations.

Only three squadrons were provided on the night of January 17th for an attack on Bremen. 311 Squadron would be joined by 214 Squadron, each providing six crews while 149 Squadron, who were still getting to grips with the Stirling provided one.

*Winter 1941/42 and conditions had improved little for the ground crews.*

---

[61] *The SYKO S.P. 02266 cipher device was a compact British cipher apparatus used for RAF aircraft communications until at least June 1944.*

## January 17th 1942 : Salmon 'B' – Bremen

| Pilot | 2nd Pilot | Serial | Code | Bomb Load | Result |
|---|---|---|---|---|---|
| Sgt J Svoboda | Sgt J Sichrovský | T2971 | KX-J | 1x1000lb+3x500lb+1x250lb | **MISSING** |
| Sgt O Jambor | F/Sgt A Jedounek | Z1098 | KX-U | 1x1000lb+3x500lb+1x250lb | Duty Carried Out |
| Sgt F Naxera | Sgt V Žežulka | R1598 | KX-C | 1x1000lb+3x500lb+1x250lb | Duty Carried Out |
| P/O F Taiber | Sgt K Pospíchal | Z1167 | KX-A | 1x1000lb+3x500lb+1x250lb | Duty Carried Out |
| Sgt M Plecitý | W/O K Weiss | Z1111 | KX-N | 1x1000lb+3x500lb+1x250lb | Duty Carried Out |
| Sgt F Bulis | Sgt V Šponar | Z1155 | KX-F | 1x1000lb+3x500lb+1x250lb | Alternative Target |

The crews encountered solid cloud over the North Sea, and this would continue almost the entire journey to Bremen. Sergeant F Bulis had a bit of bad luck on route. They were badly knocked about by flak over Osnabruck. Having veered slightly off the planned route, they found themselves over the city at 10,000ft. Held by searchlights and surrounded by bursting flak, the bombs were dropped at 19:15hrs. They had been lucky to escape, the crew landing back at R.A.F East Wretham at 21:55hrs. 3 Group would again be first over the target, but there would not be much to see. They were greeted by solid clouds plus moderate to intense heavy flak and intense light flak up to 16,000ft. The clouds thankfully rendered the searchlights ineffective. All but Pilot Officer F Taiber bombed on either ETA or flak concentrations between 19:55hrs and 19:59hrs. Pilot Officer Taiber reported on return, *'Town identified by river and docks and woods northwest. Bombed one-stick near railway line 1 mile northwest of aiming point. Bursts seen'*. They were the only crew from 3 Group who identified the target.

Sergeant J Svoboda was reported to have had an inconclusive encounter with a fighter soon after leaving the target. Almost immediately afterwards, they were hit by flak in the starboard engine and fuel tank. The crew turned for home, nursing the engine. The Wellington slowly began to lose height, and the strain on the one good engine resulted in the engine temperature rising. The pilot ordered his wireless operator to send an S.O.S. This may have been picked up by R.A.F Honington over the R/T at 22:49hrs but was dismissed as an error. The crew were ordered to their crash positions as it was apparent a North Sea crossing was out of the question. Now having to contend with a badly running port engine, the Wellington became difficult to control until finally, at 22:36hrs,[62] both pilots brought the Wellington in for a crash landing near Zandstraat, northeast of Tilburg. Almost immediately after hitting the frozen ground, the Wellington hit trees, which tore off the front turret, ripped the cockpit apart, ruptured, and tore out the fuel tanks on the starboard wing. The floor of the Wellington was almost completely ripped away on impact. What was left of the Wellington immediately burst into flames. Three of the crew were killed, the body of the pilot was found some yards from the crash, likewise the wireless operator. It is not known if they survived the crash only to succumb to their injuries or were thrown out of the Wellington on impact. The navigator's body was found on the wing badly burnt.

### Vickers Wellington Mk.IC T2971 KX-J

| Manufacturer | Vickers Armstrong | |
|---|---|---|
| Contract | 328600/39 | |
| Taken on Charge | 15/01/1941 via No.9 MU. | |
| Cat E Missing | 17/01/1942 | |
| Struck Of Charge | 01/02/1942 | |
| Total Flying Hours | 239hrs 45 minutes | |
| Take-Off Time | 17:16hrs | |
| Bomb Load | 1x1000lb+3x500lb+1x250lb | |
| | **CREW** | |
| Captain | Sergeant Jindřich **Svoboda** 787165 RAFVR. Age 24. | Plot A. Row 1. Grave 22. |
| Second Pilot | Sergeant Josef Zdeněk **Sichrovský** 787331 RAFVR | Survived / Pow |
| Navigator | Pilot Officer Jaromír **Brož** 61917 RAFVR. Age 25. | Plot A. Row 1. Grave 21 |

---

[62] *The time of the crash varies depending on what source.*

| | | |
|---|---|---|
| Wireless Operator | Sergeant Rudolf **Mašek** 787865 RAFVR. Age 23. | Plot A. Row 1. Grave 23 |
| Front Gunner | Sergeant Karel **Batelka** 787947 RAFVR | Survived / Pow |
| Rear Gunner | Sergeant Josef **Šnajdr** 787798 RAFVR | Survived / Pow |
| | | |
| Posting History | Sgt J Svoboda posted via RAF Hendon, 28/03/1941. Sgt J Sichrovský posted via No.2 S.F.T.S, 17/07/1941. | |
| Operations Flown | Sgt J Svoboda : 24 / Sgt Z Sichrovský : 4 | |
| Buried | TILBURG (GILZERBAAN) GENERAL CEMETERY | |

The front gunner was the most fortunate, a few small burns and lacerations were all he suffered, a miracle given the impact. He stumbled upon the unconscious and badly injured body of the second pilot. With the help of the injured rear gunner, they somehow pulled the unconscious body away from the flames, thus saving his life. The bodies of the crew were buried on January 21$^{st}$. The remaining crews were back at East Wretham by 23:00hrs. Snow settled over the airfield on the 18$^{th}$ and continued overnight. On the 21$^{st}$, the squadron learnt that they were to return to Bremen. Seven crews were required from 311 Squadron.

### January 21$^{st}$ 1942 : Salmon 'B' – Bremen

| Pilot | 2$^{nd}$ Pilot | Serial | Code | Bomb Load | Result |
|---|---|---|---|---|---|
| Sgt B Hradil | Sgt O Špaček | W5711 | KX-H | 2x500lb+420x4lb | Duty Carried Out |
| Sgt M Plecitý | W/O K Weiss | DV515 | KX-D | 2x500lb+420x4lb | MISSING |
| Sgt O Jambor | F/Sgt A Jedounek | Z1098 | KX-U | 1x1000lb+3x500lb+1x250lb | Duty Carried Out |
| P/O J Bala | F/Sgt J Doktor | T2962 | KX-T | 1x1000lb+3x500lb+1x250lb | Early Return |
| Sgt F Bulis | Sgt V Šponar | Z1155 | KX-F | 6x500lb | Duty Carried Out |
| Sgt M Šebela | F/Sgt J Kalenský | Z1105 | KX-R | 6x500lb | Alternative Target |
| Sgt F Dostál | F/O O Hořejší | R1802 | KX-P | 6x500lb | Duty Carried Out |

Nine squadrons would be airborne on this night from 3 Group, plus a solitary aircraft from 3 Group Training Flight. Targets ranged from Bremen, Soesterberg airfield, Emden and Boulogne. The lessons of dividing the forces obviously had not been learnt with the departure of Sir Richard Peirse. Fourteen crews would be allocated the docks at Bremen. Joining 311 would be 214 and 149 Squadron, who provided one Stirling. Crossing the coast over Haisborough, the squadron had their last sight of England. Unlike the previous visit to Bremen, the weather over the North Sea was ideal, with just a few isolated clouds to the Dutch Coast and then almost cloudless to the target. Once again, the squadron notched up an early return. Pilot Officer J Bala reported a faulty rev counter aboard Wellington T2962 KX-T. The crew jettisoned their six 500-pounders 'safe' into the sea and were back at East Wretham within an hour of take-off. Sergeant M Šebela encountered severe icing on the run into Bremen and was obliged to drop his bombload on Emden at 19:2hrs. These were seen to burst in the southeast of the town. 1 Group would open proceedings over Bremen, with 3 Group scheduled to bomb between 20:05hrs and 20:40hrs. With 1 Group having stirred up the defences, the squadron encountered a cauldron of exploding flak. The heavy flak, described as accurate and intense appeared to be working with groups of 20 to 30 searchlights. These were accompanied by a maelstrom of light flak that swept the sky above Bremen up to 16,000ft. Northwest of the town, barrage balloons were observed in the moonlight. The Czechs approached the target area between 16,000ft to 18,500ft out of reach of the vicious light flak. Conditions were ideal, but the glow of the searchlights and the ferocity of the defences meant that the crews did not hang about over the target. Sergeant B Hradil dropped his mixed load at 20:07hrs observing his incendiaries ignite and start a number of fires. Two minutes later, Sergeant J Jambor let his bombload go but was unable to observe the results as he was bracketed by flak. Up at 18,500ft the crew of Sergeant F Bulis reported his bombs burst at 20:15hrs. These were followed by a bright red explosion, brighter and bigger than the flash of a 1000-pounder, the glow of which was visible from over 40 miles away. By 23:15hrs all but one crew had returned. There was the chance that the crew of Sergeant M Plecitý had wireless trouble and diverted, there was always hope. The crew of Sergeant F Dostál reported that they heard aircraft 'D' message at 21:15hrs reporting *'In sea at position PXRH5903'*.[63]

---

[63] Miloslav Pajer book, *Wings Aimed at Germany: 311th Czechoslovak Bomber Squadron during its period of service at the RAF Bomber Command.*

By early morning, the realisation that another crew was missing slowly sunk in. A fading RDF plot had been picked up of an aircraft in the sea twenty miles southeast of Orfordness. No.16 Minesweeper Group took immediate action and two trawlers were sent out to search at 23:35hrs. As soon as it was daylight, three Wellingtons from 311 Squadron were armed, fuelled, and made ready. The first away was Pilot Officer F Taiber at 11:20hrs, followed by Sergeant Pánek. The last crew airborne was Sergeant Naxera. Despite searching for nearly six hours, nothing was found.

<u>Vickers Wellington Mk.IC DV515 KX-D</u>

| | | |
|---|---|---|
| Manufacturer | Vicker Armstrong (Chester) | |
| Contract | 124362/40 | |
| Taken on Charge | 10/01/1942 via No.46 MU. | |
| Cat E Missing | 21/02/1942 | |
| Struck Of Charge | 01/02/1942 | |
| Total Flying Hours | 9hrs 10 mins | |
| Take-Off Time | 17:32hrs | |
| Bomb Load | 2x500lb+420x4lb | |
| | **CREW** | |
| Captain | Sergeant Miroslav **Plecitý** 787392 RAFVR | Runnymede Panel 75 |
| Second Pilot | Warrant Officer Karel **Weiss** 787519 RAFVR | Runnymede Panel 73 |
| Navigator | Flying Officer Zdeněk **Skořepa** 82528 RAFVR | Runnymede Panel 66 |
| Wireless Operator | Sergeant Stanislav **Rouš** 787878 RAFVR | Runnymede Panel 92 |
| Front Gunner | Sergeant Ladislav **Němeček** 787353 RAFVR | Runnymede Panel 90 |
| Rear Gunner | Sergeant Čeněk **Král** 787854 RAFVR | Runnymede Panel 87 |
| Posting History | Sgt M Plecitý posted via RAF Dumfries, 02/11/1940. Sgt K Weiss posted via RAF Hendon, 28/03/1941. | |
| Operations Flown | Sgt M Plecitý : 25 / W/O K Weiss : 15 | |
| Remembered | RUNNYMEDE MEMORIAL | |

The exact cause of the loss is uncertain. Author Miloslav Pajer suggests that the Wellington could have been shot down by Ofw Hans Rasper of 5./NJG2. Recent research by author and historian Pavel Vančata sheds new light on this lose.

*"Wellington Mk.IC DV515/KX-D (Sgt M. Plecitý, W/O K. Weiss, F/O Z. Skořepa, Sgt S. Rouš, Sgt L. Němeček, Sgt Č. King) took-off at 17.32 for a raid on Bremen and most sources indicate that on his return Sgt Rouš sent a message that the aircraft was landing in the sea about 65 km from the English coast. The KX-D raid timetable lacks information on any message received by the base, but it does contain testimony from radio operator Sgt Josef Holub, who took part in the raid as part of Sgt František Dostál's crew on R1802/KX-P: 'Aircraft P reported that machine D called the group at 21.15' and 'aircraft at sea PXRH 5903'. The coordinates given correspond to a position of 52°59'N 02°03'E, which is much further north and much closer than previously thought - only 36 km from the coast and 48 km from the town of North Walsham, which the crews used as a landmark when crossing the coast.*
*The RAF lost seven bombers that night (5 Hampdens, 1 Wellington, 1 Whitley), three of which were sent to Emden, all of which fell victim to German night fighters. No. 455 Squadron RAAF lost two Mk.I Hampdens: AT119/UB-Y Sgt G. F. Poulton (Ofw. Siegfried Ney from 6./NJG 2) and AE352/UB-R Sgt E. H. Thompson (Uffz. Rudolf Frank from 1./NJG 3). The Wellington was the only one reported shot down that night by Ofw. Hans Rasper of 5./NJG 2 at 22.16 German time at 4,000 metres above sea level in quadrant 7571. The bottom left corner of the quadrant corresponds to the position 54°00'N 07°00'E, which was also the first planned turning point of Sgt Plecitý's aircraft on the route to the target and the crew was apparently returning the same way. In view of these facts and the fact that British time was GMT+1 at the time while German GMT+2 had been moved forward by one hour, it would appear to Ofw. Rasper as the most likely*

*cause of the loss of Sgt Plecitý's crew. The radio operator, Sgt Rouš, had managed to send at least one more distress call, which suggests that the Wellington did not crash immediately after the dog fight, but that Sgt Plecitý was trying to fly the damaged machine to the British shores".* [64]

The loss of another experienced crew was a painful reminder of the dangers of operations over Germany, experience was crucial, but luck was essential to survive. Adverse weather restricted flying for the next three days. It was not until the 26th that the sleet and snow gave way to drizzle. It was a return to Brest on the 26th, with 3 Group providing 24 Wellingtons and seven Short Stirlings courtesy of 7 Squadron. The Czechs at East Wretham would contribute eight crews.

### January 26th 1942 : 'Toads' – Brest

| Pilot | 2nd Pilot | Serial | Code | Bomb Load | Result |
|---|---|---|---|---|---|
| F/O J Stránský | F/Sgt J Doktor | T2962 | KX-T | 6x500lb+1x250lb | Duty Carried Out |
| Sgt B Hradil | Sgt O Špaček | W5711 | KX-H | 6x500lb+1x250lb | Early Return |
| Sgt F Naxera | Sgt V Žežulka | R1598 | KX-C | 6x500lb+1x250lb | Duty Carried Out |
| Sgt F Bulis | Sgt V Šponar | Z1155 | KX-F | 6x500lb+1x250lb | Duty Carried Out |
| Sgt V Pánek | F/Sgt A Jedounek | R1161 | KX-X | 6x500lb+1x250lb | Duty Carried Out |
| Sgt F Dostál | F/O O Hořejší | R1802 | KX-P | 6x500lb+1x250lb | Duty Carried Out |
| Sgt M Šebela | F/Sgt J Kalenský | Z1167 | KX-A | 6x500lb+1x250lb | Duty Carried Out |
| Sgt J Fína | Sgt A Nožička | Z1098 | KX-U | 6x500lb+1x250lb | Duty Carried Out |

The raid started badly with four aircraft failing to take-off, three of them from 7 Squadron. Those aircraft remaining, headed south, and crossed the Dorset coast in clear conditions. Predictably the squadron notched up an early return. The crew of Sergeant B Hradil jettisoned their SAP bombs 'live' at 49:30N / 03:50W due to port engine concerns. With the other Groups of Bomber Command busy over Hanover and Emden, 3 Group had Brest all to itself. The Group was scheduled to start the raid at 21:10hrs, which it did promptly. The early crews found Brest clear of clouds and smoke. Arriving at 21:10hrs was the crew of Flying Officer J Stránský, who had not operated since December 7th. The crew ran the gauntlet of intense and accurate predicted flak fired from the town and surrounding areas. One area northeast of the town seemed especially active. Also observed were flak ships moored in the Goulet de Brest. The crew's bombload was dropped from 14,000ft and seen to burst on the north side of the entrance to Port de Militaire. Interestingly, the crew reported on return, *'No ships were'*. The later arrivals had to contend with an effective smoke screen, which had slowly drifted across the docks. Sergeant J Fína reported that his bombs exploded close to the Torpedo Boat Station, while the bombs dropped by Sergeant F Bulis landed on a jetty at 22:15hrs. By 22:40hrs, the raid was over, a few isolated fires were seen through gaps in the smoke screen as the bombers headed home for their bacon and eggs. The raid inflicted only superficial damage to the harbour and the town, but importantly for the squadron it was loss free.

The recent wintery weather, rain, frost, heavy snow and then melting ice was having an impact on the grassy landing ground at East Wretham. The runways were becoming increasingly rutted and muddy. The heavily loaded Wellingtons were having difficulty taking off safely, and with the thaw it was only going to get worse. Munster and its railways was the target for 84 bombers on the 28th drawn from, 1, 3 and 5 Group, of which 27 were provided by 3 Group. Nine Wellingtons would be offered by 311 Squadron but Z1098 KX-U with crew of Sergeant J Fína's crew was later withdrawn due to warped doors of the rear turret.

### January 28th 1942 : Rudd 'B' – Munster

| Pilot | 2nd Pilot | Serial | Code | Bomb Load | Result |
|---|---|---|---|---|---|
| Sgt F Dostál | F/O O Hořejší | R1802 | KX-P | 1x1000lb+3x500lb+1x250lb | Duty Carried Out |
| Sgt V Pánek | F/Sgt A Jedounek | R1161 | KX-X | 1x1000lb+3x500lb+1x250lb | Duty Carried Out |
| Sgt M Šebela | F/Sgt J Kalenský | Z1105 | KX-R | 1x1000lb+3x500lb+1x250lb | Duty Carried Out |

---

[64] *Used with permission.*

| Sgt K Danihelka | Sgt O Hlobil | Z1167 | KX-A | 1x1000lb+3x500lb+1x250lb | Early Return |
| Sgt B Hradil | Sgt O Špaček | T2962 | KX-T | 6x500lb | Duty Carried Out |
| F/O J Stránský | Sgt V Pára | DV516 | KX-K | 6x500lb | Duty Carried Out |
| Sgt F Bulis | Sgt V Šponar | Z1155 | KX-F | 6x500lb | Duty Carried Out |
| Sgt F Naxera | Sgt V Žežulka | R1598 | KX-C | 6x500lb | Duty Carried Out |

Rain fell throughout the afternoon, making the already mud-soaked flarepath even more treacherous for the heavily ladened Wellingtons. Over the North Sea, the squadron encountered 10/10th clouds ranging in height from 2,000ft to 20,000ft. These clouds continued the entire route to Munster. Over Holland, Sergeant K Danihelka's crew encountered severe icing and jettisoned their bombs 'live' at position 52:00N / 04:45E a few miles west of Oosterbeek. The squadron arrived over a hidden Munster between 20:00hrs and 21:10hrs. None of the crews saw any ground detail and bombed on either E.T.A or the barely visible flashes of the flak batteries below the cloud. Sergeant V Pánek dropped a single bomb at 20:00hrs and, deciding not to waste the remainder, dropped the rest on the small town of Ahaus near the German – Dutch Border. The raid was a complete failure, bombs were reported dropped all over North Rhine-Westphalia. Munster reported no bombs were dropped on the city.

*This is a poor-quality photograph that graphically illustrates the harsh conditions during the winter of 1941/42. A cold-looking bunch of ground crew is seen here, huddled beside Wellington Mk.Ic R1598 KX-C. This Wellington would survive until January 23rd, 1943, when it crashed while serving with the 3 Operational Training Unit. (John Costin)*

On the morning of the 29th, it was obvious that the flare path and landing ground were in such bad condition that operations from East Wretham would have to be halted. Deep ruts, mud and large puddles were more reminiscent of the Somme than an operational airfield. Both Wing Commander Ocelka DFC and Wing Commander Batchelor were instructed via 3 Group HQ that all offensive operations would be flown from R.A.F Stradishall until further notice. Two squadrons were already operating from Stradishall, 214 Squadron equipped with the Wellington and 138 Squadron equipped with a selection of aircraft flying secret and clandestine operations. The squadron stood down on the 29th, and a planned operation on the 30th to Boulogne was cancelled. The following day, nine unloaded Wellingtons flew to R.A.F

Stradishall, where the crews would be briefed and issued maps and equipment when operating. Wing Commander Ocelka DFC and Wing Commander Batchelor had arrived at Stradishall earlier in the day in preparation of their arrival. Fuel and bombs would continue to be loaded by Czech ground crews who had the unenviable job of travelling in the backs of canvas-covered lorries or, if they were lucky, busses in freezing conditions back and forth between stations a distance of around 35 miles. On January 31st, the harbour at Brest was once again to be visited.

<u>January 31st 1942 : 'Toads' – Brest Harbour</u>

| Pilot | 2nd Pilot | Serial | Code | Bomb Load | Result |
|---|---|---|---|---|---|
| Sgt B Hradil | Sgt O Špaček | T2962 | KX-T | 6x500lb | Duty Carried Out |
| Sgt F Naxera | Sgt V Žežulka | Z1147 | KX-G | 6x500lb | Duty Carried Out |
| F/O J Stránský | Sgt V Pára | DV516 | KX-K | 6x500lb | Duty Carried Out |
| Sgt J Fína | Sgt K Mazurek | Z1098 | KX-U | 6x500lb | Duty Carried Out |
| Sgt F Bulis | Sgt V Šponar | R1532 | KX-B | 6x500lb | Duty Carried Out |
| Sgt V Pánek | F/Sgt A Jedounek | R1161 | KX-X | 6x500lb | Duty Carried Out |
| Sgt F Dostál | F/O O Hořejší | R1802 | KX-P | 6x500lb | Duty Carried Out |
| Sgt M Šebela | F/Sgt J Kalenský | Z1105 | KX-R | 6x500lb | Duty Carried Out |
| Sgt K Danihelka | Sgt O Hlobil | R1021 | KX-W | 6x500lb | Duty Carried Out |

Having taken off safely from Stradishall, the crews headed south to the departure point over Lyme Regis on the Dorset coast. The Czechs of 311 Squadron would be the only 3 Group Wellington squadron over the target on this night. Two Stirlings of 149 Squadron would join them.

An effective smoke screen had already spread across the town and docks as Sergeant F Naxera and his crew dropped the first of the squadron's bombs at 19:30hrs. These were seen to burst across La Rade Abri and into the dock area. Over the next 35 minutes, individual crews attacked from 13,000ft to 15,500ft in the face of considerable flak and searchlight activity. The effective smoke screen started to cause problems, but a few crews still managed to identify ground features. Flying Officer J Stránský and Sergeant K Danihelka both reported that their SAP bombs burst close to the Torpedo Storage station. At the same time, Sergeant F Dostál believed his bombs landed 600 yards north of Dock No.1 but was unable to confirm due to the glare of the searchlights. One of the most persistent crews on this night was Sergeant F Bulis. He spent 65 minutes over the target waiting for visibility to improve, finally at 19:56hrs he dropped his six 500-pounders 2 miles east of Port Militaire. By this time, the squadron was heading home, the two Stirlings having failed to bomb and the crews of 4 and 5 Groups were busy at work. Two aircraft were seen by the Czechs held in searchlights and surrounded by flak. The first, at 9:55hrs, was hit and burst into flames, crashing into the target area in two pieces. The second was timed at 20:10hrs. This unidentified bomber was held by searchlights and was seen in flames, diving steeply towards the ground. All the squadron's Wellington's landed safely back at R.A.F Stradishall between 23:30hrs and 00:25hrs. The month of January 1942 would see the squadron top the Group's sorties flown list for the first time with a creditable 85 sorties, dropping in the process 225,190lb of bombs over 10 raids. This sadly came at the expense of two experienced crews, both lost on Bremen. The weather had been the biggest issue during the month. Snow, frost, sleet and then rain caused various problems especially with the condition of the landing ground at East Wretham. The temporary move to R.A.F Stradishall was welcomed but came with logistical difficulties.

Two awards were gazetted, a DFC to Squadron Leader Josef J Šnajdr's and a DFM to Pilot Officer Josef Čapka both well-deserved and celebrated. There were the usual Czech awards recorded in the ORB. A total of twelve 2nd & 3rd Bar awards were won, seven War Crosses and 14 Czech Gallantry Medals. The New Zealander, Flight Lieutenant Thomas Baber MiD, Gunnery Instructor was awarded the Czech Medal of Bravery, the following letter explains why:[65]

---

[65] *From C M Hanson's 'By Such Deeds - Honours and awards in the Royal New Zealand Air Force, 1923-1999'*

> *President of the Czechoslovak Republic, Chancellery, 9 Grosvenor Place, LONDON, SW1,*
>
> *26 November 1941.*
>
> *Dear Mr Dunbar,*
>
> *The Czechoslovak Minister of National Defence has the intention of proposing to the President of the Czechoslovak Republic that Flt Lt T J D Baber, RNZAF should be decorated with the Czechoslovak Medal for Bravery.*
> *This British officer is attached to Czechoslovak Squadron 311 as instructor in aerial gunnery. From the beginning he has shown great interest in the activity of the Squadron and has voluntarily taken part in three operational flights, the last of which was especially difficult owing to bad weather and strong icing.*
> *I am writing this to enquire whether the proposed measure will meet with the approval of His Majesty's Government and remain.*
>
> *Sincerely yours J. Šejnoha, Chief of Protocol.*
> *Robert Dunbar Esq., MC., Foreign Office, LONDON.*

Flight Lieutenant Baber MiD had flown seven op's with 311 Squadron, plus 32 with 75(NZ) Squadron prior to his arrival on August 27th 1941. Sadly, this brave young Kiwi would be killed on his 39th operation on March 12th 1942 with 75(NZ) Squadron.

*1942 found many of the ground crews still living in tents on, or around the airfield. (John Costin)*

# February 1942 : The Balloon Goes Up

Snow fell almost continuously on February 1st and continued to do so for the next two days. It was not until the 4th that the snow was replaced by drizzle and then rain. The squadron was ordered to 'Stand-By' from 06:30hrs on two-hour's notice on the 4th on 'Fuller' alert. This was eventually cancelled at 15:47hrs. To add to the misery, especially for the ground crews the rain, and melted snow began to turned into ice with a dropping of the temperature towards the evening. This made the work of the armourers and mechanics even harder and more dangerous out at the dispersal. They then had to be transported back to East Wretham. The Czechs were informed that four Wellingtons would be arriving from 218 Squadron based at R.A.F Marham any day to replace some of the squadrons older veteran Wellingtons. On the 5th, a raid on Brest was planned but thankfully cancelled as snow returned, the squadron stood idle. It was not just at East Wretham, the recent heavy snowfalls rendered the majority of 3 Group airfields unserviceable. The squadron was again put on 'Fuller' Alert on the 6th and at the same time ordered to make ready seven crews for another crack at the 'Toads' at Brest. It was a small-scale effort by 3 Group, 11 Wellingtons being provided by 419(RCAF) and 311 (Czech) Squadrons, while 149 Squadron would supply three Stirlings.

### February 6th 1942 : 'Toads' – Brest Harbour

| Pilot | 2nd Pilot | Serial | Code | Bomb Load | Result |
|---|---|---|---|---|---|
| Sgt B Hradil | Sgt O Špaček | T2962 | KX-T | 6x500lb | Bombs Brought Back |
| Sgt V Pánek | F/Sgt A Jedounek | R1161 | KX-X | 6x500lb | Duty Carried Out |
| Sgt K Danihelka | Sgt O Hlobil | R1021 | KX-W | 6x500lb | Crashed on return |
| F/Sgt J Kalenský | Sgt K Pospíchal | Z1105 | KX-R | 6x500lb | Duty Carried Out |
| Sgt V Pára | F/Sgt J Doktor | DV516 | KX-K | 6x500lb | Duty Carried Out |
| Sgt F Bulis | Sgt V Šponar | Z1147 | KX-G | 6x500lb | Duty Carried Out |
| Sgt F Dostál | F/O O Hořejší | R1802 | KX-P | 6x500lb | Duty Carried Out |

The crews had departed R.A.F Stradishall by 17:36hrs and made the now-familiar flight south, skirting the defences of Greater London and heading out over the coast of Dorset. One crew did not reach the south coast. Sergeant K Danihelka's crew found themselves icing up at 3,000ft thirty minutes after take-off. The pilots managed to coax the struggling Wellington up to 5,000ft in an attempt to clear the icing levels when the port engine suddenly cut due to carburettor failure. Unwilling to jettison the bombload for fear of killing civilians, the crew turned back to Stradishall. This was quickly discounted with rapidly deteriorating weather so they decided to land at R.A.F Hunsdon with their bombs. The conditions were atrocious, a snowstorm was encountered as they came into land. With visibility down to just 400 yards in places, the two pilots touched down at 18:30hrs. The landing ground was shorter than expected and at the end of the runway the port wheel hit a snow pile, left after a recent runway clearance. The port wheel collapsed, and the Wellington swung violently around, coming to a halt in a shower of snow and ice. The only crewmember injured was the second pilot, Sergeant O Hlobil, who the Air 81 Crash Report states injured himself while exiting the Wellington and leaping on the ground! He was taken to Haymeads Hospital, Bishop Stortford.

| Date | 06/02/1942 |
|---|---|
| Mark | Mk.Ic |
| Serial | R1021 |
| Code | KX-W |
| Taken On Charge | 28/09/1940 via No.22 MU. |
| Manufacturer | Vickers (Chester) |
| Contract | B.992424/39 |
| Pilot (s) | Sergeant Karel Danihelka / Sergeant Oldřich Hlobil |
| Flight | Operations |
| Time | 18:30hrs |
| Cause | Icing – landed in snowstorm. |

No blame was attributed to the pilots, who did a tremendous job bringing the loaded Wellington in to land on an unfamiliar airfield in a snowstorm. Sadly, it was the end of the line for Wellington R1021, a veteran of 30 raids with the squadron.

Once over the sea, the remaining crews encountered 10/10th cloud which persisted the entire route and over the target. The small force from 3 Group would follow 1 Group, who opened the attack at 19:35hrs. The Czechs began their attack at 19:41hrs, encountering moderate intense flak but little light flak and, surprisingly, no searchlights. No ground features could be seen, and all but one crew bombed on E.T.A. or flak concentrations. Unable to identify the target, Sergeant B Hradil retained his bombs and landed at R.A.F Stradishall. Flight Sergeant J Kalenský landed at R.A.F Exeter on return from his first raid as captain. The raid achieved little if any damage. On the 6th, the four Wellingtons from 218 Squadron arrived, T2739, X9745, Z1070 and R1497. To the frustration of the squadron, the condition of these aircraft were not much better than the Wellingtons they would replace. The 8th found the squadron once again on 'Fuller' alert. The squadron had a welcomed early reprieve at 10:20hrs, when informed they would be operating that night, but this in turn was cancelled at 16:30hrs. The following day, the squadron again remained grounded, rain, drizzle then fog descended. A small token force was provided by 3 Group for another raid on Brest on the 10th. The squadron would provide seven crews that would once again be joined by 419 RCAF Squadron and eight Stirlings drawn from 15 and 149 Squadrons.

## February 10th 1942 : 'Toads' – Brest Harbour

| Pilot | 2nd Pilot | Serial | Code | Bomb Load | Result |
|---|---|---|---|---|---|
| Sgt B Hradil | Sgt O Špaček | Z1098 | KX-U | 6x500lb | Duty Carried Out |
| Sgt V Pára | Sgt K Pospíchal | DV516 | KX-K | 6x500lb | Duty Carried Out |
| Sgt F Naxera | Sgt K Danihelka | Z1147 | KX-G | 6x500lb | Duty Carried Out |
| Sgt F Bulis | Sgt V Šponar | Z1155 | KX-F | 6x500lb | Duty Carried Out |
| Sgt F Dostál | F/O O Hořejší | R1802 | KX-P | 6x500lb | Duty Carried Out |
| Sgt V Pánek | F/Sgt A Jedounek | R1161 | KX-X | 6x500lb | Duty Carried Out |
| F/Sgt J Kalenský | F/Sgt J Doktor | R1532 | KX-B | 6x500lb | Duty Carried Out |

The weather before departure was terrible, and the crews were warned that there was a very good possibility of being diverted on return. There was no improvement in the weather over the target, which was hidden below 10/10th cloud. The Czechs were over the target between 20:00hrs and 20:25hrs, dropping 21,000lb of S.A.P. bombs. The crews either made timed runs from Ile-de-Batz or bombed on flak concentrations. Not one crew observed the target. As predicted, the squadron was diverted, and all the crews landed at R.A.F Exeter. Having been de-briefed, fed and enjoying a night's sleep all the squadron managed to return to Stradishall the following day. Once there, they then had to be transported to East Wretham. It was a tiring business. The appearance of snow on the 11th kept the squadron on the ground, it was the lull before the storm. That night at 21:14hrs, Vice-Admiral Otto Cilliax, the Brest Group commander ordered 'Operation Cerberus' to begin. *Scharnhorst*, *Gneisenau* and *Prinz Eugen* slipped anchor, and headed into the English Channel under an escort of destroyers and E-Boats. What followed was a major embarrassment to the British government and the nation and does not need to be covered in this book other than record the activity of 311 Squadron. At 07:00hrs the squadron was put on four-hour standby, at 11:40hrs Bomber Command requested 3 Group to make ready all available aircraft for immediate take-off as *Scharnhorst*, *Gneisenau* and *Prinz Eugen* had been reported off Boulogne. 3 Group put up all its available aircraft, numbering 28 Wellingtons and 3 Stirlings to attack in the second wave between 16:00hrs and 16:30hrs at position 52-10N / 03:30E. A third wave, consisting of 16 Wellingtons and 8 Stirlings started taking off at 16:37hrs. This wave included seven Wellingtons of 311 Squadron which began taking off from Stradishall at 16:46hrs in atrocious weather.

## February 12th 1942 : 'Operation FULLER' – Third Wave

| Pilot | 2nd Pilot | Serial | Code | Bomb Load | Result |
|---|---|---|---|---|---|
| P/O J Bala | Sgt V Pára | T2962 | KX-T | 6x500lb | Early Return |
| Sgt B Hradil | Sgt O Špaček | W5711 | KX-H | 6x500lb | Early Return |
| Sgt F Dostál | F/O O Hořejší | Z1167 | KX-A | 6x500lb | Abandoned |
| Sgt V Pánek | F/Sgt A Jedounek | Z1105 | KX-R | 6x500lb | Abandoned |
| Sgt F Bulis | Sgt V Šponar | Z1155 | KX-F | 6x500lb | Abandoned |
| P/O F Taiber | F/Sgt J Doktor | R1532 | KX-B | 6x500lb | Abandoned |
| Sgt J Fína | Sgt K Mazurek | X9760 | KX-O | 6x500lb | Abandoned |

Almost immediately after taking off, the Czechs encountered rain, sleet and poor visibility. The squadron was to form-up into two formations over the airfield. The first section consisted of Wellington 'O', 'F' and 'B'. Out in the lead was Pilot Officer F Taiber in KX-B. They were quickly followed by Wellington 'H', 'R', 'T' and 'A'. This section was led by Pilot Officer J Bala. In rapidly failing light, they headed towards the English coast, from where they would head out over the North Sea to position 52-40N / 04-03E, where it was believed the battleships would be. In the appalling weather, some confusion crept in. Sergeant B Hradil turned back to base when over the sleepy village of Bungay, when he was unable to locate and join his section leader. Minutes passed, and in the darkness, Pilot Officer J Bala turned for home after a confusing message was received from the crew of Wellington 'H'. What was left of the formation headed towards the waters off Alkmaar, where, in the darkness, they searched for the battleships. Conditions could not be worse for the crews. Ice-filled clouds, rainstorms and squally conditions compounded the difficulties of finding the German fleet in pitch darkness. Eventually, the crews turned for home, jettisoning their S.A.Ps 'live' into the sea. The entire episode was a major embarrassment to the government and the R.A.F which had meticulously planned for such an event. 3 Group's efforts had been restricted by the order that no aircraft equipped with TR.1335 (Gee) was to be allowed to participate thus, at a stroke, 73 serviceable aircraft stood idle.

Luck, however, did play a part. *Scharnhorst* struck a mine in the late afternoon and started to lag behind. At 19:55hrs, a magnetic mine detonated close enough to *Gneisenau*, when sailing off Terschelling, to open a hole in the starboard side, and temporarily slow her progress also. Later still, at 21.34hrs, when passing through the same stretch of water, *Scharnhorst* hit another air-dropped mine which stopped both engines and damaged steering and fire control. After frantic repairs, the vessel got under way again at 22.23hrs using its starboard engines and making just twelve knots. *Gneisenau* and *Prinz Eugen* reached the Elbe Estuary at 07.00hrs on the 13th and docked at Brunsbüttel North Locks at 09.30hrs, while S*charnhorst* arrived at Wilhelmshaven at 10.00hrs with months-worth of damage to repair ahead. There was one positive, the distraction of the 'Toads' was finally over. Bomber Command could now throw its full weight against strategic targets in Germany. On the 13th, the squadron stood down and watched yet more falling snow then sleet from the various Mess windows. On February 14th 1942, R.A.F Bomber Command received a new directive which was to mould the significant part of its activities throughout the rest of the war. The directive was simple, to set about the destruction of German industrial cities, focusing on, *'The morale of the German civilian population in particular the industrial workers'*. In reality, Bomber Command had been doing this since the summer of 1940, but now, under this new directive, it did not have to cover up its shortcomings in accuracy. The restrictions on using TR1335 (Gee) were lifted on this date. There was considerable enthusiasm at Bomber Command H.Q about the potential of Gee. Now unleashed, H.Q Bomber Command was eager to deploy it over German cities. Mannheim was the target on the 14th. Thankfully, the ground crews did not have to endure the freezing journey to R.A.F Stradishall anymore, with operations resuming from R.A.F East Wretham. Out at the dispersals, eight Wellingtons were prepared in artic like conditions.

## February 14th 1942 : 'Chub B' – Mannheim

| Pilot | 2nd Pilot | Serial | Code | Bomb Load | Result |
|---|---|---|---|---|---|
| P/O J Bala | Sgt K Pospíchal | T2962 | KX-T | 1x1000lb+3x500lb | Duty Carried Out |
| Sgt J Fína | Sgt K Mazurek | X9760 | KX-O | 1x1000lb+3x500lb | Duty Carried Out |
| Sgt F Bulis | Sgt V Šponar | Z1155 | KX-F | 1x1000lb+3x500lb | Duty Carried Out |
| P/O F Taiber | F/Sgt J Doktor | Z1147 | KX-G | 6x500lb | Duty Carried Out |
| F/Sgt A Jedounek | Sgt V Hanzl | R1532 | KX-B | 6x500lb | Duty Carried Out |
| Sgt B Hradil | Sgt V Žežulka | W5711 | KX-H | 6x500lb | Bombs Jettisoned |
| Sgt F Dostál | F/O O Hořejší | Z1167 | KX-A | 6x500lb | Duty Carried Out |
| Sgt V Pára | Sgt K Danihelka | X9733 | KX-L | 6x500lb | Duty Carried Out |

Flight Sergeant Arnošt Jedounek would be occupying the captain's seat for the first time. Joining him in the cockpit for his first raid was Sergeant V Hanzl. Group would provide just twelve Wellingtons and three Stirlings against Mannheim on this night, the bulk of the bombers being provided by 1, 4 and 5 Groups. The squadron departed between 18:31hrs and 18:39hrs. Once airborne, they headed to Orfordness, a shingle spit on the Suffolk coast. As forecast, the entire route to Mannheim was flown above almost solid cloud. Severe icing was experienced, with temperatures recorded as low as -40. Near Courtrai, at 19:35hrs, Sergeant B Hradil's crew were involved in a sharp encounter with two Me110s. Having jettisoned the bombs 'safe' the crew turned for home. 311 Squadron were over the target between 21:00hrs and 21:46hrs. Mannheim's vicious defences did not disappoint as the Czechs dropped 19,500lbs of bombs on either E.T.A, the glow of fires or flak positions. Only one crew, Sergeant V Pára claimed to have seen the target. They reported their bombload landing on the Badische Anilin und Sodafabrik[66] chemical plant creating a large fire. Minor damage was inflicted on Mannheim, but the clouds once again got the better of the participating crews. On return, it was reported that the existing searchlight belt to the south of Mannheim appeared to have been extended by 4-5 miles. With the airfield now serviceable, 1429 COTF resumed training, with flights over Berners Heath and Langham following over the next few days. On the 18th, a concert by Belfast born, Mr Howard Ferguson and Denis Mathews, a serving member of the R.A.F was held in the Officers Mess. At 10:05hrs on the morning of the 19th, news that an enemy cruiser plus escort had been sighted in the Dover Straits had 3 Group scrambling to get a number of squadrons ready for a possible attack. Three Wellington crews drawn from 311 would be joined by six from 214, and a further three from 156, while three Stirlings from both 15 and 149 Squadrons were prepared. The crews waited patiently, until at 12:30hrs when the welcomed ordered to stand down arrived. On the 22nd, six experienced crews were back over northern Germany while five 'Freshmen' crews would visit Ostende Docks. It would be the squadron's busiest night for nearly six weeks. The night would be a baptism of fire for a number of new pilots fresh from completing their time with 1429 COTF.

## February 22nd 1942 : 'Kipper' – Wilhelmshaven

| Pilot | 2nd Pilot | Serial | Code | Bomb Load | Result |
|---|---|---|---|---|---|
| Sgt B Hradil | Sgt J Kotrch | Z1070 | KX-Y | 6x500lb+1x250lb | Duty Carried Out |
| F/Sgt J Doktor | Sgt O Špaček | R1497 | KX-D | 6x500lb+1x250lb | Duty Carried Out |
| P/O F Taiber | Sgt J Štark | DV516 | KX-K | 6x500lb+1x250lb | Duty Carried Out |
| P/O J Bala | Sgt V Výcha | T2962 | KX-T | 6x500lb+1x250lb | Duty Carried Out |
| Sgt J Fína | Sgt K Mazurek | X9760 | KX-O | 6x500lb+1x250lb | Duty Carried Out |
| F/Sgt A Jedounek | Sgt K Kodeš | R1161 | KX-X | 6x500lb+1x250lb | Early Return |

The six crews selected to attack Wilhelmshaven had departed by 18:42hrs. Slowly climbing for height they skirted Norwich and passed over Happisburgh on the Norfolk coast, and out over the North Sea. Again, 3 Group would provide a token force of 14 Wellingtons. The squadron notched up an early return when Flight Sergeant A Jedounek experienced problems with the constant speed control aboard Wellington R1161 KX-X. Unable to select the prop and engine speed, the crew wisely turned for home jettisoning 'live' the entire bombload at 19:28hrs at position 03:23N / 19:28E. They

---

[66] BASF

were back at dispersal within two hours. The first crew arrived over Wilhelmshaven at 20:28hrs. Pilot Officer J Bala found it hidden by an impenetrable layer of cloud. Flak was plentiful, heavy flak was accompanied by streams of intense tracer fire hosing the sky which would spell inevitable disaster if a crew decided to bomb below 12,000ft. Some faint glows beneath the cloud were seen courtesy of 1 and 5 Groups who opened the attack. Only one crew bombed on the glow of fires, Sergeant J Fína, at 21:02hrs from 17,000ft, the majority of the squadron having bombed on flak concentrations. The night's objective, the floating docks were undamaged, and bombs were scattered all over Lower Saxony.

*Formidable job of the ground personnel in any weather –AC2 Bartoloměj Ranofrej (Flight Mechanic A) is stretching out a cover on the Wellington elevator in biting*

## February 22nd 1942 : CC13A – Ostende

| Pilot | 2nd Pilot | Serial | Code | Bomb Load | Result |
|---|---|---|---|---|---|
| Sgt K Danihelka | Sgt V Hanzl | Z1167 | KX-A | 16x250lb | Bombs Jettisoned |
| Sgt F Naxera | Sgt O Havlík | Z1111 | KX-N | 16x250lb | Bombs Jettisoned |
| F/Sgt J Kalenský | Sgt J Hadrávek | Z1105 | KX-R | 16x250lb | Duty Carried Out |
| Sgt F Dostál | Sgt O Soukup | X9733 | KX-L | 16x250lb | Duty Carried Out |
| F/O O Hořejší | F/Lt J Štrégl | W5711 | KX-H | 16x250lb | Duty Carried Out |

The Czechs would be the only crews over Ostende on this night. They had the docks all for themselves. Like their colleagues flying towards Wilhelmshaven, they experienced solid cloud over the North Sea and considerable icing. Bombing was to commence at 19:55hrs, and to the second, the crew of Flying Officer O Hořejší dropped his bombs through a gap in the clouds from 13,000ft. These were seen to burst east of the docks and north of Nouvenu Rassin De Chasse. The welcome gaps in the clouds also allowed Sergeant F Dostál and crew to claim to have bombed the docks at 20:10hrs. They were the only success. Flight Sergeant J Kalenský also dropped his bombs at 20:10hrs but were uncertain of the exact location. Nevertheless, they believed it to be the dock area. Two crews found Oostende completely cloud covered. The bombload of Sergeant K Danihelka was jettisoned 'live' at 20:26hrs in the sea 10 miles off Oostende. Sergeant F Naxera dropped 1 x 250-pounder on flak positions at Dunkirk before jettisoning his remaining bombs in the sea 10 miles from the port. On return, the crews made landfall over the English coast at Orfordness and were safely back at R.A.F East Wretham by 22:17hrs.

Air Chief Marshal Sir Arthur Harris took up his post as the new Commander-in-Chief of Bomber Command on the 22nd. Harris had his own ideas on how the bomber offensive should be fought. He was not an advocate of small-scale raids on multiple targets dotted all over Germany as favoured by his predecessor. Harris knew that to inflict maximum damage to a target and overwhelm the defences, the maximum number of bombers needed to be involved. A determined and at times ruthless commander, Bomber Command now had at its head a leader who was totally committed to the destruction of German towns and cities. While the operational crews slept, 1429 COTF were busy over the firing range at Langham on the 23rd.

*The right man at the right time. Air Chief Marshal Sir Arthur Harris.*

The training continued the following day but was blighted by a crash during night flying practice. Wellington L7841 was airborne at 22:10hrs when problems with the port engine were experienced. The crew decided to land, but on nearing the flare path at 30ft with full flaps and the undercarriage down, it was found that the port engine could not be shut off. Aborting the landing, the crew did another circuit and tried again. On the second attempt, the starboard engine showed no revolutions, while the port engine showed 2200 per minute. With the Wellington halfway down the flare path, the pilot retracted the undercarriage in an attempt to maintain height and gain speed, but it was too late. An obstruction was seen, and in an attempt to avoid it, the port wing dropped and hit the ground. The Wellington was, at the time, flying at just 60 mph. The Wellington slammed into the ground, coming to a halt near the boundary fence where it burst into flames at 22:20hrs. Fortunately, the crew were able to get clear. The rear gunner, Sergeant Gustav Michalec was thrown out of his turret and injured his back. He was admitted to Ely Hospital. The subsequent investigation attributed the cause to the pupil pilot's mishandling of the airscrew controls. The instructor was the experienced Pilot Officer Josef Čapka DFM.

| Date | 24/02/1942 |
| --- | --- |
| Mark | Mk.Ic |
| Serial | L7841 |
| Code | KX-S |
| Taken On Charge | 13/06/1941 via 311 Squadron after being repaired by No.43 Group. |
| Manufacturer | Vickers Armstrong (Chester) |
| Contract | 992424/39 |
| Pilot (s) | Pilot Officer Josef Čapka DFM / Sergeant Ferdinand Kepka |
| Flight | Night Training Flight |
| Time | 22:20hrs |
| Cause | Mis-handling error when landing. (1429 COTF) |

The Wellington was Struck off Charge on March 4th 1942. A planned 'Freshmen' operation was cancelled at 16:17hrs on the afternoon of the 25th. The COTF lost another aircraft if but temporarily after a routine inspection. It was discovered that Wellington R3237 KX-R had suffered damage to the undercarriage probably due to its early retraction on take-off on a previous flight. The damage unusually had gone unreported! On the 26th, the morning activities were tragically shattered at 11:10hrs when Spitfire Mk.IIa P7546 of 1401 Met Flight based at nearby R.A.F Bircham Newton smashed into the ground on the aerodrome between two dispersed Wellingtons, killing the New Zealand pilot, Sergeant William McLeod RNZAF instantly. Wreckage from the Spitfire hit Wellington R1497 KX-D. Damage was classed as CAT Ac./FA and repaired on site. It would be March 7th before it returned to service. That night's operation was cancelled at 16:10hrs.

The late cancellation trend continued on the 27th when a planned 'Freshmen' operation was abandoned after the aircraft and crews had been made ready. Five 'Freshmen' crews were to operate on the last day of the month, but this was cancelled at 16:25hrs after the Wellingtons had been fuelled and bombed-up. The month had been one of disappointment and some irritation. The weather had once again played a significant part in the squadron's low operational tally of just 40 sorties over six nights, during which the squadron dropped approximately 77,750lb of bombs without loss. The use of R.A.F Honington had caused considerable hardship, especially for the ground crews who would leave East Wretham at 04:30hrs and travel to Honington in freezing conditions. After a day's work, often in blizzard-like conditions, they had to return to East Wretham. This may have been the cause of the drop in aircraft serviceability during the month. Group H.Q reports 311 Squadron had 60% serviceable aircraft throughout February, the second lowest in the Group's Wellington units. Sitting at the bottom were the Canadians of 419 RCAF Squadron, who were still finding their way at R.A.F Mildenhall, having just started operations in January. There was a flood of awards during the month with the announcement of four Third Czech War Crosses, 12 Second Czech War Crosses, 16 Czech War Crosses, one Second Czech Gallantry Medal and two Czech Gallantry Medals. The various Mess bars and local pubs would have been busy celebrating the awards. Despite the abysmal weather, the squadron was well entertained. ENSA Films and Concerts were held on the 6th, 11th, 23rd, 24th and 26th. There were also two piano concerts on the 18th and 20th by Irishman Mr Howard Ferguson and Englishman Mr Denis Mathews, both serving members of the R.A.F. Apart from the music, a new gymnasium and games room was built by converting the old Dining Hall, which doubled as a lecture room during the day. Here, the Czechs were given English lessons, but it was not just the Czechs eager to learn. The R.A.F members were taught Czech, French, and German. One of the popular classes was Russian!

# March 1942 : Germany Bound

The month started with yet another cancelled operation. Five 'Freshmen' crews were to attack the docks at Oostende, but at 21:30hrs the Group commander decided the weather was too unpredictable. A new target appeared on the Operations Board on the 2nd, the Renault Works at Billancourt located west of the centre of Paris. There was a real buzz around the aerodrome. This would have been the first occasion the Czechs had visited Paris, but frustratingly the operation was cancelled late in the day. Over at 1429 COTF an inspection of Wellington Z8854 found what appeared to be British tommy gun bullets in the mainframe. The Wellington was believed to have been taken on with the damage, having arrived via 75(NZ) on December 5th 1941. It was a strange discovery, even for the Czechs.

*The distinctly battered looking Wellington Mk.IC W5711/KX-H. The photo was taken after 12th March 1942, when the machine completed it 50th sortie with the Bomber Command. The letter "V" for "Victory" and its Morse code were painted underneath the bomb symbols. Despite the high wear-out rate, the aircraft carried out another 27 Anti-Submarine Patrols with Coastal Command between 12th August 1942 and 14th January 1943 to became 311 Squadron's Wellington with the most operational flights. (Václav Kolesa)*

On the 3rd, Op's were back on, and so was the attack on the Renault Works. This operation was a result of the Air Ministry reconsidering its previous orders not to attack industrial war-producing factories in the occupied countries. On February 5th 1942 in memo S.46368/DCAS this original directive was withdrawn. The gloves were now off, it was time to demonstrate to the French people the offensive power of R.A.F Bomber Command. The squadron would provide eight crews, plus four 'Freshmen' on Emden. 3 Group would offer 12 squadrons totalling 43 Wellingtons and 27 Stirlings. Gee-equipped aircraft were excluded, but many new records were broken. 235 aircraft were sent to one target, and a record tonnage of bombs was dropped. From a tactical point, the use of experienced crews to drop flares to illuminate the target was tested and employed.

## March 3rd 1942 : 'Herring' – Emden

| Pilot | 2nd Pilot | Serial | Code | Bomb Load | Result |
|---|---|---|---|---|---|
| Sgt V Pára | Sgt O Soukup | Z1147 | KX-G | 13x250lb | Jettisoned |
| Sgt K Danihelka | Sgt V Hanzl | Z1167 | KX-A | 13x250lb | **MISSING** |
| Sgt V Žežulka | Sgt O Havlík | Z1111 | KX-N | 13x250lb | Jettisoned |
| Sgt K Pospíchal | Sgt J Hadrávek | Z1105 | KX-R | 13x250lb | Duty Carried Out |

Bomber Command H.Q in its wisdom ordered just four Wellingtons to operate against Emden on this night. Unfortunately, the squadron chosen was 311 Squadron. Taking off between 18:50hrs and 18:53 hours, the four crews first headed to North Walsham before the long sea journey to a point at 54.00N / 07.30E over the North Sea, from where they made their run into the target. The crews were briefed to be over Emden from around 21:10hrs. In the bright moonlight, two crews fought off prowling Me110s within ten minutes of each other. At 21:10hrs, Sergeant V Pára and crew were attacked by a Me110 which, during the encounter, mortally wounding the rear gunner, Sergeant František Binder. At 23.52hrs a signal was received at East Wretham, *'R-Gnr maybe dead'*. Extensive damage was inflicted to the Wellington, and the bombload was jettisoned over the target. Excellent flying by the pilot ultimately saved the crew. At 21:20hrs, Sergeant V Žežulka was also attacked over the target. Outstanding pilot-gunner coordination prevented serious damage to the aircraft, but the bombload was again jettisoned in the target area during the combat. Sergeant K Pospíchal escaped the attention of the prowling fighters and dropped his bombs from 17,000ft at 21:30hrs. These were seen to burst across the town, but no results were observed. Sadly, not all the crews were lucky. At 21:20 hrs, Sergeant K Danihelka and crew were shot down into the sea 12 miles northwest of Terschelling by Oberfeldwebel Paul Gildner of 5./NJG 2, who was flying from Leeuwarden airfield. There were no survivors.

## Vickers Wellington Mk.IC Z1167 KX-A

| Manufacturer | Vickers Armstrong | |
|---|---|---|
| Contract | B.97887/39 | |
| Taken on Charge | 20/10/1941 via No.44 MU. | |
| Cat E Missing | 03/03/1942 | |
| Struck Of Charge | 27/03/1942 | |
| Total Flying Hours | N/K | |
| Take-Off Time | 18:51hrs | |
| Bomb Load | 13x250lb | |
| | **CREW** | |
| Captain | Sergeant Karel **Danihelka** 787172 RAFVR. Age 26. | Panel 81 |
| Second Pilot | Sergeant Vladimír **Hanzl** 787491 RAFVR. Age 22. | Panel 85 |
| Navigator | Flying Officer Ladislav **Říha** 82527 RAFVR. Age 25. | Panel 66 |
| Wireless Operator | Sergeant František **Janča** 787845 RAFVR. Age 19. | Panel 86 |
| Front Gunner | Sergeant Dobromil **Špinka** 787889 RAFVR. Age 22. | Panel 94 |
| Rear Gunner | Sergeant Adolf **Podivínský** 787589 RAFVR. Age 26. | Panel 91 |
| Posting History | Sgt K Danihelka posted via RAF Church Fenton, 12/05/1941. Sgt V Hanzl posted via 2 S.F.T.S & 1429 COTF, 02/03/1942. | |
| Operations Flown | Sgt K Danihelka : 19 / Sgt V Hanzl : 2 | |
| Remembered | RUNNYMEDE MEMORIAL | |

The first crew back at East Wretham at 00:07hrs was Sergeant V Pára, followed by Sergeant K Pospíchal at 00:12hrs and three minutes later by Sergeant V Žežulka. The participating crews and the squadron had been lucky. Whoever had

Two of the more conventional Combat Reports. These were submitted by the squadron on return from the operation to Emden on March 3rd 1942.

Left: Sergeant František Binder was 311 Squadron's only gunner who succumbed in the air to injuries received from the enemy night fighter fire while the squadron was with the Bomber Command in 1941–1942. (Anna Trehy) Above: Sergeant František Binder is being lowered into the grave at St. Ethelbert Cemetery, East Wretham, 7th March 1942. The coffin bearers are pilots Sgt Josef Horáček and Sgt Oldřich Soukup and Wireless Operators/Air Gunners Sgt Jaroslav Klvaňa, Sgt Zdeněk Donda, Sgt Ján Šimko and Sgt Pavel Tofel. (Pavel Vančata)

Tail of Wellington Mk.IC Z1147/KX-G seen damaged after return from Emden on the night of 3rd/4th March 1942. Sgt Binder suffered fatal injuries in the rear gun turret riddled by the German night fighter fire. After repairs at the Vickers factory in Weybridge, aircraft was returned to 311 Squadron on 15th July 1942 and flew 33 Anti-Submarine patrols marked with an individual letter Q until 10th April 1943. During a transfer flight from 311 Squadron on 23 May 1943 both engines failed before landing at Pembrey and aircraft crashed north of Kidwelly. (Jan Rail)

*Memorial plaque to František Binder.*

agreed to send just four 'Freshmen' crews to Emden had made a serious error of judgement. Sergeant František Binder was dead on arrival back at base. The crew had done all they could trying to save their crewmate. He was buried in East Wretham (St. Ethelbert) Churchyard, Grave 21 on March 7th, with full military honours. The sacrifice of this young airgunner was honoured on October 27th 2014 when a memorial plaque was unveiled at Hojná Voda, near České Budějovice, the birthplace of František Binder.

### March 3rd 1942 : Renault Works – Paris

| Pilot | 2nd Pilot | Serial | Code | Bomb Load | Result |
|---|---|---|---|---|---|
| Sgt B Hradil | Sgt J Kotrch | Z1070 | KX-Y | 2x1000lb+1x500lb | MISSING |
| F/Sgt J Doktor | Sgt O Špaček | Z8838 | KX-Z | 2x1000lb+1x500lb | Duty Carried Out |
| P/O F Taiber | Sgt J Štark | DV516 | KX-K | 2x1000lb+1x500lb | Duty Carried Out |
| P/O J Bala | Sgt V Výcha | T2962 | KX-T | 2x1000lb+1x500lb | Duty Carried Out |
| F/Sgt J Kalenský | Sgt K Kodeš | R1161 | KX-X | 2x1000lb+1x500lb | Duty Carried Out |
| Sgt F Naxera | Sgt K Mazurek | X9760 | KX-O | 2x1000lb+1x500lb | Duty Carried Out |
| Sgt F Dostál | Sgt H Dostál | R1802 | KX-P | 2x1000lb+1x500lb | Duty Carried Out |
| Sgt F Bulis | Sgt J Bláha | Z1098 | KX-U | 2x1000lb+1x500lb | Duty Carried Out |

The crews selected to attack the Renault works were all safely airborne by 18:40hrs. The operation would see the return of Hugo Dostál, he had not operated since June 1941 having completed 22 operations. A fluke, or by design, the two Dostál's would fly together for the first time. 3 Group's aircraft were given three routes to target, each routed over or via St Valery, Dieppe or Fecamp before making their run into the target over pre-determined times and altitudes. The times and heights are recorded in the Interception and Tactics Report.[67] These were 20:35hrs-21:59hrs, bombing height 3,000ft – 20:45hrs-21:20hrs, bombing height 3,500ft - 21:05hrs-21:49hrs, bombing height 4,000ft and finally 21:50hrs-22:10hrs had the bombing height as 6,300ft. This would indicate 3 Group would be over the target in four waves. This is at odds with the 3 Group Operational Appendices[68] which records the following:

### HIGHBALL FORCE : **Advance Force** (All Experienced Crews)

| R.A.F Mildenhall | 149 Squadron | Short Stirling |
|---|---|---|
| R.A.F Wyton | 15 Squadron | Short Stirling |
| R.A.F Oakington | 7 Squadron | Short Stirling |
| R.A.F Marham | 218 Squadron | Short Stirling |

---

[67] *Interception and Tactics Reports (Air14/33722)*
[68] *3 Group Operational Appendices (Air25/65)*

### HIGHBALL FORCE : **Main Force** (All Experienced Crews)

| R.A.F East Wretham | 311 (Czech) Squadron | Vickers Wellington |
| --- | --- | --- |
| R.A.F Honington | 9 Squadron | Vickers Wellington |
| R.A.F Feltwell | 57 Squadron | Vickers Wellington |
| R.A.F Feltwell | 75(NZ) Squadron | Vickers Wellington |
| R.A.F Stradishall | 214 Squadron | Vickers Wellington (Not TR1335) |
| R.A.F Wyton | 156 Squadron | Vickers Wellington (Not TR1335) |
| R.A.F Oakington | 101 Squadron | Vickers Wellington (Not TR1335) |
| R.A.F Mildenhall | 419 (RCAF) Squadron | Vickers Wellington (Not TR1335) |

### HIGHBALL FORCE : **Rear Force** (All Experienced Crews)

| R.A.F Stradishall | 214 Squadron | Vickers Wellington (4-432) |
| --- | --- | --- |

Regardless of the route or the number of waves, 3 Group's bombers encountered moderate flak and spasmodic fighter interference en route to the target. Flak was especially fierce around Abbeville and Dieppe and almost all towns and larger villages en route to Paris. The Renault factory complex was built on the 1 mile-long Ile Seguin island, just 2 miles downstream from the Eiffel Tower. The whole factory complex stood 20 feet above the River Seine. The factory housed its own coal-fired power station located at the western tip of the island, and in the centre was a vast 5-storey factory that accommodated chassis assembly, body construction, painting and upholstery workshops. A two-tier covered track for testing completed vehicles ran around the island's perimeter. Flight Sergeant J Kalenský skippered the first squadron crew over the target at 21:05hrs. In the bright moonlight, they bombed from between 2,000 and 2,500ft, and reported on return, *'Bombed areas E & F. Bursts seen on island east of river and on aerodrome possibly on hangars'*. Flying immediately behind him was Sergeant F Bulis at just 1,400ft. The crew reported, *'Target area brightly lit by flares and bombed across area F (Diesel assembly plant). Bomb bursts seen amongst buildings. Red glow resulting. Fire with much smoke seen on arrival north of the end of Island'*. Opposition over Paris was negligible, and in the clear conditions, the squadron pressed home their attacks with some relish. All the crews reported hitting the target. The last over the Renault works was Pilot Officer J Bala at 21:42hrs. He dropped his three bombs from 1,400ft on the Assembly plant. Back at East Wretham, he reported, *'Fires seen on arrival near Power Station and on west bank of River. Dropped on 'C'. Buildings certainly hit, bursts seen, no other results'*.

The squadron turned for home leaving the island of Seguin covered in smoke and flames. It had been an exhilarating operation, low-level in good visibility and no opposition. By 23:48hrs, all but one crew was back at East Wretham. An aircraft had been seen to crash, but little did the squadron realise it was from their squadron. Wellington Z1070 KX-Y was caught by searchlights on the return flight and shot down by flak, crashing in flames near Verneul-en-Halatte on the main road between Creil and Pont Saint Maxence, killing all the crew. The charred remains were removed from the burnt-out wreckage, and only the body of Sergeant Alois Tolar, the navigator, could be positively identified. He was buried on March 6th in Creil Cemetery, Grave 341. That same day, the rest of the crew were buried in five coffins as 'Unknown'. Post-war, their bodies were exhumed, and efforts to positively identify the remains were undertaken.

## Vickers Wellington Mk.IC Z1070 KX-Y

| Manufacturer | Vickers Armstrong | |
|---|---|---|
| Contract | B.97887/39 | |
| Taken on Charge | 06/02/1942 via 218 Squadron. | |
| Cat E Missing | 03/03/1942 | |
| Struck Of Charge | 27/03/1942 | |
| Total Flying Hours | N/K | |
| Take-Off Time | 18:19hrs | |
| Bomb Load | 2x1000lb+1x500lb | |
| | **CREW** | |
| Captain | Sergeant Bohuslav **Hradil** 787578 RAFVR. Age 26. | Plot 2. Coll. grave 341-345 |
| Second Pilot | Sergeant Jan **Kotrch** 787532 RAFVR. Age 24. | Plot 2. Coll. grave 341-345 |
| Navigator | Sergeant Alois **Tolar** 788183 RAFVR. Age 27. | Plot 2. Coll. grave 341-345 |
| Wireless Operator | Sergeant Josef **Svoboda** 787881 RAFVR. Age 30. | Plot 2. Coll. grave 341-345 |
| Front Gunner | Sergeant Imrich **Kormanovič** 787440 RAFVR. Age 40. | Plot 2. Coll. grave 341-345 |
| Rear Gunner | Sergeant Přibyslav **Strachoň** 787569 RAFVR. Age 22. | Plot 2. Coll. grave 341-345 |
| Posting History | Sgt B Hradil posted via Wilmslow Depot, 28/03/1941. Sgt J Kotrch posted via 2 S.F.T.S, 17/07/1941. | |
| Operations Flown | Sgt B Hradil : 26 / Sgt J Kotrch : 1 | |
| Buried | CREIL COMMUNAL CEMETERY | |

The euphoria of the raid was quickly lost with the realisation that the squadron had lost two experienced crews. An estimated 300 bombs fell on the factory complex destroying approximately 40% of the buildings. This resulted in a halt of production of almost a month, with further repairs taking 6 – 7 months. With the success came the inevitable civilian casualties, 367 were killed, with hundreds more injured, and over 9,000 people are reported to have lost their homes. This, however, seems a remarkably high figure. Reports from Paris confirmed a highly destructive raid, and surprisingly, the same report remarked that the Parisians were overjoyed with the operation's success.

Heavy rain fell all day on the 4th, cancelling any planned operations throughout the Group. The rain was followed by a severe frost on the 5th that rendered the airfield unserviceable for the next two days. To add to the problems, snow returned on the night of March 6th and 7th. On the 7th, Air Vice Marshal Karel Janoušek visited East Wretham, to decorate members of the squadron. Inclement weather and poor landing ground conditions again resulted in a cancelled operation that night.

On the 8th, the squadron learnt that it was to return to Essen, a target it had not visited in over nine months. Seven experienced crews would be detailed and briefed, while four 'Freshmen' crews would bomb the docks at Le Havre. A change in tactic was deployed on this night. 3 Group would open the attack using twenty of its most experienced TR1335-equipped crews. Each would carry 12 bundles of flares, plus bombs and would illuminate the aiming point. They would be followed by a 54-strong incendiary force that would bomb on the flares, setting ablaze the target. Finally, the main force would bomb on the fires. The Czechs of 311 Squadron would be part of this wave of 34 bombers.

 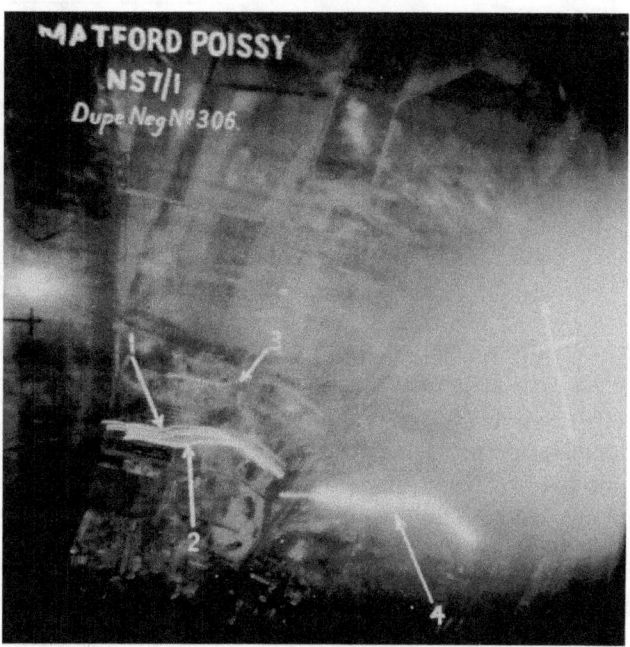

*Two target photographs taken by Vickers Wellingtons of No.3 Group over Matford Poissy, March 3rd 1942.*

*Two pilots who made the supreme sacrifice. Sergeant Jan Kotrch and Sergeant Oldřich Helma. Fate played a hand in their time on the squadron. Left: Jan Kotrch completed just two operations, while Oldřich Helma completed 18.*

## March 8th 1942 : Krupps Works – Essen

| Pilot | 2nd Pilot | Serial | Code | Bomb Load | Result |
|---|---|---|---|---|---|
| F/Sgt J Doktor | Sgt O Špaček | R1161 | KX-X | 2x1000kb+2x500lb | Early Return |
| Sgt J Fína | Sgt J Štark | Z1098 | KX-U | 2x1000kb+2x500lb | Early Return |
| P/O J Bala | Sgt V Výcha | T2962 | KX-T | 2x1000kb+2x500lb | Duty Carried Out |
| F/Sgt J Kalenský | Sgt K Kodeš | Z1105 | KX-R | 2x1000kb+2x500lb | Duty Carried Out |
| Sgt F Naxera | Sgt K Mazurek | X9760 | KX-O | 2x1000kb+2x500lb | Early Return |
| Sgt F Dostál | Sgt H Dostál | R1802 | KX-P | 2x1000kb+2x500lb | Duty Carried Out |
| Sgt F Bulis | Sgt J Bláha | Z1155 | KX-F | 2x1000kb+2x500lb | Duty Carried Out |

The squadron departed R.A.F East Wretham between 00:57hrs and 01:08hrs and within nine minutes the first of the night's early returns was back in the circuit. Flight Sergeant J Doktor landed with instrument failure aboard Wellington R1161 KX-X. He was closely followed by Sergeant J Fína, who failed to gain sufficient height on retracting the undercarriage on take-off. The Wellington struck barbed wire fencing damaging the airscrews and ripping the fabric bomb doors. The aircraft completed a wide circuit while the remaining Wellingtons took off. Once complete, the crew gingerly came in for a landing, their operation having lasted just 13 minutes. It was not the best of starts.

| Date | 08/03/1942 |
|---|---|
| Mark | Mk.Ic |
| Serial | Z1098 |
| Code | KX-T |
| Taken On Charge | 13/11/1941 via No.18 MU |
| Manufacturer | Vickers Armstrong (Chester) |
| Contract | 124362/40 |
| Pilot (s) | Sergeant Jiří Fína / Sergeant Ján Štark |
| Flight | Operational |
| Time | 01:08hrs |
| Cause | Pilot Error. |

The remaining crews headed for the departure point over Southwold and then out over the North Sea. The Group's aircraft would make landfall at various points along the Dutch and Belgian coast, from where they would converge over Essen. The squadron's third early return of the night occurred over the North Sea. Sergeant F Naxera reported problems with the port engine aboard Wellington X9760 KX-O, which resulted in the crew jettisoning their bombs into the sea and turning back. It was a dreadful night for the squadron, which would not have gone unnoticed at Exning.

The crews of 3 Group would be over Essen between 02:07hrs and 03:30hrs. Heavy flak was intense over the Krupps works and easily made up for the lack of light flak, which seemed subdued. Considerable searchlight opposition was encountered, and some beams were seen to be laid horizontally over the industrial haze, making it look like a white fog. Also reported were the dreaded balloons seen dotted around Essen ranging in height between 4,000ft and 12,000ft. One of the earliest crews over Essen was Flight Sergeant J Kalenský. He arrived to find three prominent fires at the northern end of the Krupps works. They bombed at 02:07hrs, one bomb seen bursting 600 yards north of the aiming point. Forty minutes later, Pilot Officer J Bala was making his bomb run. Dense smoke and haze prevented visual identification of the aiming point. Despite this, the crew were confident that their bombs landed just west of the target. Three minutes later at 02:50hrs, Sergeant F Dostál approached Essen. He reported on return, *'Two fairly large fires seen in target area on arrival'*. Flying at 11,200ft the crew found themselves held by searchlights and pounded by flak. They dropped their bombload northwest of the town and made their escape. Finally, at 03:12hrs, Sergeant F Bulis and crew arrived to add his 3,000lbs of bombs. They dropped their bombload one mile west of the aiming point, producing a few small fires. The returning crews were enthusiastic about the raid. However Group H.Q were less so. They reported that the operation was not as successful as they had hoped. There were some positives, the flare force had performed well. However, the incendiary force's bombing accuracy was woeful. This resulted in a scattered attack and the concentration hoped for never materialised. Local sources in Essen reported a light raid with modest damage in the southern districts.

### March 8th 1942 : CC24A – Le Havre

| Pilot | 2nd Pilot | Serial | Code | Bomb Load | Result |
|---|---|---|---|---|---|
| Sgt V Šponar | F/Lt J Strégl | W5711 | KX-H | 16x250lb | Duty Carried Out |
| Sgt V Pára | Sgt O Soukup | T2739 | KX-V | 16x250lb | Duty Carried Out |
| Sgt V Žežulka | Sgt O Havlík | X9745 | KX-S | 16x250lb | Duty Carried Out |
| Sgt K Pospíchal | Sgt J Hadrávek | R1532 | KX-B | 16x250lb | Duty Carried Out |

Only 12 Wellingtons and a solitary Stirling were detailed to attack the docks at Le Havre, 3 Group provided all of them. The squadron departed over Littlehampton and unlike the Essen crews, reported no early returns. The raid was scheduled to start at 05:35hrs and finish at 06:18hrs. As was becoming a custom, a Czech crew was one of the first over the target. Sergeant K Pospíchal watched as his bombs exploded across Dock 7 and Bassin-de-Maree. Five minutes later at 05:40hrs it was the turn of Sergeant V Šponar, who dropped his entire load in one stick on the fires created by Sergeant Pospíchal. Bombing at the same time was the crew of Sergeant V Žežulka. They watched as their 16x250-pounders straddled docks 8 and 9. The last crew over the target was that of Sergeant V Pára at 05:50hrs. This crew bombed in one stick from 15,000ft on the fires raging between Dock 7 and Bassin-de-Maree.

The following day, the squadron was informed it was going back to Essen along with nearly 200 other aircraft, as there was unfinished work. The squadron would detail and brief seven crews on what would turn out to be another poor night with mechanical defects becoming increasingly common.

### March 9th 1942 : Krupps Works – Essen

| Pilot | 2nd Pilot | Serial | Code | Bomb Load | Result |
|---|---|---|---|---|---|
| Sgt J Fína | Sgt J Štark | DV516 | KX-K | 2x1000kb+2x500lb | Jettisoned |
| P/O J Bala | Sgt V Výcha | T2962 | KX-T | 2x1000kb+2x500lb | Early Return |
| Sgt V Šponar | F/Lt J Strégl | W5711 | KX-H | 2x1000kb+2x500lb | Jettisoned |
| Sgt V Pára | Sgt O Soukup | Z8838 | KX-Z | 2x1000kb+2x500lb | Alternative Target |
| Sgt F Dostál | Sgt H Dostál | R1802 | KX-P | 2x1000kb+2x500lb | Early Return |
| Sgt F Bulis | Sgt J Bláha | Z1155 | KX-F | 2x1000kb+2x500lb | Duty Carried Out |
| Sgt V Žežulka | Sgt O Havlík | X9745 | KX-S | 2x1000kb+2x500lb | Jettisoned |
| F/Sgt J Doktor | Sgt O Špaček | R1161 | KX-X | 2x1000kb+2x500lb | Withdrawn |

Group H.Q had requested nine crews for the operation. Eight were prepared but only seven took off for Essen. Wellington R1161 KX-X skippered by Sergeant Doktor was withdrawn at the last moment. Departing over Southwold in good visibility, the same tactics would be employed as the previous night. The Flare Force would illuminate the Krupps works, and the Incendiary Force would follow up, covering the target area in thousands of four-pound incendiaries. The coup de grâce would be the main force carrying mainly high explosives. The first of the night's aborts involved Pilot Officer J Bala and crew. The cockpit escape hatch aboard Wellington T2962 KX-T blew open, and despite attempts to close it, impossible in the slipstream, they aborted. The bombs were jettisoned into the sea. They were back over the airfield within an hour. Over Holland, the crew of Sergeant J Fína were flying at 11,000ft and passing north of the small village of Hellouw when they were attacked at 21:41hrs by a Bf109. The crew managed to fend off a determined attack but, in doing so, were obliged to jettison 'live' the entire bombload. Damage to Wellington DV516 KX-K was extensive. Once again excellent pilot and gunner coordination prevailed, and the crew claimed the Bf109 as damaged. The crew landed back at East Wretham at 23:18hrs. Two crews did not reach the German frontier. Sergeant F Dostál at the controls of Wellington R1802 KX-P, was unable to climb to a reasonable altitude due to issues with the engines. They were heavily engaged by Amsterdam's flak and searchlight defences. They elected to drop their bombs on the flare path of Schiphol airfield at 22:09hrs from 9,000ft. German aircraft were reported to have been observed taking off soon after the bombing. Searchlights got the better of Sergeant V Šponar and crew. Held in the beams, the crew took violent evasive action and in doing so lost considerable height. Unable to climb, the bombs were jettisoned 10 miles south of Utrecht at 22:25hrs. The crew of Sergeant V Pára got as far as Duisburg. Unable to gain sufficient altitude, the crew dropped their bombs from 12,000ft. These landed between a railway and a canal northeast of the city at 22:45hrs. There is some confusion regarding the crew of Sergeant V Žežulka aboard Wellington X9745 KX-S. The Operations Record Book reports that the crew saw numerous fires on arrival over the target and were then hit by flak, requiring the bombload to be jettisoned. This is at odds with two other reports, No.3 Group Operational Appendices[69] and the Form 'E' Report.[70] These documents state that the crew were unable to pinpoint the target and bombed Schiphol airfield at 22:31hrs. Only one crew appears to have successfully reached and then bombed Essen, and they had to drop one 1000-pounder on Emmerich to gain height. Sergeant F Bulis and crew reported numerous incendiaries on the run into the target, but due to evasive action, the remaining bombs were dropped on the northernmost district of Altenessen at 22:25hrs. All the crews are reported to have returned to East Wretham, some more badly damaged than others. Wellington DV516 was extensively damaged in the fighter attack. The rear turret was hit, as was the undercarriage, bomb-doors, hydraulics and fuselage. The pilots brought the Wellington in for a belly landing on return. None of the crew were injured. It would take over a month to repair the damage. Also showing signs of flak damage was Wellington X9745 KX-S.

*The Combat Report submitted by the crew of Sergeant J Fína on return from the operation against Essen. Right Sergeant Jiří Fína in his Czechoslovak Air Force uniform. Sadly, the crews luck would run out tantalisingly close to completing their operational tour. (John Costin)*

---

[69] *No 3 Group Operational Appendices (Air25/65)*
[70] *No 3 Group: summaries of Form E reports of bombs dropped on targets in France, Low Countries and Germany(Air14/3165)*

The following morning the Czechs took stock, it had been a dreadful night for the squadron. Something had gone very wrong, crews withdrawn before take-off, Wellingtons unable to climb, mechanical malfunctions and bad luck had blighted the whole operation. There was no consolation in the bombing results. Poor visibility caused the bombing to be widely scattered and spread across towns and cities in the Ruhr region. Some fires remained visible for eighty miles into the return flight and gave the impression of a successful raid. The 3 Group ORB rather optimistically reported, *'The operation on Essen was again on quite a considerable scale and was most successful. Large fires were left burning in the target, visible for many miles of the return journey. The result was definitely a 100% improvement on the previous night, the attack being much more concentrated in the target area'*. This confidence was also echoed by 1, 4 and even 5 Group.

The squadron had a lot to do on the 10$^{th}$. Questions had to be asked and reports submitted to Group H.Q. There was also the matter of inspecting the faulty Wellingtons. Thankfully, the squadron sat out another raid that night, once again the Krupps works were the target. Group contributed 12 Stirlings and 18 Wellingtons from R.A.F Marham, Mildenhall and Wyton. Results were once again disappointing with widespread scattered bombing. Bomber Command turned its attention northwards towards Kiel's Deutsche Werke and dock area on the 12$^{th}$. At R.A.F East Wretham the squadron prepared six crews for the night's raid.

### March 12$^{th}$ 1942 : GR3588 – Deutsche Werke, Kiel AG

| Pilot | 2$^{nd}$ Pilot | Serial | Code | Bomb Load | Result |
|---|---|---|---|---|---|
| Sgt V Šponar | Sgt O Špaček | W5711 | KX-H | 5x500lb | Duty Carried Out |
| Sgt J Fína | Sgt O Soukup | R1802 | KX-P | 5x500lb | MISSING |
| P/O J Bala | Sgt V Výcha | T2962 | KX-T | 5x500lb | Duty Carried Out |
| F/Sgt A Jedounek | Sgt H Dostál | R1532 | KX-B | 5x500lb | Duty Carried Out |
| F/Sgt J Kalenský | Sgt K Kodeš | Z1105 | KX-R | 5x500lb | Early Return |
| Sgt F Naxera | Sgt K Mazurek | X9760 | KX-O | 5x500lb | Duty Carried Out |

Group would detail and brief a total of 44 Wellingtons, which would be followed over the target by Wellingtons of 1 Group. There was a reasonably early take-off for the squadron, who were all safely airborne by 19:17hrs. It was another disappointing night for the Group. Two Wellingtons failed to take-off due to mechanical issues, while seven crews aborted for various reasons. Four were from Huntington's 9 Squadron and one from 311 Squadron. Flight Sergeant J Kalenský encountered severe icing conditions en route. The starboard engine overheated, and the engine's oil pressure fell, resulting in the Wellington, which was covered in ice rapidly losing height. The bombs were jettisoned 'live' into the sea, and the crew safely landed back at R.A.F East Wretham. Another crew in trouble was Sergeant J Fína. They encountered the same conditions, their Wellington also being covered in ice, resulting in loss of speed and altitude. The ORB reports that at 20:53hrs an S.O.S. was transmitted, which was also reported to have been picked up by other squadron wireless operators. The crew were obviously in trouble and returning to base having jettisoned their bombload. A further message was received at 21:19hrs indicating that the aircraft *'had landed in the sea'*. The Wireless Operator, Pilot Officer Josef Cibulka, continued wireless communication until the very last moment. The reported location of the ditching was 5317N 0307E.

The raid was to start promptly at 21:55hrs and predictably a Czech crew was at the forefront. The gutsy crew of Pilot Officer J Bala were the first over Kiel flying at 16,000ft. Flak was intense and working with zones of searchlights numbering between 12 to 30. Also highly visible were numerous balloons flying between 5,000 and 6,000 ft, these strangely were being illuminated by searchlights. In excellent visibility, the crew reported their bombs bursting in the area of the aiming point, producing a huge flash. It would be thirty minutes before the next Czech crew bombed. Sergeant F Naxera was over the target at 22:25hrs, and they dropped their bombs on the east bank of the Kieler Hafen. Observation of the results was impossible due to the intense flak and searchlights. Such was the ferocity of the defences that the crew of Sergeant V Šponar were unable to locate the aiming point, so they opted to bomb the flak and searchlight positions at 22:35hrs. Bomb bursts were seen but not the results. The Die Hörn commercial docks were on the receiving end of the bombload of Flight Sergeant A Jedounek at 22:52 hours. Six explosions were observed, but no other details.

It had been a successful raid by a small force. The Deutsche Werke, the Group's primary target, sustained damage, as did the naval dockyard and the town. The use of flares which were accurately dropped proved highly effective. What squadron dropped them is unclear. All apart from Wellington, R1802 KX-P landed back at R.A.F East Wretham. The squadron had already started planning for an early morning sea search. The Czechs would have been confident of a recovery. With the ditching position known, there was an excellent chance either Air Sea Rescue or Coastal Command would orchestrate a recovery. On the morning of the 13th, four Wellington crews stood by to take-off. The weather was far from ideal, high wind and rough seas prevented the Air Sea Rescue launches from participating. The No.3 Group H.Q Records Book[71] reports that a dinghy was located by a Lockheed Hudson, and a fixed was obtained. but it goes on to say that further searches were fruitless. This may mistakenly be referring to a sighting by Hudson 'A' of 279 Squadron flown by Flight Sergeant Heywood and crew on March 11th.

Search flights were flown throughout the 13th, but for some reason the four 311 Squadron Wellingtons remained grounded. On the 15th, three Hudson's of 279 Squadron found floating wreckage and a large patch of oil which may have been from Wellington R1802. Sadly, nothing was ever found, the sea claiming yet another brave crew.

<u>Vickers Wellington Mk.IC R1802 KX-P</u>

| Manufacturer | Vickers Armstrong | |
|---|---|---|
| Contract | 992424/39 | |
| Taken on Charge | 11/12/1941 via 149 Squadron. | |
| Cat E Missing | 12/03/1942 | |
| Struck Of Charge | 27/03/1941 | |
| Total Flying Hours | N/A | |
| Take-Off Time | 19:16hrs | |
| Bomb Load | 5x500lb | |
| | | |
| | **CREW** | |
| Captain | Sergeant Jiří **Fína** 787403 RAFVR. Age 28. | Runnymede Panel 83 |
| Second Pilot | Sergeant Oldřich **Soukup** 787489 RAFVR. Age 21. | Runnymede Panel 94 |
| Navigator | Flying Officer Jaroslav **Kula** 82615 RAFVR. Age 27. | Runnymede Panel 66 |
| Wireless Operator | Pilot Officer Josef František **Cibulka** 66491 RAFVR. Age 28. | Runnymede Panel 68 |
| Front Gunner | Sergeant Alois **Mezník** 787866 RAFVR. Age 30. | Runnymede Panel 89 |
| Rear Gunner | Sergeant František **Raiskup** 787876 RAFVR. Age 28. | Runnymede Panel 92 |
| | | |
| Posting History | Sgt J Fína posted via Cosford Depot, 11/04/1941. Sgt O Soukup posted via 2 S.F.T.S, 17/07/1941. | |
| Operations Flown | Sgt J Fína : 28 / Sgt O Soukup : 4 | |
| Remembered | RUNNYMEDE MEMORIAL | |

The loss of Sergeant Jiří Fína and crew was a tremendous blow to the squadron. Experienced and nearing the end of their tour, the failure to locate and rescue the crew in what should have been a risky but routine Air-Sea-Rescue operation made the loss much harder to except. The Czechs bad luck with ditched crews seemed to have struck again.

R.A.F East Wretham was visited by an International ENSA Group on the 14th, the '*Les Comediens Francais Libres*' who performed two one-act comedies. Training occupied the squadron on the 16th, cross-country flights and local training being undertaken. The same day, two Airspeed Oxfords were delivered to 1429 COTF to augment the existing flight's aircraft. The following day the Training Flight was again airborne with crews over Berners Heath, these encountered some turbulent conditions. It was the prelude of some bad weather that kept the squadron grounded. Reports of Wellington T2739 flying slightly right wing down prompted an inspection on the 19th. Surprisingly, previously

---

[71] *Air25/52*

*A moment of rest with an accordion at East Wretham, 3rd March 1942. Standing L-R: W/O Jaroslav Kamarád, Sergeants Otto Jebáček, Karel Kodeš, Adolf Mužík, Flight Sergeant Bohumil Hájek (Fitter I). Sitting L-R: Sergeants Hugo Dostál, Jaroslav Doktor, Jiří Osolsobě, Richard Květ, Bohuslav Hradil, and Rostislav Kaňovský. With the exception of Flight Sergeant Hájek and Sergeant Mužík the rest took part in the evening sortie to Emden or Paris. All but Sergeant Hradil returned. (Jaroslava Rozumová)*

unnoticed damage to the engine nacelle edge was found, the fixing rivets appearing to have sheared off. This was one of the Wellingtons supplied by 218 Squadron. On the 21st, the squadron found itself training over Holkham Bay, north Norfolk. Finally after a break of nine days the news that op's were back on arrived at East Wretham, this was quickly cancelled at 6:10hrs. The On-Off preparations continued for the next few days, interspersed by the usual training flown in between ever-changeable weather. It was on the 25th that the Training Flight suffered another accident. Wellington R3234 was to practice a loaded climb with dummy bombs. On taking off, the Wellington bounced heavily several times, damaging the undercarriage in the process. The pilot eventually managed to get airborne and continue with the flight.

On returning to East Wretham, it was discovered that the undercarriage would not lower, reports suggest ruptured hydraulics. The ageing Wellington was brought in for a belly landing at 12:30hrs.

| Date | 25/03/1942 |
| --- | --- |
| Mark | Mk.Ic |
| Serial | R3234 |
| Code | KX-N |
| Taken On Charge | 09/08/1940 via No.22 MU. |
| Manufacturer | Vickers Armstrong (Weybridge) |

| Contract | B3919/39 |
|---|---|
| Pilot (s) | Flight Sergeant Jaroslav Hájek Instructor / Pupil N/K |
| Flight | Training |
| Time | 12:30hrs |
| Cause | Undercarriage damaged on take-off (1429 COTF) |

The aircraft was damaged beyond even the skill of 311 Squadron groundcrews and was transported away to be repaired, it never returned to the squadron. Finally on the 25th, with a break from operations of 13 days eight crews assembled in the briefing room for a return visit to Essen.

## March 25th 1942 : Essen

| Pilot | 2nd Pilot | Serial | Code | Bomb Load | Result |
|---|---|---|---|---|---|
| F/Sgt J Doktor | Sgt O Havlík | Z1098 | KX-U | 2x1000lb+1x500lb+1x250lb | Dropped in clouds |
| Sgt K Pospíchal | Sgt J Hadrávek | Z8838 | KX-Z | 2x1000lb+1x500lb+1x250lb | Jettisoned |
| P/O F Taiber | Sgt K Mazurek | T2962 | KX-T | 2x1000lb+1x500lb+1x250lb | Jettisoned |
| Sgt F Dostál | Sgt J Štark | R1497 | KX-D | 2x1000lb+1x500lb+1x250lb | Alternative target |
| F/Sgt A Jedounek | Sgt H Dostál | R1161 | KX-X | 2x1000lb+1x500lb+1x250lb | Duty Carried Out |
| F/Sgt J Kalenský | Sgt K Kodeš | Z1105 | KX-R | 2x1000lb+1x500lb+1x250lb | Alternative target |
| F/O O Hořejší | F/Lt J Štrégl | X9745 | KX-S | 2x1000lb+1x500lb+1x250lb | Jettisoned |
| Sgt F Bulis | Sgt J Bláha | Z1155 | KX-F | 2x1000lb+1x500lb+1x250lb | Alternative target |

3 Group assembled a force of 93 Wellingtons and 30 Stirlings for the raid. It would follow the now familiar routine of selecting experienced crews who would be assigned to the flare, incendiary or main strike force. The raid's success would be determined by the accuracy and diligence of the Gee set operators within the group. The first of the squadron crews departed at 19:50hrs on what would turn out to be one of the squadron's blackest and most disappointing nights. Four squadron crews reported that they were unable to climb to altitude and either jettisoned or bombed alternative targets. The first crew back at 22:42hrs was Sergeant K Pospíchal. They stated they could not climb above 6,000ft. The bombs were jettisoned 'live' in position 5153N/0326E. Port engine trouble aboard Wellington T2962 KX-T forced the crew of Pilot Officer F Taiber to jettison his bombs at position 5140N/0338E. They landed at 23:00hrs. Five minutes later, Flight Sergeant J Kalenský touched down at East Wretham. They were unable to climb but bombed the flarepath at Haamstede airfield, which was showing lights at 21:43hrs. This crew landed at 23:05hrs. Another crew unable to climb was Flying Officer O Hořejší, they jettisoned their bombs at position 5155N/0335E, landing back at base at 23:24hrs.

Contrary to the ORB entry, Flight Sergeant J Doktor did not bomb Essen. Once again, problems with reaching operational height resulted in the bombs being dropped through the clouds at position 5114N/0630E. The crew landed back at East Wretham at 00:19hrs. Flight Sergeant A Jedounek were over the Krupps Werkes at 22:22hrs at just 12,000ft. They reported that fires were almost extinguished on arrival, yet despite the vicious flak they bombed in one stick before heading home, landing at 00:35hrs. Sergeant F Bulis and crew found themselves behind schedule and unlikely to arrive over the target on time. They opted to bomb Duisburg dock from 13,500ft at 23:05hrs. This is another crew the ORB claims to have bombed the target. Sergeant F Bulis landed at 00:44hrs. The last crew to touch down at East Wretham was Sergeant F Dostál at 01:10hrs. They, like Sergeant Bulis, would have been late over the target, so they decided to bomb Gladbach at 22:21hrs. The only positive from the squadron's activities this night was that all the crews returned safely. The squadron had a visitor in the shape of a Short Stirling Mk.I of 218 (Gold Coast) Squadron, which landed at East Wretham on return from an abortive trip to Essen. The TR.1335 Gee aboard Stirling N6077 HA-V failing, so the pilot, Sergeant John Webber, jettisoned. Exactly why the crew diverted is unclear.

What had caused the issues with the Wellington's reluctance to climb would be investigated. In contrast, the crew's inability to keep on time was something for the commanding officer and the navigation leader. It was not a good night for 311 Squadron's reputation and would have raised eyebrows at Group and Command level. Both the Station Commander, Group Captain Ommaney and Wing Commander Batchelor DFC, R.A.F East Wretham Station

*Former 311 Squadron pilots Sergeant Jaroslav Doktor (44 sorties) and Flight Sergeant Josef Filler (38 sorties) in spring 1942 when they served as Pilot-Instructors at 1429 Czech Operational Training Flight. F/Sgt Filler was a well-known "troublemaker" and Sgt Doktor ably seconded him in carrying out all sorts of "boy pranks", although both were already in their thirties. The attentive reader must have noticed the discrepancy in rank of Jaroslav Doktor as he completed the tour of operations in the rank of F/Sgt. He was reported for low flying between 100 and 150 feet during an air test of Wellington Mk.IC X9733/KX-L over Bury St. Edmunds on 25th March 1942 and he was reduced in rank to Sergeant for unauthorized low flying on 25th May 1942. (Jaroslav Popelka)*

Commander were not slow in voicing their opinions about the squadron's performance. Their reports would eventually reach 3 Group H.Q and what followed was both unfavourable and unforgivable and would ultimately leave the Czechs feeling somewhat irritated. On the 26th, Group Captain Alois Kubita, the Czech Liaison Officer in London was summoned to 3 Group H.Q to meet AVM J.E.A Baldwin CB, OBE, DSO to discuss the reports from Honington and East Wretham and specifically 311 Squadrons performance the previous night.

Hardly had the dust settled than the squadron was ordered to prepare for a return to Essen. The squadron would supply ten crews.

<u>March 26th 1942 : Essen</u>

| Pilot | 2nd Pilot | Serial | Code | Bomb Load | Result |
|---|---|---|---|---|---|
| F/Sgt J Doktor | Sgt O Havlík | R1497 | KX-D | 2x1000lb+1x500lb+1x250lb | Duty Carried Out |
| Sgt K Pospíchal | Sgt J Hadrávek | Z1098 | KX-U | 2x1000lb+1x500lb+1x250lb | Duty Carried Out |
| P/O J Bala | Sgt V Vycha | X9760 | KX-O | 2x1000lb+1x500lb+1x250lb | Duty Carried Out |

| Sgt F Dostál | Sgt K Mazurek | X9880 | KX-C | 2x1000lb+1x500lb+1x250lb | Alternative target |
| Sgt V Žežulka | Sgt O Špaček | X9733 | KX-L | 2x1000lb+1x500lb+1x250lb | Duty Carried Out |
| F/Sgt A Jedounek | Sgt H Dostál | R1161 | KX-X | 2x1000lb+1x500lb+1x250lb | Duty Carried Out |
| P/O F Taiber | Sgt J Štark | R1532 | KX-B | 2x1000lb+1x500lb+1x250lb | Jettisoned |
| F/Sgt J Kalenský | Sgt K Kodeš | X9787 | KX-G | 2x1000lb+1x500lb+1x250lb | Alternative target |
| F/O O Hořejší | Sgt V Pára | Z1111 | KX-N | 2x1000lb+1x500lb+1x250lb | Duty Carried Out |
| Sgt F Bulis | Sgt J Bláha | X9871 | KX-A | 2x1000lb+1x500lb+1x250lb | Duty Carried Out |

It was a much smaller force that returned to the Ruhr on this night. Group cobbled together 78 Wellingtons from seven of its squadrons and twelve Stirlings from 7 and 218 Squadrons. They would again be joined by 1 Group and just six Wellingtons from 4 Group. A flare force was once again used, selected crews from 9, 57, 75(NZ), 101, 115 and 218 Squadrons would act as target finders and illuminate the aiming point prior to the main force arrival. Flying conditions were ideal, scattered cloud, excellent visibility and very little haze was encountered as the squadron crossed the coast of Holland. It was no surprise given the conditions that night fighter activity was observed soon after crossing the coast. Flying ten miles east of Haamstede at 13,000ft was the crew of Pilot Officer F Taiber. They were involved in a desperate battle of survival when attacked by a Bf109 at 21:11hrs. Crew discipline and steady nerves and a good rear gunner would see the crew fend-off three attacks. During the encounter, the bombs were jettisoned.

Two crews were unable to reach Essen, both reporting their Wellingtons were incapable of climbing to bombing height. One of them was Flight Sergeant J Kalenský, he stated that he could not climb above 10,500ft. On nearing Essen, the crew were alarmed to see balloons at 11,000ft and above. Unable to get safely above them the crew headed back towards Wesel where they bombed the town at 22:15hrs. Sergeant F Dostál also reported his aircraft struggling to reach heights above 10,600ft. Unable, or unwilling to penetrate the Ruhr's formidable searchlight belt, they too opted to bomb Wesel at 22:29hrs. The first squadron crew over the target was Sergeant V Žežulka at 22:12hrs. They bombed the western part of Essen from 12,500ft experiencing considerable opposition on the bomb run. Within six minutes four crews had delivered their bombloads. Flight Sergeant J Doktor and Sergeant K Pospíchal bombed the western part of the town with the former's bombs falling near the marshalling yards with the aid of flares. Flying Officer O Hořejší dropped his bombs in one stick in the centre of Essen, while Sergeant A Bulis bombed Essen's northern suburbs at 22:17hrs. The last two squadron crews were over Essen between 22:30hrs and 22:32hrs. Pilot Officer J Bala bombed first from 16,000ft. His bombs were not seen to burst as he was engaged by flak and held by searchlights. The only crew to report seeing the Krupps Werkes was Flight Sergeant A Jedounek. They reported on return, *'Fires in northern part of Krupps works observed on arrival over the target. Own bombs fell on town area south of marshalling yards, burst seen'*. All the squadron crews returned safely but agreed that the flak and searchlight defences had dramatically increased in both numbers and volume. Exning House seemed pleased with the result, TR1335 had worked well and the flares were considered an excellent aid to identifying the target.

The new AOC was not just an outstanding leader, he was also extremely shrewd. He was acutely aware that his command needed to produce a devastating attack on a German town or city that would reverberate horror throughout the Third Reich but simultaneously capture the imagination of the British public and finally quell the snipers in the Air Ministry and Admiralty. The recent operations directed against targets in the Ruhr had proved that TR1335 still had limitations over densely populated areas. He and his trusted staff at Bomber Command H.Q devised a simple but risky solution. Harris understood the limitations of his crews. Their bravery was never questioned, but traditional navigation aids had advanced little since 1939. Weather, blacked-out cities, and decoy fires heaped numerous problems on the navigators and crews. Unlike a target in the Ruhr, a target located on Germany's northern coast could be located reasonably simply by following the coastal features, a bay, a peninsular, and even islands. A target like a large port could be found and bombed, and with the aid of TR1335, it was now possible. The target chosen by the AOC and his staff was Lübeck, a busy port and home to several shipyards, including the Lübecker Flender-Werke. Lübeck is located a few miles inland from the Bay of Lübeck and sits beside the River Trave. Lübeck is rather unique in that its old town is an island entirely surrounded by the river. Its narrow streets, half-timbered buildings and importantly light defences made it a tempting target.

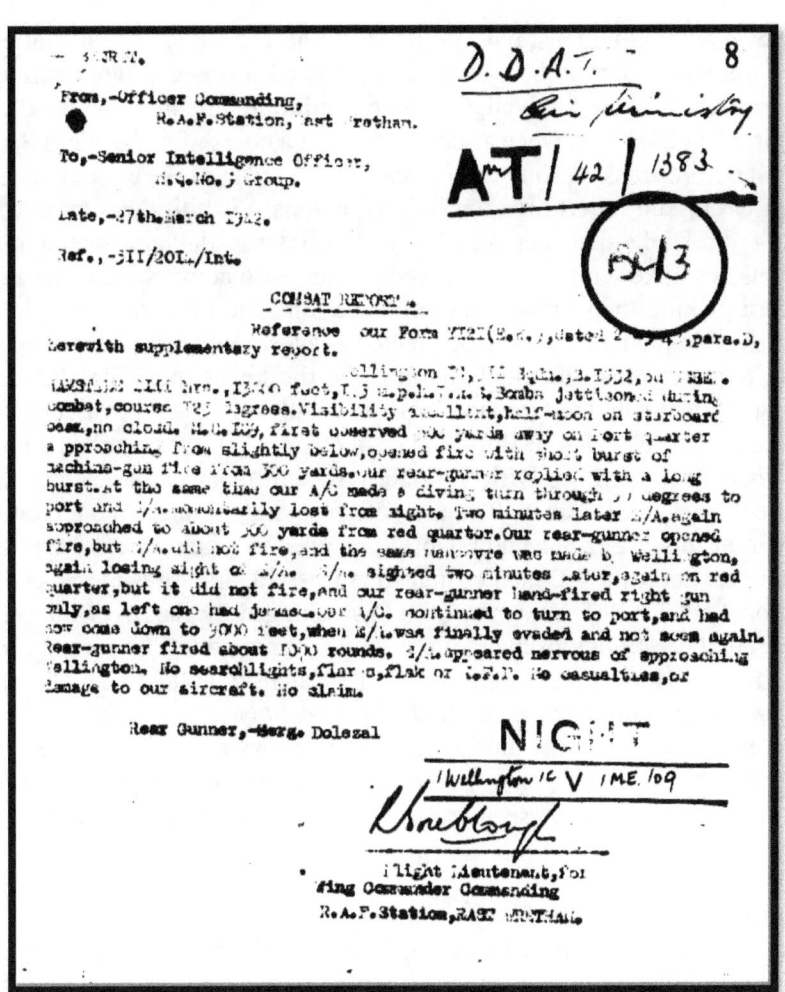

Left : Combat report submitted by Sergeant František Taiber and crew. Above: Sergeant Taiber.

In his book, 'Bomber Offensive' A.V.M Harris wrote, *'The nature of its buildings easier than most cities to set on fire. It was a city of moderate size, of some importance as a port, and with some submarine building yards of moderate size not far from it. It was not a vital target but seemed to me better to destroy an industrial town of moderate importance that to fail to destroy a large industrial city. However, the main objective of the attack was to learn to what extent a first wave of aircraft could guide a second to the aiming point by starting a conflagration'.*

Throughout the day the squadron's groundcrews prepared eight Wellingtons for the trip to northern Germany. Each would carry a full load of incendiaries. Bomber Commands intentions were very clear.

March 28th 1942 : Mackerel 'A' – Lübeck

| Pilot | 2nd Pilot | Serial | Code | Bomb Load | Result |
|---|---|---|---|---|---|
| P/O J Bala | Sgt V Výcha | T2962 | KX-T | 540x4lb | Duty Carried Out |
| Sgt K Pospíchal | Sgt J Hadrávek | Z1098 | KX-U | 540x4lb | Duty Carried Out |
| F/O O Hořejší | Sgt V Pára | Z1111 | KX-N | 540x4lb | Duty Carried Out |
| Sgt V Žežulka | Sgt O Špaček | X9733 | KX-L | 540x4lb | Duty Carried Out |
| Sgt F Bulis | Sgt J Bláha | Z1155 | KX-F | 540x4lb | Duty Carried Out |
| P/O F Taiber | Sgt J Štark | DV507 | KX-W | 540x4lb | Duty Carried Out |
| F/Sgt J Doktor | Sgt O Havlík | R1497 | KX-D | 540x4lb | Duty Carried Out |
| Sgt F Dostál | Sgt K Mazurek | X9880 | KX-C | 540x4lb | Duty Carried Out |
| F/Sgt J Kalenský | Sgt K Kodeš | Z1105 | KX-R | 540x4lb | Duty Carried Out |

The Group would detail and brief a total of 112 crews, 95 Wellingtons, plus a further 26 Short Stirlings. The operation would be conducted as those employed against Paris and Essen. Lübeck would be attacked in three waves opening proceedings at 22:25hrs were hand-picked crews from Honington, Feltwell, Oakington and Marham, dropping bundles of flares. All 311 Squadron's Wellingtons had departed by 20:37hrs. Once airborne, they set a course for the North Sea, aiming to make landfall on Denmark's western coast before crossing southern Jutland. To the credit of the groundcrews, there were no early returns on this night, although Group did report nine for various reasons. Visibility was excellent along the entire route to the target. Both Hamburg and Kiel's defences were fully alert, intense flak and searchlight activity was reported. Such were the clear conditions, a smoke screen was observed at Kiel. Also active were a number of night fighters from NJG2 and NJG3 which were making their unwelcome presence felt. The first wave, the flare force, which included two crews of 9 Squadron, arrived in the target area to be greeted by clear skies, bright moonlight and just a handful of searchlights. The AOC had requested that the attack be made at the lowest possible altitude, depending on the scale of defences. The Czechs would make up the Group's final wave and were over the target area between 23:50hrs and 00:10hrs. Three crews, Sergeants K Pospíchal, F Bulis and Pilot Officer F Taiber bombed a few minutes before midnight on what they described as an inferno. Pilot Officer Taiber reported his incendiaries landing just south of the railway station, producing a red fire that was still visible 40 minutes after leaving the target. Four crews, Pilot Officer J Bala, Flight Sergeant J Doktor and Sergeants V Žežulka and F Dostál, bombed in quick succession after midnight, each crew reporting the town a sea of flames. Sergeant F Dostál estimated at least 40 large fires in the town. Pilot Officer J Bala was unfortunate, his Wellington T2962 KX-T was one of the few damaged by flak over Lübeck. Thankfully, the crew and engines were undamaged, but the Wellingtons hydraulic systems were ruptured. The last crews from 311 Squadron over the target were Flying Officer O Hořejší and Flight Sergeant J Kalenský at 00:10hrs, Hořejší reporting on his return, *'Observed a mass of red flames raging in the town, individual fires could not be counted'*. Flight Sergeant J Kalenský had this to say at debriefing and summed up the amazement of the crews at the inferno. *'Fires visible from Kiel, whole town in flames on arrival. Own bombs dropped northwest of harbour and railway station. Fires started and large bluish yellow fire in harbour before aircraft left. Fires seen at Westerhaven on return and smoke blown as far as the Kiel Canal.'*

All the crews landed back at East Wretham. Pilot Officer J Bala made a belly landing at 03:40hrs due to hydraulic failure. The aircraft would be repaired on site but then transferred to 21 O.T.U on April 24th. Flight Sergeant J Doktor and crew fell foul of flak and searchlights on the return flight. To escape, the Wellington was put into a steep dive. The manoeuvre worked but it put a tremendous amount of stress on the fabric skinned Wellington. On return to base it was discovered that sections of fabric were torn away presumably in the near vertical dive. There was a real buzz around the crew room on return, never had there been a raid of this intensity and destructive power, the crews already knew that it was a complete success. It would be revealed that approximately 190 acres of Lübeck's old town had been destroyed, primarily by fire, and this amounted to 30% of the city's built-up area. A total of 1,425 buildings had been destroyed, and almost two thousand others were seriously damaged. Only a matter of weeks under his leadership, Bomber Command had its first significant success implementing the new area bombing policy. Bomber Command and Harris had delivered.

On the 28th, Group Captain Alois Kubita submitted a detailed two-page document to the AOC clarifying the cause and reasons for the early returns on March 25th. The report also went on to explain in some detail the availability of the Czech crews and answer issues regarding sickness. This shows that there were other concerns raised by either Group Captain Ommaney or Wing Commander Batchelor DFC other than the Essen operation. On the same date, Group Captain Alois Kubita sent a letter to the C.I.G. in London, updating them on the ongoing situation and the displeasure and upset it had created. Lübeck was the last raid of March for the Czechs, and it was an outstanding success and was just the boost the squadron needed after a disappointing month. The squadron had flown on seven nights, flying a total of 64 sorties carrying aloft 174,400lb of bombs, not all of which reached their intended targets. Three experienced crews were lost, each nearing the end of their tours, all irreplaceable. Group H.Q stopped publishing the Group's serviceability rates in February, it was counterproductive. It was initially hoped that producing a Group pool would motivate each squadron, after all, squadron pride was at stake, but it had the opposite effect.

To date, 1942 had not been a good year for the Czechs, serviceability and reliability being a major issue. Almost every operation would witness early returns, the majority due to mechanical problems. The main reason was the unreliability of the ageing Wellington Mk.Ic's then on strength. They had given valiant service but too many were ready for replacement. When replacement aircraft did arrive, some had already seen considerable action with previous squadrons. The four Wellingtons supplied by 218 Squadron were a prime example, Wellington T2739 had been in service since 1940 with 149, 99 and 218 Squadron. These old 'Wimpey's' did not inspire confidence. The unreliability and poor performance of the aging Wellingtons had been the cause of a dip in morale from January onwards. This demonstrated itself in an alarming fashion on three occasions in March. The operations on the 8th, 9th and 25th were particularly bad. These failures were out of character of the Czechs, who were renowned for pressing home their attacks. Unlike the previous year, strong characters seemed to be lacking on the squadron. The likes of Šnajdr, Breitcetl, Šedivý, Fencl, Čapka and half a dozen more were the very heartbeat of the squadron, its inner strength. Their departure seemed to leave a noticeable void. They were big shoes to be fill, and in early 1942, it would appear that they hadn't been. The squadron, or specifically 1429 COTF had a quiet month. It lost Squadron Leader Murrey Peter-Smith to 57 Squadron March 16th, he would replace Wing Commander Mackenzie Southwell MiD, who arrived at Honington the same day to join 311 (Czech) Squadron.

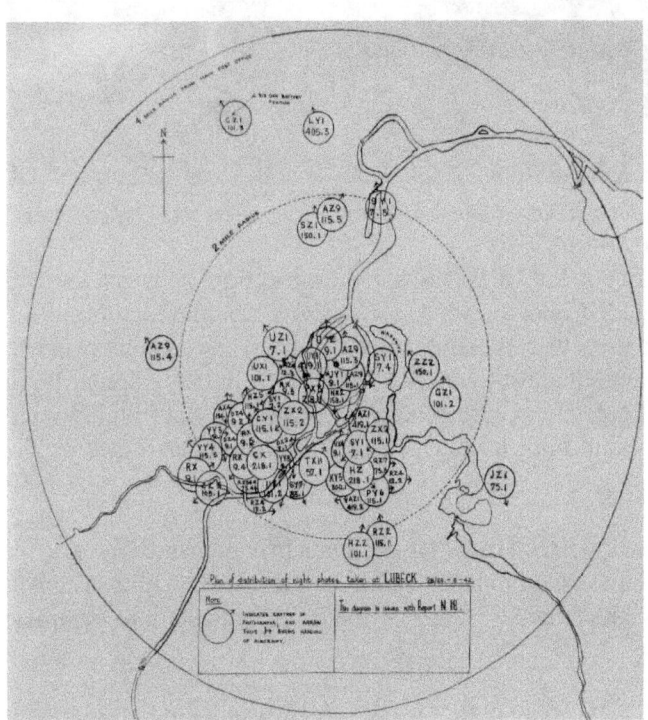 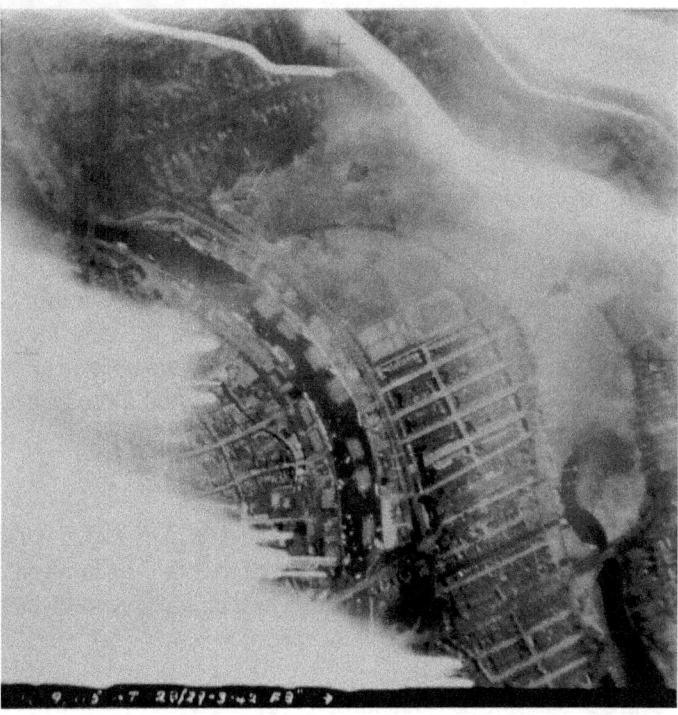

*Left: This Bomber Command HQ diagram graphically records the operation's accuracy and identifies the position of known aiming point photographs. Right: The clear conditions are seen in this 9 Squadron bombing photograph. Numerous features can be identified.*

# April 1942 : The Beginning of the End

The month started with a concerted push on training on both the squadron and Training Flight. Cross country flights were flown and both the COTF Advanced crews and Initial crews were active in between occasional rain showers. Thankfully, the squadron was not involved in the disastrous Operation *"Lineshoot"* an ill-conceived operation at low-level on the railway installations at Hanau. 3 Group detailed and briefed 36 crews. Twelve Wellingtons failed to return from the primary target, seven from 214 Squadron and five from 57 Squadron.

The recent accusations had left the squadron unhappy, the squadron's reputation and Czech honour had been called into question. Perhaps sensing this, Group Captain Alois Kubita telephoned the C.I.G. on April 1st to arrange a visit to the squadron by President Dr. Beneš. There was a deep-rooted mistrust among Czech servicemen about Czechoslovak high-ranking officers, but Dr Beneš was the exception, he was respected and popular. His visit, when it happened, would be a much-needed boost to morale and show his appreciation of the squadron's efforts. It could also help smooth the way for a change in command. In late March, discussions between Group Captain Kubita and C.I.G. had already taken place regarding the posting in mid-April of the hugely popular Wing Commander Ocelka DFC. His replacement had already been agreed, Squadron Leader J Šnajdr, another highly rated and respected officer would take over. This transition, which would not be easy, would be eased if endorsed by the President.

*A posed photograph showing a 'made-up crew'. Identified are, Valach, Plzák, Bernát, Konštacký, Horák, Filler.*

From his office, at Harraton House, the AOC, AVM J.E.A Baldwin CB, OBE, DSO penned a private handwritten letter dated April 1st. The letter was addressed to R.A.F Honington's Station Commander. Baldwin wrote that he fully accepted the explanations given by Group Captain Alois Kubita in his letter dated March 28th and declared he had complete confidence in the Czechs.

During this period, the squadron produced a list of its own. Perhaps prompted by Group Captain Kubita, the list outlined a number of problems. Number one on the list was the reliability of the ageing Wellington Mk.Ic they were operating, many of which had already seen considerable service. There was also a chronic lack of trained ground crews and a shortage of trained aircrew. Another issue was the basic facilities. Both operational and domestic sites at R.A.F East Wretham, had hardly improved since the squadron arrived. There were no hangars, some ground crews were still living in tents, and all servicing had to be carried out in the open, exposed to the elements. Any major servicing had to be undertaken at R.A.F Honington, which meant travelling back and forth, with ground crews already stretched. Dividing the already limited resources between two stations was proving both difficult and demoralising. Hopes were high that something would at last be done to address the problems. It was a return to Paris on the 2nd, or more specifically the Matford Works at Poissy located in the north western suburbs of Paris. The Poissy plant was commissioned by Ford France in 1937 and opened in 1940 a few weeks before the German invasion. 4 Group had visited the plant the previous night without much success. It was now up to 1 and 3 Groups to finish off the job. 3 Group detailed and briefed 13

Wellington crews, of which 311 would provide three who would be joined by ten Stirling drawn from 15 and 218 (Gold Coast) Squadron.

### April 2$^{nd}$ 1942 : Z.166 – Matford Works – Poissy

| Pilot | 2$^{nd}$ Pilot | Serial | Code | Bomb Load | Result |
|---|---|---|---|---|---|
| P/O F Taiber | Sgt J Štark | DV507 | KX-W | 2x1000lb+2x500lb+1x250lb / Leaflets | Duty Carried Out |
| Sgt F Bulis | Sgt J Bláha | X9871 | KX-A | 2x1000lb+2x500lb+1x250lb / Leaflets | Duty Carried Out |
| Sgt H Dostál | Sgt J Lenc | X9760 | KX-O | 2x1000lb+2x500lb+1x250lb / Leaflets | Duty Carried Out |

The selected crews were airborne by 20:02hrs. Thankfully conditions were excellent, with no clouds and moderate visibility en-route and over the target area. The Wellingtons of 1 Group would be the first over the target experiencing intense light flak. The Czechs were over the target between 22:22hrs and 22:40hrs. Upon arrival, five good fires were observed in the northern part of the complex. Bombing between 1,500ft and 3,800ft, the crews had an excellent view of the bombing and the plant. Pilot Officer F Taiber claimed his bombs burst in the western quarter of the Matford factory. Both Sergeants Bulis and Dostál carried out shallow dive-bombing attacks, the latter claiming direct hits on the factory. Sergeant Bulis dived on the target from 4,000ft dropping his bombs in a salvo on the northwestern edge of the factory. On pulling out of his dive at just 1,500 feet he was engaged by light flak which thankfully was inaccurate. Three significant fires were reported at 22:43hrs in the southwest corner and north of the factory complex, with smoke enveloping the whole area. Returning crews reported one aircraft on fire while on its bomb run and believed it crashed 1 mile west of the target. The last of the Group's aircraft bombed at 23:05hrs, leaving the whole factory complex hidden by smoke. The raid was considered a success.

On the 3$^{rd}$, AVM J.E.A Baldwin CB, OBE DSO arrived at R.A.F East Wretham to decorate Flying Officer Karel Bečvář with the DFC for his service with the squadron as navigator. The squadron personnel were paraded for the ceremony. On conclusion, the AOC inspected the canteen, washrooms, and other facilities, no doubt due to the squadrons raised concerns. During the visit, the AOC discussed, amongst other things, the squadron's immediate future with Group Captain Kubita. A number of topics were covered. The following day, Group Captain Kubita wrote to C.I.G about the visit and his talk with Baldwin, he also wrote privately to AVM Janoušek KCB. Two points in particular are worth noting and may have influenced the squadrons future. Below is an extract from the letter from Group Captain Kubita to AVM Janoušek KCB;

*'1) Before leaving, he (AOC) took me aside and asked what we would say about the disbandment of the squadron. He claimed it was suggested to him by Bomber Command Head Quarters that each flight would be assigned to one British squadron. I replied that there is no reason for that and that I doubted we would agree to that. To that he said that he would regret it himself, when the squadron has been existing for almost two years, but that we still have personnel shortages. He asked me to help Šnajdr to organize the service like at a British squadron. It is said that our people must realize that they are in England and that they must therefore also behave like Englishmen. I have the impression that the reason for this remark may be political influences and also denigration by local officials. After returning from Group H.Q last week, the station commander was in an even worse mood than usual. He wasn't even in the operating room and he only asked to be informed about the situation. The next day he called Batchelor and the intelligence officer, who then told me that he should go on leave. I guess he got scolded for misreporting the operation of 311 Squadron on March 25$^{th}$ and now they are finding fault where they can. Serviceability is now satisfactory but he still has some comments every day. Otherwise, he had no reason to complain.*

*2) The situation will be critical in the near future, when 4 to 5 crews will finish their tour, it would be very soon due to the present high speed of operations and the gap in training will disband the squadron automatically.'* [72]

---

[72] *VÚA Praha, ČSL VB, sign. 107/CI-2ab/7/27, Korespondence tajná a důvěrná, čj. 1601–1650, 1942.*

It is easy to conclude from the above that the apparent issues with 311 (Czech) Squadron had reached H.Q Bomber Command and the new Commander-in-Chief. Also that a frosty relationship with Honington Station Commander had developed since March 25th. Perhaps exacerbated by Wing Commander Batchelor and East Wretham Adjutant, Flight Lieutenant Herbert Maurice Jaffe.

A raid on Stuttgart was planned for the 3rd, but this was cancelled at 18:20hrs due to unpredictable weather. On the 5th, Bomber Command was out in force with Cologne city centre the primary target for 144 medium bombers and 40 heavies. An additional 29 crews would be given the Humboldt Deutz Motoren AG Werkes (GN.3777) as the primary target. 311 Squadron would detail ten crews for a return visit to the Cathedral city.

April 5th 1942 : Cologne 'A'

| Pilot | 2nd Pilot | Serial | Code | Bomb Load | Result |
|---|---|---|---|---|---|
| F/O O Hořejší | Sgt V Pára | Z1111 | KX-N | 2x1000lb+1x500lb+1x250lb | Duty Carried Out |
| W/O F Bulis | Sgt J Bláha | Z1155 | KX-F | 2x1000lb+1x500lb+1x250lb | Duty Carried Out |
| P/O J Bala | F/Sgt J Doktor | Z1090 | KX-Q | 2x1000lb+1x500lb+1x250lb | Duty Carried Out |
| Sgt K Pospíchal | Sgt J Hadrávek | Z1098 | KX-U | 2x1000lb+1x500lb+1x250lb | Duty Carried Out |
| Sgt V Žežulka | Sgt O Špaček | X9871 | KX-A | 2x1000lb+1x500lb+1x250lb | Duty Carried Out |
| Sgt F Naxera | Sgt K Mazurek | R1497 | KX-D | 2x1000lb+1x500lb+1x250lb | Duty Carried Out |
| Sgt V Šponar | F/Lt J Strégl | R1532 | KX-B | 2x1000lb+1x500lb+1x250lb | Duty Carried Out |
| Sgt H Dostál | Sgt O Havlík | X9760 | KX-O | 2x1000lb+1x500lb+1x250lb | Duty Carried Out |
| P/O F Taiber | Sgt J Štark | DV507 | KX-W | 2x1000lb+1x500lb+1x250lb | Duty Carried Out |
| F/Sgt J Kalenský | Sgt F Dostál | X9787 | KX-G | 2x1000lb+1x500lb+1x250lb | Early Return |

The squadron had not visited Cologne in over six months. Nothing had changed apart from the increase in flak and searchlights. The attacking force would be divided into a Flare, Incendiary and Striking Force. 3 Group would provide the bulk of the bombers over Cologne with 110[73] crews detailed, scheduled to be over the city from 01:30hrs. The Czechs of 311 Squadron would bomb during the latter stages part of the Striking Force. The weather conditions over the North Sea did not bode well, with cloud cover between 8/10th to 10/10th. Fortunately, on reaching the German border and to the relief of the Flare Force, the cloud suddenly started thinning with large sways of ground visible. One crew did not reach the German frontier. Flight Sergeant J Kalenský and crew had just passed south of Charleville-Mézières (49.45N/04.45E) at 10,000ft when a Me110 tore in for an attack. The rear gunner returned fire, and a vicious battle ensued. Four attacks were made, but each time repulsed by accurate fire from the rear turret. Damage was sustained, especially to the wiring which resulted in only one 1000-pounder being jettisoned. Finally, to the relief of the Czechs the Me110 sheared off, the crew claiming hits on the fighter. They landed safely back at RAF East Wretham.

By the time the first crews of 311 Squadron were over Cologne at 03:00hrs, conditions had deteriorated, and smoke was drifting over the city. This caused some problems with crews unable to positively identify the aiming point. Flying Officer O Hořejší identified Bonn and used the River Rhine, clearly seen shimmering in the bright moonlight, to reach Cologne, where he dropped his bombs from 16,000ft on a previously pinpointed concentration of flak. Sergeant V Šponar and crew had no such problems. They reported, *'River and town clearly seen, Bombs burst 1000 yards north of Aiming Point. Two good red fires in west part of town when arriving'*. The last four crews were over the target between 03:35hrs and 03:45hrs and gave conflicting accounts of the raid. Sergeant F Naxera, aboard Wellington R1497 KX-D, only managed to identify the general target area by the explosions of his own bombs, which he believed landed close to the railway tracks south of the town. Five minutes later at 03:40hrs it was the turn of the crew of Sergeant V Žežulka. They appeared more confident stating, *'Bomb bursts 1000 yards south of town centre in Marshalling Yards area'*. Two crews bombed simultaneously at 03:45hrs. The recently promoted Warrant Officer F Bulis was unsure where his bombs fell in the town but reported fires seen below the cloud. Equally uncertain was Sergeant K Pospíchal, who bombed from 15,000ft having been unable to identify the target owing to cloud cover. They bombed on flak positions. All the crews

---

[73] *Numbers differ depending on what document you use, between 110 and 128 are known to have been detailed.*

*Pilot Sergeant František Bulis and Air Gunner of his crew Sergeant Ján Timko seen by a propeller of a Wellington which sports 13 mission symbols. It is believed it is Bulis' personal Wellington Mk.IC Z1155/KX-F. If so, the picture was taken in early March 1942. (Marie Lambourne)*

returned safely from what looked like a successful operation. However, there were some concerns raised about the effectiveness of the raid. R.A.F Feltwell, Marham and Oakington were convinced that the bombing was scattered. Contrary to the doubts raised to Group H.Q, R.A.F Honington, Mildenhall, Stradishall, and Wyton were confident that their bombs had landed in the city. A local report subsequently proved that the doubters were correct. Scattered bombing on both sides of the Rhine was reported, with one industrial building hit and ninety houses destroyed.

The AOC obviously still had a point to prove with Essen. On the 6th, the squadron was ordered to prepare six crews, on what would turn out to be a disastrous night for the Czechs and 3 Group. The weather on the 6th started out promising but towards the afternoon heavy thunderstorms were making their presents felt. A cross-country exercise was undertaken by 1429 COTF in the morning and to the dismay of the whole squadron news began to filter in that a Wellington, possibly from the Czech Flight had crashed in Wales killing the entire crew. The tragic news was confirmed later that day when it was learnt that Sergeant Aldis Keda had descended through cloud and crashed into the hillside at 13:12hrs near Pystell Gwyn Llanymawddwy, Wales. The bodies of the crew were recovered from this inhospitable spot and initially taken to R.A.F Llandbedr. On April 10th, they were transported by train to Honington, where on April 13th the six man crew were buried at East Wretham (St. Ethelbert) Churchyard. Although 1429 COTF were independent of 311 (Czech) Squadron, both were still very much connected. Most of the flying instructors were season veterans from 311 and that close bond was still evident.

## Vickers Wellington Mk.Ic P9299 KX-A No.1429 COTF

| Manufacturer | Vickers Armstrong (Weybridge) | |
|---|---|---|
| Contract | 549268/36 | |
| Taken on Charge | 16/06/1941 via 218 Squadron (Czech Training Flight) | |
| Cat FA/E | 07/04/1942 | |
| Struck Of Charge | 07/04/1942 | |
| Total Flying Hours | - | |
| Take-Off Time | 11:15hrs | |
| Bomb Load | - | |
| | | |
| | **CREW** | |
| Captain | Sergeant Aldis **Keda** 788087 RAFVR. Age 27. | Grave 28/42 |
| Second Pilot | Sergeant Rudolf **Vokurka** 787584 RAFVR. Age 25. | Grave 23/42 |
| Navigator | Pilot Officer Jan **Štefek** 100023 RAFVR. Age 27. | Grave 25/42 |
| Wireless Operator | Sergeant Rudolf **Grimm** 787831 RAFVR. Age 30. | Grave 27/42 |
| Front Gunner | Sergeant Jan **Stanovský** 787880 RAFVR. Age 27. | Grave 26/42 |
| Rear Gunner | Sergeant Jindřich **Hořínek** 787929 RAFVR. Age 30. | Grave 24/42 |
| | | |
| Posting History | Unknown | |
| Operations Flown | Nil | |
| Buried | EAST WRETHAM (ST. ETHELBERT) CHURCHYARD | |

## April 6th 1942 : 'Stoat 'B' - Essen

| Pilot | 2nd Pilot | Serial | Code | Bomb Load | Result |
|---|---|---|---|---|---|
| P/O J Bala | F/Sgt J Doktor | Z1090 | KX-Q | 2x1000lb+2x500lb | Jettisoned |
| Sgt K Pospíchal | Sgt J Hadrávek | Z1098 | KX-U | 2x1000lb+2x500lb | Jettisoned |
| Sgt V Žežulka | Sgt O Špaček | X9871 | KX-A | 2x1000lb+2x500lb | Jettisoned |
| Sgt H Dostál | Sgt O Havlík | X9745 | KX-S | 2x1000lb+2x500lb | Jettisoned |
| F/O O Hořejší | Sgt J Štark | DV507 | KX-W | 2x1000lb+2x500lb | Jettisoned |
| W/O F Bulis | Sgt J Bláha | Z1155 | KX-F | 2x1000lb+2x500lb | Jettisoned |

It was a late take-off for the squadron, with the crews departing a few minutes after midnight. Once airborne, they headed towards the departure point over Orfordness. There was already some doubt about the conditions before take-off, but little did the crews of 3 Group realise what they were flying into. A total of 77 Wellingtons and Stirlings were detailed and briefed, of which only 63 managed to get airborne. Over the North Sea, the weather took an ominous turn for the worse. On reaching the Dutch coast, the crews were confronted with 10/10th ice-bearing clouds and electrical storms. Over Holland, crew after crew either jettisoned and turned for home or identified targets of opportunity, icing and storms, making it almost impossible to continue. The first of the squadron crews were back in the circuit within 71 minutes. Pilot Officer Bala jettisoned his bombs 10 miles off Orfordness. Twenty minutes later, Sergeant K Pospíchal landed, having jettisoned his load at 51.23N/02.04E. By 02:25hrs all six crews were back at East Wretham, all having jettisoned. They were not alone. 3 Group reported 43 aborts due to the weather and a further five due to mechanical issues. The Czechs were not alone in all their aircraft returning. Both 7 and 101 Squadrons had their entire force return. Predictably, the operation was a failure, local sources confirming that only a few bombs had found their way into the city, damage was negligible.

On the 8th, 3 H.Q ordered 13 of its squadrons to begin preparations for a significant raid directed against Hamburg that night. 112 crews were involved, 88 Wellingtons, plus 24 Stirlings, with the Czechs of 311 Squadron providing nine.

### April 8th 1942 : 'Dace' – Hamburg

| Pilot | 2nd Pilot | Serial | Code | Bomb Load | Result |
|---|---|---|---|---|---|
| Sgt K Pospíchal | Sgt J Hadrávek | X9745 | KX-S | 540x4lb | Duty Carried Out |
| Sgt H Dostál | Sgt O Havlík | Z8838 | KX-Z | 540x4lb | Alternative Target |
| F/O J Bala | Sgt J Štark | DV507 | KX-W | 540x4lb | Early Return |
| Sgt V Žežulka | Sgt O Špaček | X9880 | KX-C | 540x4lb | Duty Carried Out |
| F/Sgt J Kalenský | Sgt K Kodeš | Z1105 | KX-R | 540x4lb | Early Return |
| W/O F Bulis | Sgt J Bláha | Z1155 | KX-F | 540x4lb | Duty Carried Out |
| W/O F Naxera | Sgt F Dostál | R1497 | KX-D | 2x1000lb | Early Return |
| Sgt V Šponar | F/Lt J Štrégl | R1532 | KX-B | 2x1000lb | Early Return |
| F/O O Hořejší | Sgt J Štark | Z1111 | KX-N | 2x1000lb | Duty Carried Out |

The majority of the Group would carry incendiaries to Hamburg, with the aiming point being the centre of Hamburg's old town, the Altstadt. Similar to Lübeck, the old town was predominantly an area of closely packed warehouses, narrow streets and timber buildings, the AOC's intentions were clear. It was hoped that the weather would be better than Essen, surely the meteorologists could not make another balls-up on the scale they did two days prior. The squadron began taking off at 21:21hrs with the Met officer's announcement that Hamburg would be clear of clouds still fresh in their minds. Having passed over Cromer, the bombers headed out over the North Sea to be confronted with gigantic columns of cumulonimbus cloud rising to 20,000ft and containing violent electrical storms and ice-ladened clouds which continued ominously over the Schleswig-Holstein coast and on to Hamburg. The first of the night's aborts landed at 22:15hrs. Flying Officer J Bala jettisoned his incendiaries into the sea 'safe' and was back at his dispersal within 55 minutes. At 23:31hrs, Flight Sergeant J Kalenský requested permission to land. Sergeant V Šponar and Warrant Officer F Naxera followed him in quick succession, each having encountered severe icing and violent electrical storms.

The crews of 3 Group were scheduled to bomb between 00:08hrs and 01:57hrs. The squadron was to lose yet another crew. Sergeant H Dostál bombed the island of Nordeney at 23:42hrs, having observed lights on the ground and flak bursts. Those crews that battled the weather conditions were met with solid clouds over Hamburg despite the Met officer's promise of clear skies. Sergeant V Žežulka skippered the first crew over Hamburg at 00:08hrs. They bombed from the dangerously low altitude of 5,000ft, reporting that Hamburg was hidden below a blanket of clouds. Annoyingly, icing prevented some of their incendiaries from dropping. The flak defences were considered light, and due to the cloud, searchlight interference was practically nil. At 00:35hrs, it was the turn of Flying Officer O Hořejší and crew, who

*The lucky ones. Four members of Bulis' crew posing at the side window with covered machine gun mounting of Wellington Mk.IC Z1155/KX-F at East Wretham in March 1942. L-R: Rear Gunner Sergeant Ján Timko, Skipper Sergeant František Bulis, Navigator Flight Lieutenant Miroslav Cígler and Wireless Operator Sergeant Jiří Osolsobě. (Milan Šindler)*

bombed from 18,000ft, dropping their two 1000-pounders on the faint glow of fires. Warrant Officer F Bulis arrived within minutes, and they were confronted with scattered fires over a vast area much larger than the target. They dropped their incendiaries in one stick from 16,000ft. It was not until 00:53hrs that the crew of Sergeant K Pospíchal were over the target. They had battled their way to Hamburg only to be disappointed. Like most of the crews, they bombed on the glow of fires below and turned for home. The raid was another failure made only more palatable by the exceptionally low losses. Little if any damage was inflicted on Hamburg. Bremen reported that incendiaries landed on the Vulkan shipyards, damaging buildings and U-boats under construction. Thirty-two bombers from 3 Group had returned early, 16 due to icing or storms and 16 with mechanical defects, 25% of the Group's effort. The unpopular Wing Commander Batchelor DFC stepped down from his role as British Liaison Officer and East Wretham Station Commander on the 10th, he was replaced by Wing Commander Mackenzie Southwell MiD, who had arrived in mid-March to learn the ropes under the guidance of Ken Batchelor. It had been a baptism of fire since his arrival, embroiled in the agitation between 311 Squadron, Wing Commander Batchelor, Group Captain Ommaney and Group H.Q. It was no wonder he requested posting back to 57 Squadron. A pre-war veteran, he arrived on the squadron with extensive operational and leadership experience behind him. He had trained initially in Egypt in 1933 and served with distinction as a flight commander with 75(NZ) Squadron before being rewarded for his service with the command of 57 Squadron between May 1941 and March 1942.

Essen was the objective for the Group on April 10th, with Cologne an alternative target. 3 Group would detail and brief 102 Wellingtons and Stirling crews with 311 Squadron offering nine. The now familiar three-phase attack would again be used. The TR1335 equipped Flare Force, carrying 2,259lb of flares would begin dropping from 23:45hrs onwards.

### April 10th 1942 : 'Stoat B' – Essen

| Pilot | 2nd Pilot | Serial | Code | Bomb Load | Result |
|---|---|---|---|---|---|
| F/Sgt J Doktor | Sgt O Havlík | X9745 | KX-S | 1x1000lb+5x500lb | Duty Carried Out |
| F/O J Bala | Sgt V Výcha | Z1090 | KX-Q | 1x1000lb+5x500lb | Jettisoned |
| Sgt K Pospíchal | Sgt J Hadrávek | Z1098 | KX-U | 1x1000lb+5x500lb | Duty Carried Out |
| Sgt V Žežulka | Sgt O Špaček | X9880 | KX-C | 1x1000lb+5x500lb | Alternative target |
| Sgt V Šponar | F/Lt J Štrégl | R1532 | KX-B | 1x1000lb+5x500lb | Duty Carried Out |
| P/O F Taiber | Sgt J Stark | DV507 | KX-W | 1x1000lb+5x500lb | Duty Carried Out |
| F/Sgt J Kalenský | Sgt K Kodeš | Z8838 | KX-Z | 1x1000lb+5x500lb | MISSING |
| F/O O Hořejší | Sgt V Pára | Z1111 | KX-N | 1x1000lb+5x500lb | Jettisoned |
| W/O F Bulis | Sgt J Bláha | Z1155 | KX-F | 1x1000lb+5x500lb | Jettisoned |

On crossing the English coast above Southwold, the crews slowly climbed to their operational height before reaching Knokke on the Belgian coast. It was while over the sea that the crew of Warrant Officer F Bulis were amazed to see the unmistakeable shape of a He111 on its way to bomb a target in England.[74] The first of yet another frustrating night of early returns began shortly after take-off. Engine trouble aboard Wellington Z1111 KX-N resulted in Flying Officer O Hořejší jettisoning all but one bomb, which hung up at position 52.23N/02.30E. An electrical fire in the wing which filled the fuselage with acrid smoke and starboard engine problems aboard Wellington X9880 KX-C led to Sergeant V Žežulka dropping his bombs on flak positions north of Ijmuiden at 23:19hrs. At 00:08hrs, a message was received back at R.A.F East Wretham from Wellington Z8838 KX-Z stating that they had developed engine trouble and were returning to base. The next crew to return to East Wretham was Flying Officer Bala at 01:10hrs. They also reported engine trouble and abnormal petrol consumption aboard their Wellington. Flak and searchlights very nearly claimed the crew of Warrant Officer F Bulis. They had to carry out violent evasive action to escape, but in doing so, they jettisoned their bombs at 00:26hrs northwest of Gladbach on searchlight and flak positions. Once clear of danger, they turned for home. It had not been a promising start to the operation, and Essen had not even been reached. Sergeant V Šponar was compelled to jettison part of his bombload near Ijmuiden to gain sufficient height when a vibrating starboard engine came close to claiming yet another early return. Showing remarkable courage, the crew selected to continue the operation, which was a bold decision. Having run the gauntlet of flak and fighters, the first of the squadron's crews were over Essen at 00:18hrs. Opposition was intense, both heavy and light flak were especially vicious. Crews reported light flak reaching up to 20,000ft, well within the reach of the Wellingtons. Searchlights were also active and working in groups. Also evident were balloons seen flying between 9,000 and 10,000ft. The pre-raid forecast of clear skies was again wildly inaccurate. Essen lay beneath a broken layer of cloud. At 00:18hrs, Flight Sergeant J Doktor and crew dropped their bombs on fires on the eastern edges of Essen from 15,000ft. Two minutes later, Pilot Officer F Taiber identified the target by the light of numerous flares. He dropped his bombs on the northwestern suburbs. Sergeant K Pospíchal found himself being subjected to intense flak as he made his bomb run from 14,000ft at 00:25hrs. Such was the intensity of the flak, they were unsure exactly where their bombs dropped but were just thankful to have escaped unmolested. The last Czech crew over the target at 00:55hrs was plucky Sergeant V Šponar. They arrived over Essen at just 12,500ft with the engine still underperforming. The crew's navigator, Flight Lieutenant Jaromír Foretník, dropped their remaining 1000-pounder and 2x500lb on the western edge of the target. All the crews had to do now was get back to East Wretham. At 03:40hrs what turned out to be the last crew touched down. Sergeant K Pospíchal had been airborne for 5 hours 35 minutes. One crew was still unaccounted for. Flight Sergeant J Kalenský had last been heard at 00:08hrs. Even with a defective engine, they should have been home. The now standard overdue aircraft procedure was implemented, but sadly, it would prove fruitless. The crew of Wellington Z8838 KX-W were intercepted and shot down

---

[74] *This is probably a misidentification. There were no He111 based in Belgium / Holland at this time. The only Luftwaffe unit flying the He111 in the West at the time of were III/KG.40, who were based in the south of France. The aircraft was probably a Junker Ju88.*

at 00:29hrs by Oberleutnant Helmut Lent of the Stab II./NJG, crashing near Molenweg, Middenmeer, Noord-Holland killing all onboard. It was Oberleutnant Helmut Lent's 35th victory. The encounter may have been witnessed by a 5 Group crew who reported on return, *'Enkhuizen 00:30hrs, 12,000ft. A small red glow which was probably an aircraft at about 3,000ft. Also red tracer being fired into the glow a few seconds later. The presumed aircraft was seen to hit the ground and burn.'* Enkhuizen is only six miles southeast of the crash site.

## Vickers Wellington Mk.IC Z8838 KX-Z

| Manufacturer | Vickers Armstrong | |
|---|---|---|
| Contract | B.71441/40 | |
| Taken on Charge | 09/12/1941 via 149 Squadron. | |
| Cat E Missing | 10/04/1942 | |
| Struck Of Charge | N/K | |
| Total Flying Hours | N/K | |
| Take-Off Time | 22:01hrs | |
| Bomb Load | 1x1000lb+5x500lb | |
| | **CREW** | |
| Captain | Flight Sergeant Josef **Kalenský** 787177 RAFVR. Age 25. | Coll. grave 31. A. 5-6. |
| Second Pilot | Sergeant Karel **Kodeš** 787261 RAFVR. Age 21. | Coll. grave 31. A. 5-6. |
| Navigator | Flying Officer Karel **Rychnovský** 82632 RAFVR. Age 23. | Coll. grave 31. A. 5-6. |
| Wireless Operator | Sergeant Josef **Politzer** 787875 RAFVR. Age 31. | Grave 31. A. 11 |
| Front Gunner | Sergeant Jan **Peprníček** 787872 RAFVR. Age 22. | Coll. grave 31. A. 5-6. |
| Rear Gunner | Flight Sergeant Josef **Hrdina** 787310 RAFVR. Age 30. | Coll. grave 31. A. 5-6. |
| | | |
| Posting History | F/Sgt J Kalenský posted via 11 O.T.U, 11/10/1941. Sgt K Kodeš posted via 2 S.F.T.S, 06/08/1941. | |
| Operations Flown | F/Sgt J Kalenský : 20 / Sgt K Kodeš : 8 | |
| Buried | BERGEN-OP-ZOOM WAR CEMETERY | |

*Left : Flight Sergeant Josef Kalenský. Right His 2nd pilot, Sergeant Karel Kodeš. (J Costin)*

The bodies of the crew were eventually recovered by the Germans, but only Sergeant Josef Politzer was positively identified. It is reported that his body was found away from the main wreckage. He was buried with full military honours on April 24th 1942, in Huisduinen General Cemetery (Den-Helder) Grave 121. The remaining crew were buried as 'unknowns' on the same day, two in Grave 122 and three in Grave 136. Post-war, in an effort to identify the 'unknowns'. The bodies were exhumed to establish positive identification. Unfortunately, the severity of the crash and ensuing years made this impossible even for the dedicated teams from the M.R.E Units. The bodies were taken and reburied in the Bergen op Zoom War Cemetery in 1947. In 2004, the Royal Netherland Air Force while salvaging the crash site of Wellington Z8838 found *'a fair amount of human remains, uniforms and equipment'* still in the wreckage. The remains of the crew were placed in a small coffin and re-interned into the collective grave at Bergen op Zoom during a low-key ceremony attended by members of the Czech Ministry of Defence on April 11$^{th}$ 2005, the 63rd anniversary of the crew's tragic death. In October 2021, a memorial pole was erected near the crash site in honour of the crew. Wreaths and flowers were laid by Mark Versteeg, deputy mayor of the Municipality of Hollands Kroon, Mr Čeněk Hajný from the Czech Embassy, Mr. Jozef Kušlita and Mrs Silvia Swan from the Slovak Embassy, the Veterans, the Wieringermeer Historical Society and Willy de Bruin.

The raid was yet another disappointment, local sources in Essen confirming the operation to have been another miserable failure, which destroyed only twelve houses in Essen and caused no industrial damage. The failure of four crews to reach the target brought yet more unwanted attention to the squadron at a time when it was already under the microscope. Of the four early returns, three Wellingtons had seen previous service on other squadrons, Z1090 with 101, 99 and 156 Squadron, X9880 had arrived via 149, then 156 Squadron, finally Z8838, this Wellington had two spells with 149 and a period with 40 Squadron. There was some justification in the squadrons complaints they were getting cast-offs. On the 12$^{th}$, orders were received across all four of Bomber Command's Groups to prepare for a return to Essen that night. 3 Group would offer 124 crews, of which 31 would be Short Stirlings. The AOC was anxious that as many aircraft as possible should carry cameras. He had been told that the *'weather was especially suitable for night photography'* by the less than dependable meteorologists at command H.Q. At R.A.F East Wretham the Czechs prepared seven Wellingtons. Little did the squadron know that the operation would turn out to be a disaster on par with Essen on March 25$^{th}$.

*The memorial to the crew of Wellington Z8838 KX-Z. (Photo via Stichting Herdenkingspalen Hollands Kroon)*

### April 12$^{th}$ 1942 : 'Stoat B' – Essen

| Pilot | 2$^{nd}$ Pilot | Serial | Code | Bomb Load | Result |
|---|---|---|---|---|---|
| F/O J Bala | Sgt V Výcha | X9745 | KX-S | 1x1000lb+4x500lb | Alternative target |
| Sgt V Šponar | Sgt J Hadrávek | R1532 | KX-B | 1x1000lb+4x500lb | Jettisoned |
| Sgt V Žežulka | Sgt O Špaček | X9880 | KX-C | 1x1000lb+4x500lb | Jettisoned |
| Sgt H Dostál | Sgt J Lenc | X9760 | KX-O | 1x1000lb+4x500lb | Duty Carried Out |
| P/O F Taiber | Sgt J Štark | DV507 | KX-W | 1x1000lb+4x500lb | Alternative target |
| Sgt V Pára | Sgt O Havlík | Z1105 | KX-R | 1x1000lb+4x500lb | Jettisoned |
| W/O F Bulis | F/Lt O Hořejší | Z1111 | KX-N | 1x1000lb+4x500lb | Jettisoned |

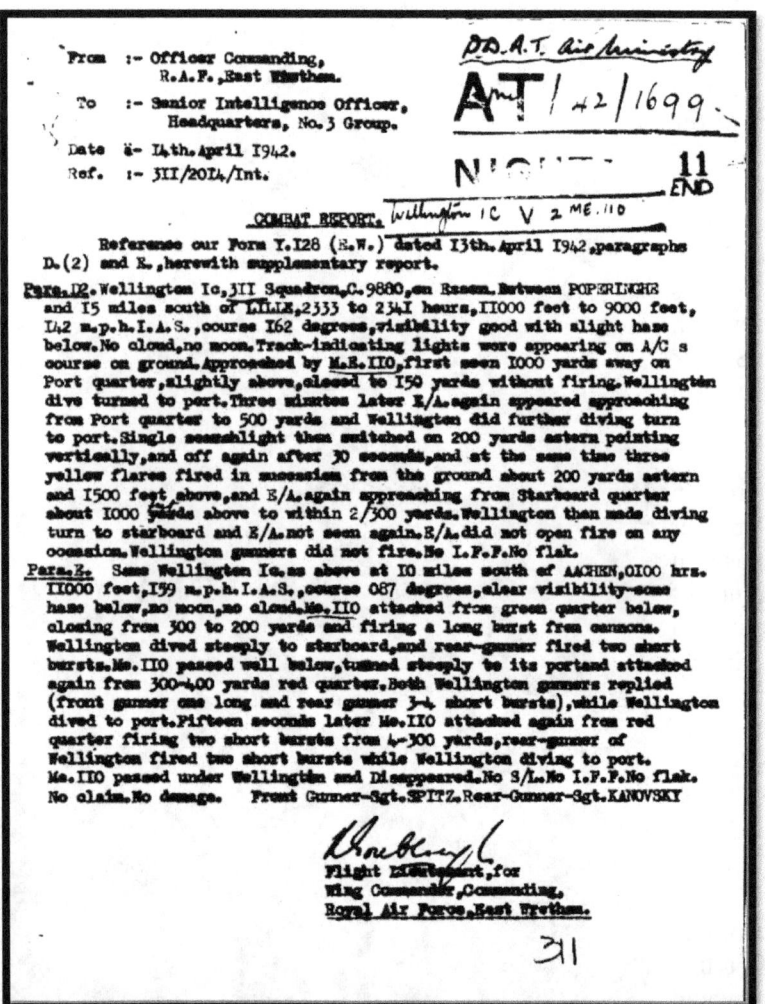

There was no indication as the crews trundled down East Wretham's flarepath of the problems about to unfold for the squadron. Once safely aloft, the seaside town of Orfordness passed below as the crews struggled for altitude. Only five aircraft failed to take-off throughout the Group, a low number even for 3 Group. All seemed well until, after nearly 3 hours, the first of the night's six early returns requested permission to land. Warrant Officer F Bulis had reached mid-channel at position 51.20N/02.05E before he jettisoned his bombload due to an overheating engine and an inability to climb.[75] Within nine minutes, it was the turn of Sergeant V Šponar at the controls of Wellington R1532 KX-B. The crew had jettisoned their bombs 'safe' in the sea when unable to climb to altitude. Only 48 hours prior, the crew had staunchly pressed on to Essen when R1532 suffered engine vibration. The remaining crews found the route relatively cloud-free over northern France and Belgium, but worryingly, the first signs of enemy fighter activity were observed. At 23:33hrs, Sergeant V Žežulka had the first of two clashes with Me110s. While at 9,000ft over Poperinge, West Flanders, the crew had an inconclusive encounter with a Me110. Neither crew managed to open fire or gain the advantage. The crew, having lost the fighter continued to Essen. Having got back on track and gained altitude, the crew were 10 miles south of Aachen when, at 01:00hrs, they were again attacked. This time, the Me110 pilot appeared more aggressive, carrying out a series of attacks, both front and rear gunners aboard X9880 KX-C engaged the fighter with accurate

*The Combat report submitted by the crew of Sergeant V Žežulka. Below: A Messerschmitt Bf 110 of Gruppe Stab II/NG1 Spring 1942.*

return fire. During this encounter, the bombs were jettisoned, and soon after, the crew turned for home, landing at 02:45hrs. The fourth abort of the night landed at 02:57hrs. Sergeant V Pára and crew reported that they encountered icing near the Belgian city of Durbuy, some 45 miles south of Liege and almost 130 miles southwest of Essen. The crew jettisoned their bombs at position 51.10N / 01.40E from 7000ft at 02:10hrs in the English Channel. Of the 20 crews that aborted this night, this was one of only two incidents involving icing. The port inner engine aboard Wellington DV507 KX-W began cutting en route to Essen. Unwilling to risk his crew flying into the Ruhr with a misfiring engine, Pilot Officer F Taiber decided to bomb the area between the town of Eukirchen and position 50.40N/06.50E from 16,000ft at 00:55hrs. They landed back at R.A.F East Wretham at 03:01hrs. Another crew who bombed Eukirchen was flying

---

[75] Conflicting reports, one records only 1000-pounder jettisoned.

Officer J Bala, who bombed at 01:17hrs. They were the second crew reporting issues due to icing. They landed back at base at 04:24hrs. Of the seven crews that took off from East Wretham only one reached Essen. Sergeant H Dostál and crews had dodged the flak, the fighters and the icing. They arrived over Essen at 01:41hrs to be greeted by a mass of bursting flak and searchlights. Almost 100 searchlights were reported and a cauldron of flak that exceeded any previous raid on Essen. Despite the promise of clear skies the crew were unable to identify the aiming point and eagerly dropped their bombs from 16,000ft. The only positives from a disastrous night was all the 311 crews returned. The night's performance would not go unnoticed and given the already frosty relationship between the Czechs and Honington's Station Commander, Group Captain Ommaney, questions would be asked and answers expected. As for the raid, it was a partial failure. Some useful damage had been inflicted. Local sources confirmed that five high explosive bombs and two hundred incendiaries had hit the Krupp complex, causing a large fire, and eighty houses were destroyed or seriously damaged, but it was a poor return for the size of the force. The following day, the squadron carried out Air-Sea gunnery practice over Holkham Bay, while two Wellingtons were ferried to Sealand, while another was collected from Hawarden, the home of 48 MU where were Vickers Armstrong shadow factory was located at Broughton.

Form B.781 arrived at R.A.F East Wretham during the early hours of the 14th. The target for that night was Dortmund. 14 of the Group's squadrons would be involved. 311 Squadron would provide seven crews.

### April 14th 1942 : 'Sprat' – Dortmund

| Pilot | 2nd Pilot | Serial | Code | Bomb Load | Result |
|---|---|---|---|---|---|
| W/O F Naxera | F/Lt O Hořejší | R1497 | KX-D | 1x1000lb+3x500lb+1x250lb | Duty Carried Out |
| Sgt V Šponar | Sgt V Výcha | DV757 | KX-X | 1x1000lb+3x500lb+1x250lb | Duty Carried Out |
| Sgt V Žežulka | Sgt O Špaček | X9880 | KX-C | 1x1000lb+3x500lb+1x250lb | Duty Carried Out |
| Sgt H Dostál | Sgt J Lenc | X9760 | KX-O | 1x1000lb+3x500lb+1x250lb | Duty Carried Out |
| Sgt F Dostál | Sgt J Štark | X9745 | KX-S | 1x1000lb+3x500lb+1x250lb | Duty Carried Out |
| Sgt V Pára | Sgt O Havlík | Z1098 | KX-U | 1x1000lb+3x500lb+1x250lb | MISSING |
| W/O F Bulis | F/Lt J Štrégl | Z1111 | KX-N | 1x1000lb+3x500lb+1x250lb | Duty Carried Out |

Dortmund, the largest city in the Ruhr Valley and a vast industrial centre boasted more than twenty important war producing factories, as well as collieries, coking, chemical and explosives plants, iron foundries and steel works, all of which were powered by ten modern power stations. It was also a communications and transportation hub with six extensive railway marshalling yards and an inland harbour connected by canal to the North Sea. 3 Group would detail and brief 86 Wellington crews from nine squadrons, plus 22 Stirlings. All four of Bomber Command's Groups would be involved, with 3 Group providing the bulk of the bombers. 311 Squadron departed between 22.05hrs and 22:25hrs and made their way to the now familiar departure point of Orfordness. Conditions were again ideal, with a cloudless sky and visibility between 5 and 10 miles. The Group's contribution was reduced when 11 bombers aborted, none from 311 Squadron. The Group was scheduled to be over Dortmund between 01:30hrs and 02:58hrs, with 311 Squadron arriving during the middle and latter stages. The first over the target was Sergeant V Šponar and crew at 02:02hrs. Industrial haze was made worse by the glare of continuous salvos of ground-fired flares over the city making identification of the aiming point difficult. Flak was intense and appeared to be working in conjunction with searchlights which seemed to be accurately stalking individual bombers. Also cooperating closely with the searchlights was a murderous array of light flak. If this was not enough, night fighters were observed in some numbers. Bombing from just 13,500ft the crew were unable to identify the aiming point so bombed on E.T.A. Three minutes later it was the turn of Warrant Officer F Naxera. They also had trouble identifying the aiming point, which would be a reoccurring problem for the Czech crews on this night. Flying over the city at 16,500ft was the crew of Sergeant H Dostál on his 34th operation. Dortmund was identified, but haze and smoke prevented the crew from positively identifying the aiming point. The bombs were dropped, and bursts were seen, but no other details. Two crews followed within five minutes of each other. Sergeant V Žežulka dropped his bombs from 16.000ft on what he believed to be the town. Sergeant F Dostál, flying his 32nd operation, deposited his bombload at 02:30hrs, remarking, *'One big red fire emitting heavy black smoke'*.

It would be 25 minutes before Warrant Officer F Bulis ran the gauntlet of flak and searchlights over Dortmund. They managed to drag Wellington Z1111 KX-N up to a respectable 18,000ft, a marvellous effort considering it was unable to climb two days earlier. Ominously, while running into the target, the crew reported seven large fires at Hagen, six miles south of the centre of Dortmund. This was a clear indication of what would be a scattered raid. Fires were observed over the target, but the crew were not entirely sure where their bombs landed. It is understood that Sergeant V Pára's crew had successfully reached and bombed the target and met their end on the return flight. The Wellington had been damaged by flak over Dortmund, rendering an engine useless and injuring the navigator. It was while limping home the crew were shot down at 04:10hrs by Uffz Walter Schienbein of 2./NJG1. It is believed that the pilots tried to crash land the barely flyable Wellington near the village of Boschoven near Weert, Limburg, the blazing wreck coming to a halt near some conifer trees close to a children's play area. Three of the crew survived, including the pilot, Sergeant Vladimír Pára, who was pulled out of the wreckage unconscious by the Navigator, Flying Officer Milan Zapletal. The two injured survivors were taken to the St Jans – Ziekenhuis Hospital. The bodies of those killed were recovered and buried on April 16th.

<u>Vickers Wellington Mk.IC Z1098 KX-U</u>

| Manufacturer | Vickers Armstrong (Chester) | |
|---|---|---|
| Contract | B.97887/39 | |
| Taken on Charge | 13/11/1941 via No. 18 MU. | |
| Cat E Missing | 14/04/1942 | |
| Struck Of Charge | N/K | |
| Total Flying Hours | N/K | |
| Take-Off Time | 22:25hrs | |
| Bomb Load | 1x1000lb+3x500lb+1x250lb | |
| | **CREW** | |
| Captain | Sergeant Vladimír **Pára** 787616 RAFVR | PoW |
| Second Pilot | Sergeant Oldřich **Havlík** 787521 RAFVR. Age 25. | Plot JJ. Grave 48. |
| Navigator | Flying Officer Milan František **Zapletal** 82535 RAFVR | PoW |
| Wireless Operator | Sergeant Josef **Taláb** 787892 RAFVR. Age 32. | Plot JJ. Grave 49. |
| Front Gunner | Sergeant Pavel **Varjan** 787901 RAFVR. Age 30. | Plot JJ. Grave 50 |
| Rear Gunner | Sergeant Jaroslav **Klvaňa** 787271 RAFVR | PoW |
| Posting History | Sgt V Pára posted via Wilmslow Depot, 22/05/1941. Sgt O Havlík posted via 2 S.F.T.S, 10/08/1941. | |
| Operations Flown | Sgt V Pára : 31 / Sgt O Havlík : 12 | |
| Buried | EINDHOVEN (WOENSEL) GENERAL CEMETERY | |

The raid was yet another disappointment. Post raid analysis established that bombs fell across a forty mile stretch of the Ruhr, damage to Dortmund was minor. Back at R.A.F East Wretham the already fragile morale of the squadron was knocked once again by the loss of an experienced aircrew, but more importantly, friends. To date, 1942 had not been kind to the Czechs, seven crews had *'Got the chop'*. The loss of close friends was tough on the Czechs. Relatively few in number, they had become close, and their absence was more apparent. Once familiar faces in the mess or the local pub vanished, leaving an empty void that broke the camaraderie's close bond. The concerns about the numbers of replacement crews from the Training Flight keeping pace with operational losses were becoming a reality. Bomber Command H.Q would have been fully aware that the raid did not achieve the desired results, so it was surprise that the following night the target was again Dortmund. 3 Group H.Q detailed and briefed 78 crews for the raid, just three of which were supplied by 311 Squadron.

## April 15th 1942 : 'Sprat' – Dortmund

| Pilot | 2nd Pilot | Serial | Code | Bomb Load | Result |
|---|---|---|---|---|---|
| W/O F Naxera | F/Lt O Hořejší | R1497 | KX-D | 1x1000lb+3x500lb | Duty Carried Out |
| Sgt V Šponar | Sgt V Výcha | DV757 | KX-X | 1x1000lb+3x500lb | Early Return |
| Sgt F Dostál | Sgt J Štark | X9871 | KX-A | 1x1000lb+3x500lb | Early Return |

The three selected crews had all flown the previous night, highlighting the serious shortage of crews within the squadron. All three were safely away by 23:16hrs and heading for Orfordness. No sooner had the drone of the engines fallen silent when the first of the night's aborts requested permission to land. Sergeant V Šponar and crew had not even reached the Norfolk coast before port engine trouble brought the operation to an early conclusion. They had been airborne for just 25 minutes. Within the hour, it was the turn of Sergeant F Dostál and crew, who also reported port engine failure. Honour was saved by the crew of Warrant Officer Naxera. They reached the target at 03:05hrs, identifying a cloud-covered Dortmund by a combination of E.T.A, flak and flares. They bombed from 18,000ft, bursts were observed, but once again, scattered fires were reported, this time in Witten, six miles southwest of the centre of Dortmund. The crew landed at 05:46hrs, it was František Naxera's 30th operation.

*A lovely photograph. Five experienced operational airmen of the 311Squadron – four navigators with a pilot in the middle – at East Wretham airfield in early spring 1942. L-R: Flight Lieutenants Lubomír Svátek, Miroslav Cígler, Warrant Officer František Bulis, Flight Lieutenants Antonín Hruška, and Flying Officer Josef Doubek. (Pavel Vančata)*

The squadron stood down on the 16th, it was a brief reprieve. H.Q Bomber Command turned its attention northwards, away from the industrial Ruhr to an old favourite, Hamburg. A force of 173 aircraft was made ready for Hamburg on the 17th, of which 3 Group provided 73 Wellingtons and Stirlings. 311 Squadron would provide four crews.

### April 17th 1942 : Dace 'D' – Hamburg

| Pilot | 2nd Pilot | Serial | Code | Bomb Load | Result |
|---|---|---|---|---|---|
| W/O F Naxera | Sgt K Mazurek | R1497 | KX-D | 1x1000lb+2x500lb+1x250lb | Early Return |
| Sgt V Šponar | Sgt J Lenc | DV757 | KX-X | 1x1000lb+2x500lb+1x250lb | Duty Carried Out |
| Sgt F Dostál | F/Lt O Hořejší | DV516 | KX-K | 1x1000lb+2x500lb+1x250lb | Duty Carried Out |
| Sgt H Dostál | Sgt O Špaček | X9760 | KX-O | 1x1000lb+2x500lb+1x250lb | Duty Carried Out |

*The squadron's third Commanding Officer A/W/Cdr Ocelka DFC*

The four crews left R.A.F East Wretham between 23:00hrs and 23:04hrs and made their way to the coastal town of Happisburgh before the long sea crossing. The now almost routine early return occurred midway between the English coast and the Frisian Islands. Warrant Officer F Naxera and crew suffered an engine problem aboard the usually dependable R1497 KX-D. The bombs were jettisoned 'safe' and the crew was back at R.A.F East Wretham by 01:15hrs. The remaining crews headed north to a point off the Amrum Island from where they would make the run into target. Weather conditions en route were clear, affording the navigators the luxury of map-reading their way to the target. Arriving north of Hamburg, the crews were greeted with intense flak of every variety and calibre. Sergeant H Dostál skippered the first Czech crew over the target at 03:00hrs. On arrival over Hamburg at 17,000ft, the crew observed one large fire burning in the town but were unable to make out the aiming point due to haze. Both Sergeants V Šponar and F Dostál arrived over Hamburg within a minute of each other. Sergeant F Dostál bombed from 18,000ft, the crew observed their own bombs burst, but little else other than one large fire south of the town. At 03:16hrs, it was the turn of Sergeant Vladimír Šponar, whose crew bombed on E.T.A., having established their position by the Elbe Estuary. The three were safely back at base by 06:40hrs. Despite the crew's favourable reports, 3 H.Q. deemed the operation only *'moderately successful'*. However, local sources suggested differently. More damage had been inflicted than estimated, a total of 75 fires were started, of which 33 were classed as significant. The following day, Saturday 18th, the Czechoslovakian President Edvard Beneš and chancellor Dr. Jaromír Smutný and senior Czech officials arrived at R.A.F East Wretham. There to meet them was Air Vice Marshal J.E.A Baldwin CB, CBE, DSO, Officer Commanding No.3 Group, Group Captain M.D Ommaney, Station Commander R.A.F Honington and Wing Commander J.M Southwell MiD, the newly appointed Station commander of R.A.F East Wretham.

The squadron aircrew, all Czech groundcrews including the hardworking servicing flight based at Honington plus 1429 COTF personnel were paraded. On completion of the inspection, all the Czechs were assembled in the Airmen's Dining Hall. Here the President informed all present that the squadron would, after a short rest period, be leaving R.A.F Bomber

Command and would re-train for operations with R.A.F Coastal Command. The crews were given no explanation, the news of the transfer came as a complete shock. They knew that the battle of the Atlantic was in full swing and Coastal Command were short of aircraft and crews, but that was general knowledge. The question was, *'why the transfer?'* For the time being, however, operations would continue. On the 20th, the popular and respected Wing Commander Ocelka DFC stepped down as 311 (Czech) Squadron Commanding Officer. He would be replaced by Wing Commander Josef Šnajdr DFC. The loss of Wing Commander Ocelka DFC was a bittersweet posting. Josef Ocelka DFC had commanded the squadron since June 1941 and had gently steered the squadron through some challenging times. Courageous and fair, he was an exemplary commanding officer. For the next three days the squadron had time to take in the loss of the commanding officer and come to terms with the move to Coastal Command.

On the 22nd, the squadron was called to action. The bulk of 3 Group would attack Cologne while a small force of 16 bombers, including five from 311 Squadron would visit an old stamping ground, the Docks at Le Havre. The inclusion of 311 Squadron on what was classed as a 'Freshmen' operation was no doubt deliberate. The possible loss of a trained crew and Wellington over Germany just prior to transfer would not have been appreciated at Coastal Command H.Q.

April 22nd 1942 : CC24A – Le Havre

| Pilot | 2nd Pilot | Serial | Code | Bomb Load | Result |
|---|---|---|---|---|---|
| Sgt F Dostál | Sgt F Kepka | DV779 | KX-L | 15x250lb | Duty Carried Out |
| Sgt K Pospíchal | Sgt J Šotola | DV516 | KX-K | 15x250lb | Duty Carried Out |
| Sgt V Žežulka | Sgt M Červinka | DV757 | KX-X | 15x250lb | Duty Carried Out |
| W/O F Bulis | F/O J Nývlt | DV507 | KX-W | 15x250lb | Duty Carried Out |
| F/Lt O Hořejší | F/Lt B Eichler | Z1111 | KX-N | 15x250lb | Duty Carried Out |

Finally, after what seemed an eternity, 1429 COTF offered five recently qualified pilots, two of whom only completed their training on the 19th. The last pilots to be posted onto the squadron while operational with R.A.F Bomber Command were Sergeant Miroslav Červinka, Sergeant Ferdinand Kepka, Sergeant Josef Šotola and Flying Officer Josef Nývlt. The crews departed just after 21:00hrs and made their way south to the coastal town of Littlehampton on the fortified south coast of England. There were no early returns, and the crews enjoyed favourable weather the entire route. 4 Group would open proceedings, with the Czechs over the target between 23:10hrs and 23:50hrs. Heavy flak appeared concentrated on either side of the town and to the north while light flak was particularly bothersome, this appeared to be concentrated in the centre of the town. Searchlights were busy and aggressive in the clear conditions. Flight Lieutenant O Hořejší was first over the target at 23:10hrs. The crew bombed from 12,000ft and watched with some satisfaction as the 15 x 250-pounders burst across the Bassin De Maree. The bombs of Sergeant V Žežulka were dropped in one stick from 13,000ft and seen to explode across the outer edge of Bassin De Maree. Within eight minutes, it was the turn of Sergeant Dostál and crew. He too, dropped on Bassin De Maree and the smaller Bassin Bellot, their bombs straggling both. Following close behind and at 14,400ft was the crew of Sergeant K Pospíchal. They observed seven reddish fires on arrival. Their bombs were aimed at the dock area, but details were unobserved due to haze. Warrant Officer F Bulis was the last over Le Havre at 23:50hrs. They stated their bombs were seen to burst in or around Bassin Bellot.

The only trouble with what was an excellent operation was on the return to R.A.F East Wretham. There are two conflicting reports on the crew of Sergeant V Žežulka, who crash landed on return. The ORB states that the crew suffered hydraulic failure aboard Wellington DV757 KX-X, resulting in a belly landing at 01:55hrs. The Accident Card (Form 1180), however, reports that the 2nd pilot, Sergeant M Červinka, misjudged his height on landing due to bad visibility, bounced and then stalled from 20 feet. On hitting the flarepath, the undercarriage collapsed. The report says that the 2nd pilot was *'comparatively inexperienced at night landing, completed his O.T.U. a week beforehand'*. This was signed off by the Officer Commanding. The crash prevented Warrant Officer F Bulis and crew from landing forcing them to divert to R.A.F Honington, where they landed at 02:27hrs.

| Date | 22/04/1942 |
|---|---|
| Mark | Mk.Ic |
| Serial | DV757 |
| Code | KX-X |
| Taken On Charge | 08/04/1942 via No.46 MU. |
| Manufacturer | Vickers Armstrong (Chester) |
| Contract | 124363/40 |
| Pilot (s) | Sergeant Vratislav Žežulka / Sergeant Miroslav Červinka |
| Flight | Operations |
| Time | 01:55hrs |
| Cause | Misjudge landing, inexperience. |

Thankfully, none of the crew were injured but damage to the brand new Wellington meant months of repair, it never returned to squadron service. Despite the expected departure to Coastal Command training continued. On the 23rd, bombing practice was flown over the ranges at Lakenheath. At the same time, strenuous efforts were being made to crate and pack materials, stores and equipment for the impending move. On the 24th, the squadron was ordered to prepare for another operation on the channel ports. Five crews would again be detailed and briefed. During the day, Group Captain Kubita, Wing Commander Southwell MiD, Flight Lieutenant Fantl, Squadron Operations Officer and a selection of other officers flew to R.A.F Aldergrove, Northern Ireland to make arrangements for the squadron's move.

## April 24th 1942 : CC25 – Dunkirk

| Pilot | 2nd Pilot | Serial | Code | Bomb Load | Result |
|---|---|---|---|---|---|
| Sgt F Dostál | Sgt J Neradil | DV779 | KX-L | 16x250lb | Duty Carried Out |
| Sgt K Pospíchal | Sgt J Sapák | DV516 | KX-K | 16x250lb | Duty Carried Out |
| Sgt V Žežulka | Sgt R Pancíř | DV716 | KX-Z | 16x250lb | Duty Carried Out |
| W/O F Bulis | Sgt S Petrášek | Z1155 | KX-F | 16x250lb | Duty Carried Out |
| F/Lt O Hořejší | Sgt J Říha | R1497 | KX-D | 16x250lb | Duty Carried Out |

The weather on take-off was ideal, and this thankfully continued until the target was reached. The flak was less abundant than Le Havre, the most troublesome opposition was from two flak ships positioned 2 to 3 miles northwest of the port. These kept up a steady barrage throughout the raid. The ever-dependable Sergeant F Dostál was the first to bomb at 23:32hrs, his bombs falling on the north-western part of the docks. Just before midnight, it was the turn of Sergeant V Žežulka. The crew could not accurately observe their results due to flares being fired from the ground, which along with haze prevented accurate observation. Sergeant K Pospíchal was at 13,500ft when at 00:01hrs, the crew dropped their all High Explosive load. Again, haze and the glare from the flares made observation difficult but were confident that they fell in the dock area. The area between the docks and the railway station was on the receiving end of Warrant Officer F Bulis's bombs. Six minutes later, Flight Lieutenant O Hořejší and crew deposited their entire load in one stick in the dock area at 00:29hrs. The crew reported a dummy fire west of the town on return. It had been another relatively straightforward operation without loss, or incident.

Despite some of the senior crews being in Northern Ireland, and the squadron's impending move, it would appear that both 3 Group and Bomber Command H.Q were determined to maximise the squadron to the very end. On the 25th, the squadron was called again to provide five crews to attack Dunkirk. While the Wellingtons were being prepared, training continued, air-to-air gunnery practice being completed over Holkham Bay.

## April 25th 1942 : CC25 – Dunkirk

| Pilot | 2nd Pilot | Serial | Code | Bomb Load | Result |
|---|---|---|---|---|---|
| Sgt F Dostál | Sgt F Kepka | DV779 | KX-L | 16x250lb | Duty Carried Out |
| Sgt K Pospíchal | Sgt J Šotola | DV516 | KX-K | 16x250lb | Duty Carried Out |
| Sgt V Žežulka | Sgt M Červinka | DV716 | KX-Z | 16x250lb | Duty Carried Out |
| W/O F Bulis | F/O J Nývlt | Z1155 | KX-F | 16x250lb | Duty Carried Out |
| F/Lt O Hořejší | F/Lt B Eichler | R1497 | KX-D | 16x250lb | Duty Not Carried Out |

The squadron again enjoyed good visibility on take-off and over the English Channel, but it did not last. 3 Group were timed over Dunkirk between 23:25hrs and 00:10hrs. Unfortunately, by the time the squadron arrived, broken clouds and haze had made an appearance. Once again, Sergeant F Dostál was the first over the target on his 37th operation. He and his crew were the only ones to positively identify the target, his bombs dropping on docks 7 & 8 at 23:55hrs. When the remaining crews arrived, clouds had almost covered the target. Sergeant V Žežulka dropped his 250-pounders in the general area of the docks from 12,000ft. The results were unobserved. Warrant Officer F Bulis, on his 36th operation, bombed at 00:08hrs, they managed to glimpse the docks through a gap in the clouds. The bombs were dropped, but the results were unnoticed. Preferring to attack from 16,000ft were the crew of Sergeant K Pospíchal. They reported bursts in the docks at 00:10hrs. They would be the last bombs dropped by 311 (Czech) Squadron with R.A.F Bomber Command. Unable to identify the target, Flight Lieutenant O Hořejší retained his bombs and turned for home. The last crew to land was Warrant Officer F Bulis at 01:35hrs. The squadron's bombing war with 3 Group was over.

On the 25th, the squadron was paraded, aircrew and groundcrews gathered together for the last time as members of 3 Group and Bomber Command. The parade stood in silence as the rain gently fell, two commanding officers past and present, Wing Commander Ocelka DFC and Wing Commander Josef Šnajdr DFC stood under both the Czech Flag and the Royal Air Force flag and addressed the assembled men. It was an emotional speech. Also in attendance was the AOC of 3 Group, AVM. J.E.A Baldwin CB, OBE, DSO. He thanked the squadron for their service, commitment and, above all, bravery. He also told the men that he was proud to have been their commanding officer and hated to say goodbye to them. On the 26th, the squadron, or what was left of it, put all its energy preparing for the move. Two Wellington crews left during the day taking essential equipment. Other than night flying practice flown on the 27th, the only activity was packing. On the 28th, the day of the move, the Czechs were still carrying out bombing practice over the Lakenheath ranges. A group of between 170 to 180 men of the main party departed R.A.F East Wretham for Northern Ireland. On the 29th, a frantic effort was made by all left behind. Everything had to be loaded, crated, recorded and signed for, everything from bicycles to engines needed to be ready for the following day. April 30th, 1942, found the Czechs over Holkham Bay on air firing practice. It would be their last visit. The main air party of around 100 personnel and 15 Vickers Wellingtons, DV664 KX-A, DV665 KX-B, DV738 KX-C, R1497 KX-D, Z1155 KX-F, DV516 KX-K, DV779 KX-L, Z1111 KX-N, Z1090 KX-O, X3178 KX-P, Z1105 KX-R, X9745 KX-S, T2564 KX-T, DV507 KX-W, DV716 KX-Z and finally took off for a new adventure. The rear party of around 60 men boarded the lorries and set off for the docks and the short sea crossing to Northern Ireland.

The squadron's short but remarkable association with 3 Group was at an end. The only Czechoslovak bomber squadron in Bomber Command was rather unceremoniously gifted to Coastal Command. The losses on the squadron had taken their toll, and this is often reported as the main reason behind its departure. The pool of men to draw upon to serve on the squadron was not extensive, and perhaps, fearing that operational losses could not be replaced despite the COTF finally producing crews on a regular basis, it looked at least to be a prudent transfer on paper. However, the recent issues of early returns and the frosty and, at times, strained relationship with Group Captain M.D Ommaney and Wing Commander Batchelor DFC, had reached the corridors of H.Q Bomber Command. Could this have also influenced the decision to part ways with 311 (Czech) Squadron and its transfer to Coastal Command being an ideal opportunity. We will probably never know.

# Ground Crews: The Role and Dangers They Faced

The life of a member of the ground crews was hard, and dangerous. Long tiring days, and nights were the norm, with the crews often working 15-hour days in all weather conditions.

To keep the Wellingtons operational there were a multitude of trades, each highly skilled, motivated and devoted to keeping their aircraft in 'tip-top' condition. Inspections, maintenance, fuelling and arming were inevitably undertaken out at dispersal in changeable and often severe weather.

*311 (Czech) Ground crew. A mixed bag looking every inch ground crew with a mixture of uniforms, muddy wellington boots and leather gilets.*

Behind the scenes, and usually under the guidance of the Adjutant, the Armaments Officers, Engineering Officers and Signals Officers ensured everything ran smoothly. The responsibility would be shared with their operational counterparts, each of the aircrew trades had their own specific 'Boss', these positions were usually Gunnery, Bombing, Navigation and Signals Leader. Together they would each be responsible for a number of skilled men and women. NCOs - Sergeants and Flight Sergeants - and "Rankers" - Corporals, Leading Aircraftsmen, and Aircraftsmen 1 and 2 (AC1 and AC2). All had a vital role. Away from the aircraft were the Supplies Officers, Intelligence Officers, Catering Officers, and Medical Officers, not as demanding, or dangerous, but equally important.

311 Squadron initially had one operational flight, plus the Training Flight. However, this soon changed to two operational Flights. The ground crews would be designated to service specific aircraft within each flight, which could range from 6-9 aircraft. Ideally, one set of ground crews would be responsible for one Wellington, but this was often not possible, and the crews would frequently work on two or three aircraft. The squadron often requested more trained Ground crews to join the existing Czechs and Slovaks. This was true mainly when 311 Squadron operated two operational flights, plus the training flight. A lack of trained Czech 'Erks' meant the shortfall was met by the posting of their R.A.F. counterparts. Regardless of nationality, these men took great pride in their aircraft. It was often felt and said that the ground crews "owned" the aircraft and "lent" them to aircrew, provided they brought them back in one piece!

On a typical day when "ops were on" for that night, the preparations would begin early in the morning with the Daily Inspection (DI). The crew given a particular aircraft to operate or had last flown the aircraft on ops would come "down the flights" to discuss any equipment or performance problems encountered on the aircraft with the ground crew. These would be noted on the "snag sheet" (Form 700). As long as the ground crew had the Form 700, the aircraft belonged to them and was their responsibility. Daily inspections included checking all engine controls and parts, all hydraulics and pneumatics, and all electrical systems, radios, navigation aids, lights, brakes, etc. It also included repairing minor battle damage and attention to anything reported by the aircrew on the "snag list". Ground running of engines to check for "mag drop", fuel leaks, oil burning, sticking sleeve valves plus a hundred other checks were also part of the Daily Inspection. If 'Ops' were on that night, the Wellington would be fuelled and bombed-up. The oxygen systems would be filled, and the intercom systems would be checked. Only the ground crew could certify an aircraft as serviceable for ops and no one regardless of any rank could order an Erk to sign off the Form 700. It was only when all the "snags" had been checked off and the aircraft's captain was happy would he signed the Form 700, from that moment on the responsibility for the aircraft passed to the pilot.

Erks came in many forms. "Fitters" were generally responsible for the engines and other mechanical parts of the aircraft. "Riggers" were responsible for the airframe and the related components, including fuelling. "Armourers" were responsible for arming the aircraft, including bombs and defensive guns. After the checks had been completed, it was not unusual for an aircraft, especially after repair or performance issues, to be Air-Tested. Often, a ground crew member would join the crew, sometimes in an official capacity, but other times simply for the thrill of flying.

The role of the Armourer was particularly hazardous, especially during the winter months. The bombs were usually stored well away from the aircraft and the buildings. Each raid, or practice flight the bombs and occasional flares had to be transported to the aircraft dispersals from the bomb dump. It was here the fuses and detonators were installed, and safety wires were attached. It was not uncommon for the squadrons aircraft to carry mixed bomb loads for an op. Some might carry 250lb or the bigger 500-pounders. These came in various types, Medium Capacity (MC) High Explosive (HE) or Armour Piercing (AP). Others may carry SBCs full of 4lb incendiaries or combinations of bombs and incendiaries.

*Grave of AC 2<sup>nd</sup> Class P.B Bray*

## AIRMAN'S TRAGIC DEATH.

### Walked into Revolving Propeller.

A verdict of death by "Misadventure" was returned at an inquest at an East Anglian aerodrome on Wednesday night on 2nd Aircraftsman Percival Blewitt Bray (37), who accidentally walked into the propeller of a bomber which was standing with the engines running ready to take off. Deceased had just had his first flight in the machine.

L.-A.-C. S. Figures, an armourer, said that while the propellers of the stationary aircraft were ticking over he heard a bump and saw deceased on the ground. Witness thought that deceased was coming to speak to him and that contrary to orders he went on the outside of the bomb doors instead of inside. An officer had pointed out these orders that morning.

A Czech pilot-sergt. said they had landed from a practice flight and he had left the airscrews ticking over as they were going to take off again shortly. He thought that deceased walked straight from the steps in the nose of the machine into the airscrews.

An R.A.F. medical officer said that Bray was killed at once.

*How the local papers reported LAC Brays death.*

## Amateur Comedian Killed in An Aerodrome Tragedy

### SAD DEATH OF "BILLY" BRAY

### Stepped Into Airscrew After Flight

THE DEATH occurred on Tuesday at an R.A.F. station, following an accident, of Ac2 Percival Blewitt Bray (known to his his friends as "Billy"), of Myra Lodge, Wootton-rd., Gaywood. Mr. Bray had recently completed his training as an armourer. He was within two days of his 37th birthday.

"Billy" Bray had a personality which made him a great success on the stage and brought him a host of friends off it. In amateur theatricals he had shown a consistent improvement which reached its peak when the Operatic Society started its war-time concert party for the Services. Billy was very soon the life and soul of that entertaining little band. His "star turn" was "Black-out Bertha," an hilariously funny impersonation. Among many service units "Bertha" was a more familiar name for him than Billy.

When he joined the R.A.F. in April the concert party—already depleted by the loss of other members joining the Forces—broke up.

### THE INQUEST

A verdict of death by "Misadventure" was returned at the inquest held at an East Anglian aerodrome on Wednesday night. It was stated that Bray accidentally walked into the airscrew of a bomber which was standing with the engines running ready to take off. Deceased had just had his first flight in the machine.

L/Ac. S. Figures, an armourer, said that while the propellers of the stationary aircraft were ticking over he heard a bump and saw Ac2 Bray on the ground. He thought Bray was coming to speak to him and that contrary to orders he went on the outside of the bomb doors instead of inside. An officer had pointed out these orders that morning.

A Czech sergt.-pilot said they had landed from a practice flight and he had left the airscrews ticking over as they were going to take off again shortly. He thought that Bray walked straight from the steps in the nose of the machine into the airscrews.

An R.A.F. medical officer said that Bray was killed at once.

*Photo: Goodchild, Lynn.*
"BILLY" BRAY

(Long Sutton), sisters: Mr. S. Goodale, father-in-law; Mr. and Mrs. T. Barnes, uncle and aunt; Mrs. H. Bunn, Mrs. Boardman, Mr. and Mrs. E. Tolliday, Mr. T. Barnes, Mrs. Denton, Mrs. Whitehead, Mrs. Watson, Miss Watson, cousins; Mr. and Mrs. J. Feetham, brother-in-law and sister; Mr. E. Goodale, Miss C. Goodale, uncle and aunt; Mr. and Mrs. C. Carter, Mr. and Mrs. Youngs (Gaywood), Mr. and Mrs. A. L. Marsters (Gaywood), R.A.F. officer and N.C.O.'s, Mr. and Mrs. W. Portass (Lutton), Mr. and Mrs. F. Donnell (Long Sutton), Mr. and Mrs. G. Holmes (Lutton), Mr. and Mrs. H. Kitchen, Mr. S. Barnett, Mrs. Burwell, Mrs. Simpson, Mrs. Bulter, Messrs. G. and J. Bulter, Mr. A. Chapman, Mr. and Mrs. Reeder, Mr. and Mrs. S. Cole, Mrs. Watkins, Mr. C. Archer, Mr. Jefferies, Mr. W. Wilkinson, Mr. S. Kitchen, Miss Parsons and many others.

The wreaths were inscribed: "With all my love, from yours ever Mammy." "Love to Daddy, from Helen." "To our Brother, from Annie and Kate." "To Uncle, from Joan and Ron." "Uncle Billy, from little John." "Deepest sympathy, from father-in-law." "Maurice and Maisie, Lutton." "Reg., Ivy, Aunt Maud, Lutton." "Erny, Eliza and family, Lutton." "Aunt Jennie and Eliza, and George, Canada." "Jennie, Ada, Kate and Flo." "Ivy, Mary and Jess." "Uncle Erny, Chrissie

The dangers experienced by the Armourers is tragically illustrated by the death of Englishman A.C.2 P Bray an Armourer (Bombs) under probation at R.A.F East Wretham on August 12$^{th}$ 1941. Airborne on a practice bombing programme, Vickers Wellington R1523 had completed its first exercise landing at 12:15 hours. Rolling to a stop at the designated bombing up area, the pilot, Sergeant Karel Danihelka applied the airbrakes and as was customary and ordered by the Bombing Leader/Armoury Leader he allowed the engines to tick-over while the bomb-carriers were positioned under the Wellington for re-loading. The only airman authorised to go under the Wellington was L.A.C S Figuires.

While the loading of the bombs was taking place the pilot vacated the Wellington to talk with Flying Officer Albert Roman RAFVR who was both the Bombing Leader and Armoury Leader. The pilot was informed that once reloading had been completed the crew would continue with the training. It was during this discussion that A.C.2 Percival Bray vacated the Wellington and moved forward to engage with L.A.C Figuires who was standing forward of the bomb bay. Sadly in doing so, the inexperienced Bray was struck on the head by a revolving port engine airscrew and killed almost instantly. The engines were immediately shut down and the Station Doctor was ordered to attend. Flying Officer N Novak found L.A.C Bray face down, there was nothing that could be done for the unfortunate Percy Bray. There followed a lengthy and thorough investigation into the fatal accident initially presided over by Flight Lieutenant Wilfred Williams D.F.C, then Wing Commander Ken Batchelor. During the proceedings, Flying Officer Roman was questioned on a number of occasions as were others involved. Finally, the investigation concluded that Flying Officer Roman's orders to keep the engines ticking over were contrary to Orders laid down in A.P 124, para IV, chapter 1, section 5, para 5, sub-section 3 and also that the unfortunate L.A.C Bray was negligent.

Following the fatal accident, all bombing-up procedures on the squadron and the Training Flights would be carried out with the engines shut down. L.A.C Bray had no reason to fly on the Wellington on that fateful morning, but his eagerness to experience a trip in a Wellington and the bombing up process proved just too strong, sadly he would pay for his enthusiasm with his life. Born in Lutton Gowts, Lincolnshire, on August 14th 1904 'Billy' as he was known was the only son of Mr & Mrs Bray, proprietors of the 'Chequers' inn in Lutton. Pre-war, he worked for the local Grocery & Bakery store, where he eventually married the owner's daughter Elsie; the couple had a daughter in 1936. Keen on amateur dramatics and a liking for slapstick and comedy, he was well-known in the area. Percival Bray, just days away from his 38th birthday, is one of the hundreds of Ground crews killed with Bomber Command. Unknown to the vast majority, they deserve recognition and respect as much as their Aircrew colleagues. Without them, Bomber Command would have never been able to achieve the success it did.

# 311 Squadron Prisoners of War

| Rank | Name | Date | No. | Camp |
|---|---|---|---|---|
| Sgt | Karel BATELKA 787947 RAFVR | 17/01/1942 | 24771 | Oflag O4C Saalhaus Colditz |
| Sgt | Vilem BUFKA 787572 RAFVR | 23/06/1941 | 39160 | Stalag Luft L3 Sagan and Belaria |
| P/O | Emil BUSINA 82588 RAFVR | 06/02/1941 | 401 | Oflag O4C Saalhaus Colditz |
| F/Lt | Otakar CERNY 82590 RAFVR | 16/07/1941 | 3663 | Oflag O4C Saalhaus Colditz |
| P/O | Frantisek CIGOS 82541 RAFVR | 06/02/1941 | 402 | Oflag O4C Saalhaus Colditz |
| P/O | Vaclav KILIAN 82606 RAFVR | 25/09/1940 | 3771 | Stalag Luft L3 Sagan and Belaria |
| Sgt | Jaroslav KLVANA 787271 RAFVR | 14/04/1942 | 244 | Stalag 4B Muhlberg Elbe |
| Sgt | Frantisek KNAP 787396 RAFVR | 17/07/1941 | 39285 | Stalag Luft L3 Sagan and Belaria |
| Sgt | Frantisek KNOTEK 787548 RAFVR | 25/09/1940 | 315 | Stalag 4B Muhlberg Elbe |
| Sgt | Gustav KOPAL 787232 RAFVR | 06/02/1941 | 441 | Stalag 4B Muhlberg Elbe |
| P/O | Karel KRIZEK 82903 RAFVR | 06/02/1941 | 407 | Stalag Luft L3 Sagan and Belaria |
| Sgt | Emanuel NOVOTNY 787250 RAFVR | 16/10/1940 | 395 | Stalag Luft L3 Sagan and Belaria |
| Sgt | Jaroslav NYC 787166 RAFVR | 17/07/1941 | 39284 | Stalag 357 Kopernikus |
| Sgt | Vladimir PARA 787616 RAFVR | 14/04/1942 | 6446 | Stalag 357 Kopernikus |
| Sgt | Frantisek PETR 787175 RAFVR | 21/10/1941 | 24448 | Stalag 4B Muhlberg Elbe |
| P/O | Zdenek PROCHAZKA 82628 RAFVR | 25/09/1940 | 3770 | Stalag Luft L3 Sagan and Belaria |
| Sgt | Vaclav PROCHAZKA 787193 RAFVR | 21/10/1941 | 24472 | Stalag 4C Wistritz bei Teplitz |
| P/O | Josef SCERBA 82633 RAFVR | 28/12/1941 | ? | Stalag Luft L3 Sagan and Belaria |
| Sgt | Augustin SESTAK 787153 RAFVR | 16/10/1940 | 396 | Stalag Luft L6 Heydekrug |
| Sgt | Zdenek Joseph SICHROVSKY 787331 RAFVR | 17/01/1942 | 12673 | Stalag 4B Muhlberg Elbe |
| Sgt | Alois SISKA 787493 RAFVR | 28/12/1941 | 39654 | Stalag 4C Wistritz bei Teplitz |
| Sgt | Josef SNAJDR 787798 RAFVR | 17/01/1942 | 181 | Stalag Luft L1 Barth Vogelsang |
| Sgt | Karel STASTNY 787170 RAFVR | 17/07/1941 | 39287 | Stalag 4B Muhlberg Elbe |
| Sgt | Josef SUSA 787036 RAFVR | 21/10/1941 | 24446 | Stalag Luft L3 Sagan and Belaria |
| Sgt | Pavel SVOBODA 787399 RAFVR | 28/12/1941 | 24976 | Stalag 344 Lamsdorf |
| P/O | Karel Joseph TROJACEK 82580 RAFVR | 24/09/1940 | 3769 | Stalag Luft L3 Sagan and Belaria |
| Sgt | Petr URUBA 787198 RAFVR | 06/02/1941 | 450 | Stalag 4C Wistritz bei Teplitz |
| P/O | Arnost VALENTA 82532 RAFVR | 06/02/1941 | 415 | Stalag Luft III (*Murdered*) |
| Sgt | Bedrich VALNER 787899 RAFVR | 20/10/1941 | 24441 | Stalag 4B Muhlberg Elbe |
| F/L | Erazim VESELY 82534 RAFVR | 21/10/1941 | 662 | Oflag O4C Saalhaus Colditz |
| Sgt | Arnost ZABRS 787225 RAFVR | 24/09/1940 | 18350 | Stalag Luft L3 Sagan and Belaria |
| P/O | Jaroslav ZAFOUK 82217 RAFVR | 16/07/1941 | 3661 | Oflag O4C Saalhaus Colditz |
| F/O | Milan Frantisek ZAPLETAL 82535 RAFVR | 14/04/1942 | 6445 | Stalag Luft L3 Sagan and Belaria |
| Sgt | Josef ZVOLENSKY 787902 RAFVR | 21/10/1941 | 24449 | Stalag Luft L1 Barth Vogelsang |

# 311 Squadron Roll of Honour

| Rank | | DoD | Buried |
|---|---|---|---|
| Sergeant | Karel KUNKA 787252 RAFVR | 25/09/1940 | The Hague Westduin General Cemetery |
| Sergeant | Frantisek KOUKOL 787615 RAFVR | 01/10/1940 | Peterborough Eastfield Cemetery |
| Sergeant | George Owen POWIS 627228 RAFVR | 01/10/1940 | Derby Nottingham Road Cemetery |
| Pilot Officer | Jaroslav SKUTIL 82529 RAFVR | 01/10/1940 | Peterborough Eastfield Cemetery |
| Pilot Officer | Josef SLOVAK 82638 RAFVR | 01/10/1940 | Peterborough Eastfield Cemetery |
| Sergeant | Oskar VALOSEK 787244 RAFVR | 01/10/1940 | Peterborough Eastfield Cemetery |
| Pilot Officer | Hubert JAROSEK 82605 RAFVR | 16/10/1940 | Oosterwolde General Cemetery, |
| Sergeant | Otto JIRSAK 787141 RAFVR | 16/10/1940 | Oosterwolde General Cemetery, |
| Sergeant | Karl KLIMT 787547 RAFVR | 16/10/1940 | Oosterwolde General Cemetery, |
| Pilot Officer | Bohumil LANDA 82557 RAFVR | 16/10/1940 | Oosterwolde General Cemetery, |
| Sergeant | Josef ALBRECHT 787410 RAFVR | 16/10/1940 | Pinner Cemetery |
| Pilot Officer | Jaroslav MATOUSEK 82524 RAFVR | 16/10/1940 | Pinner Cemetery |
| Pilot Officer | Jaroslav SLABY 82637 RAFVR | 16/10/1940 | Pinner Cemetery |
| Squadron Leader | Jan VESELY 82582 RAFVR | 16/10/1940 | Pinner Cemetery |
| Sergeant | Frant ZAPLETAL 787242 RAFVR | 16/10/1940 | Pinner Cemetery |
| Sergeant | Karel LANG 787416 RAFVR | 17/10/1940 | Honington All Saints Churchyard |
| Sergeant | Oldrich TOSOVSKY 787237 | 17/10/1940 | Honington All Saints Churchyard |
| Pilot Officer | Miloslav VEJRAZKA 82533 RAFVR | 17/10/1940 | Honington All Saints Churchyard |
| Sergeant | Jiri JANOUSEK 787545 RAFVR | 16/12/1940 | Honington All Saints Churchyard |
| Sergeant | Jan KRIVDA 787209 RAFVR | 16/12/1940 | Honington All Saints Churchyard |
| Pilot Officer | Jaromir TOUL 82642 RAFVR | 16/12/1940 | Honington All Saints Churchyard |
| Aircraftman 2nd Class | Jindrich LIEBOLD 787284 RAFVR | 05/01/1941 | Honington All Saints Churchyard |
| Pilot Officer | Jindrich LESKAUER 82617 RAFVR | 15/01/1941 | Runnymede Memorial |
| Sergeant | Bohumil BAUMRUK 787186 RAFVR | 16/01/1941 | Runnymede Memorial |
| Sergeant | Rudolf BOLFIK 787583 RAFVR | 16/01/1941 | Runnymede Memorial |
| Flying Officer | Josef HUDEC 82604 RAFVR | 16/01/1941 | Runnymede Memorial |
| Pilot Officer | Jaromir Oldrich KRAL RAFVR | 16/01/1941 | Runnymede Memorial |
| Flying Officer | Antonin KUBIZNAK 82556 RAFVR | 16/01/1941 | Runnymede Memorial |
| Pilot Officer | Vaclav KOSULIC 82609 RAFVR | 17/04/1941 | Jonkerbos War Cemetery |
| Sergeant | Frantisek KRACMER 787244 RAFVR | 17/04/1941 | Jonkerbos War Cemetery |
| Flying Officer | Vladimir KUBICEK 82613 RAFVR | 17/04/1941 | Jonkerbos War Cemetery |
| Sergeant | Rudolf LIFCIC 787533 RAFVR | 17/04/1941 | Jonkerbos War Cemetery |
| Flying Officer | Frank SIXTA 82574 RAFVR | 17/04/1941 | Jonkerbos War Cemetery |
| Sergeant | Vaclav STETKA 787497 RAFVR | 17/04/1941 | Jonkerbos War Cemetery |

| Rank | Name | Date | Burial |
|---|---|---|---|
| Sergeant | Frantisek DUSEK 787825 RAFVR | 25/05/1941 | East Wretham St. Ethelbert C |
| Sergeant | Maxmilian STOCEK 787890 RAFVR | 25/05/1941 | East Wretham St. Ethelbert C |
| Pilot Officer | Stanislav ZEINERT 82648 RAFVR | 26/05/1941 | East Wretham St. Ethelbert C |
| Pilot Officer | Miloslav SVIC 82578 RAFVR | 04/06/1941 | East Wretham St. Ethelbert C |
| Flight Sergeant | Jan HEJNA 787204 RAFVR | 22/06/1941 | Runnymede Memorial |
| Flight Sergeant | Alois ROZUM 787169 RAFVR | 22/06/1941 | Runnymede Memorial |
| Pilot Officer | Leonard SMRCEK 82639 RAFVR | 22/06/1941 | Runnymede Memorial |
| Flight Sergeant | Karel VALACH 787551 RAFVR | 22/06/1941 | Runnymede Memorial |
| Pilot Officer | Vilem KONSTACKY 82608 RAFVR | 23/06/1941 | Runnymede Memorial |
| Sergeant | Adolf DOLEJS 787820 RAFVR | 02/07/1941 | Salisbury Devizes Road Cemetery |
| Pilot Officer | Richard HAPALA 82603 RAFVR | 02/07/1941 | Salisbury Devizes Road Cemetery |
| Sergeant | Oldrich HELMA 787190 RAFVR | 02/07/1941 | Salisbury Devizes Road Cemetery |
| Sergeant | Jaroslav LANCIK 787859 RAFVR | 02/07/1941 | The Maidenhead Register |
| Sergeant | Jaroslav PETRUCHA 787873 | 02/07/1941 | Salisbury Devizes Road Cemetery |
| Sergeant | Antonin PLOCEK 787355 RAFVR | 02/07/1941 | Salisbury Devizes Road Cemetery |
| Sergeant | Jiri MARES 787393 RAFVR | 17/07/1941 | Lemsterland Lemmer General C |
| Sergeant | Miroslav JINDRA 788021 RAFVR | 19/07/1941 | Uithuizermeeden General Cemetery |
| Sergeant | Pavel BABACEK 787148 RAFVR | 19/07/1941 | Runnymede Memorial |
| Sergeant | Jan CTVRTLIK 787433 RAFVR | 19/07/1941 | Runnymede Memorial |
| Sergeant | Vaclav VALES 787417 | 19/07/1941 | Runnymede Memorial |
| Pilot Officer | Jaroslav PARTYK 82627 RAFVR | 20/07/1941 | Sage War Cemetery |
| Sergeant | Vaclav NETIK 787211 RAFVR | 20/07/1941 | Oldebroek General Cemetery |
| Aircraftman 2nd Class | Percival Blewitt BRAY 1249720 RAF | 12/08/1941 | Lutton Cemetery |
| Sergeant | Zdenek BABICEK 787505 RAFVR | 15/09/1941 | Reichswald Forest War Cemetery |
| Sergeant | Alois JARNOT 787394 RAFVR | 15/09/1941 | Reichswald Forest War Cemetery |
| Sergeant | Jan MIKLOSEK 787339 RAFVR | 15/09/1941 | Reichswald Forest War Cemetery |
| Flying Officer | Mojmir SEDLACEK 82634 RAFVR | 15/09/1941 | Reichswald Forest War Cemetery |
| Sergeant | Vilem SOUKUP 787246 RAFVR | 15/09/1941 | Reichswald Forest War Cemetery |
| Pilot Officer | Antonin ZIMMER 82649 RAFVR | 15/09/1941 | Reichswald Forest War Cemetery |
| Pilot Officer | Frantisek DITTRICH 87616 | 23/10/1941 | Pontypridd Crematorium |
| Sergeant | Karel HURT 787557 RAFVR | 23/10/1941 | Runnymede Memorial |
| Sergeant | Otakar JANUJ 787846 RAFVR | 23/10/1941 | Runnymede Memorial |
| Sergeant | Jaroslav POLEDNIK 787593 RAFVR | 23/10/1941 | Runnymede Memorial |
| Sergeant | Jaroslav ROLENC 787686 RAFVR | 23/10/1941 | Runnymede Memorial |
| Sergeant | Stanislav LINKA 787395 RAFVR | 16/11/1941 | Runnymede Memorial |
| Flying Officer | Jan Frantiser PAROLEK RAFVR | 16/11/1941 | Runnymede Memorial |
| Sergeant | Pavel SKUTEK 787888 RAFVR | 16/11/1941 | Runnymede Memorial |
| Sergeant | Arnost VACLAVEK 787479 RAFVR | 16/11/1941 | Runnymede Memorial |
| Corporal | Robert BREEZE 540886 RAF | 23/11/1941 | Liverpool West Derby Cemetery |
| Flying Officer | Josef MOHR 82622 RAFVR | 28/12/1941 | Bergen General Cemetery |
| Sergeant | Rudolf SKALICKY 787283 RAFVR | 28/12/1941 | Runnymede Memorial |

| Rank | Name | Date | Location |
|---|---|---|---|
| Sergeant | Josef TOMANEK 787501 RAFVR | 28/12/1941 | Runnymede Memorial |
| Pilot Officer | Jaromir BROZ 61917 RAFVR | 17/01/1942 | Tilburg Gilzerbaan General C |
| Sergeant | Rudolf MASEK 787865 RAFVR | 17/01/1942 | Tilburg Gilzerbaan General C |
| Sergeant | Jindrich SVOBODA 787165 RAFVR | 17/01/1942 | Tilburg Gilzerbaan General C |
| Sergeant | Cenek KRAL 787854 RAFVR | 21/01/1942 | Runnymede Memorial |
| Sergeant | Ladislav NEMECEK 787353 | 21/01/1942 | Runnymede Memorial |
| Flight Sergeant | Miroslav PLECITY 787392 | 21/01/1942 | Runnymede Memorial |
| Sergeant | Stanislav ROUS 787878 | 21/01/1942 | Runnymede Memorial |
| Flight Lieutenant | Zdenek SKOREPA 82528 | 21/01/1942 | Runnymede Memorial |
| Warrant Officer | Karel WEISS 787519 | 21/01/1942 | Runnymede Memorial |
| Sergeant | Bohuslav HRADIL 787578 | 03/03/1942 | Creil Communal Cemetery |
| Sergeant | Imrich KORMANOVIC 787440 | 03/03/1942 | Creil Communal Cemetery |
| Sergeant | Jan KOTRCH 787532 | 03/03/1942 | Creil Communal Cemetery |
| Sergeant | Pribyslav STRACHON 787569 | 03/03/1942 | Creil Communal Cemetery |
| Sergeant | Josef SVOBODA 787881 | 03/03/1942 | Creil Communal Cemetery |
| Sergeant | Alois TOLAR 788183 | 03/03/1942 | Creil Communal Cemetery |
| Sergeant | Karel DANIHELKA 787172 | 03/03/1942 | Runnymede Memorial |
| Sergeant | Vladimir HANZL 787491 | 03/03/1942 | Runnymede Memorial |
| Sergeant | Frantisek JANCA 787845 | 03/03/1942 | Runnymede Memorial |
| Sergeant | Adolf PODIVINSKY 787589 | 03/03/1942 | Runnymede Memorial |
| Flight Lieutenant | Ladislav RIHA 82527 | 03/03/1942 | Runnymede Memorial |
| Sergeant | Dobromil SPINKA 787889 | 03/03/1942 | Runnymede Memorial |
| Sergeant | Frantisek BINDER 787400 | 04/03/1942 | East Wretham St. Ethelbert C |
| Pilot Officer | Josef Frantisek CIBULKA 66491 | 12/03/1942 | Runnymede Memorial |
| Sergeant | Jiri FINA 787403 | 12/03/1942 | Runnymede Memorial |
| Flight Lieutenant | Jaroslav KULA 82615 | 12/03/1942 | Runnymede Memorial |
| Sergeant | Alois MEZNIK 787866 | 12/03/1942 | Runnymede Memorial |
| Sergeant | Frantisek RAISKUP 787876 | 12/03/1942 | Runnymede Memorial |
| Sergeant | Oldrich SOUKOP 787489 | 12/03/1942 | Runnymede Memorial |
| Leading Aircraftman | Jan BAMBUSEK 787715 | 04/04/1942 | East Wretham St. Ethelbert C |
| Flight Sergeant | Josef HRDINA 787310 | 11/04/1942 | Bergen-Op-Zoom War Cemetery |
| Flight Sergeant | Josef KALENSKY 787177 | 11/04/1942 | Bergen-Op-Zoom War Cemetery |
| Sergeant | Karel KODES 787261 | 11/04/1942 | Bergen-Op-Zoom War Cemetery |
| Sergeant | Jan PEPRNICEK 787872 | 11/04/1942 | Bergen-Op-Zoom War Cemetery |
| Sergeant | Josef POLITZER 787875 | 11/04/1942 | Bergen-Op-Zoom War Cemetery |
| Flying Officer | Karel RYCHNOVSKY 82632 | 11/04/1942 | Bergen-Op-Zoom War Cemetery |
| Sergeant | Oldrich HAVLIK 787521 | 14/04/1942 | Eindhoven Woensel General Ce |
| Sergeant | Josef TALAB 787892 | 14/04/1942 | Eindhoven Woensel General Ce |
| Sergeant | Paval VARJAN 787901 | 14/04/1942 | Eindhoven Woensel General Ce |

Remembering...

# REMEMBERING

# REMEMBERING

# REMEMBERING

# REMEMBERING

# REMEMBERING

# REMEMBERING

# REMEMBERING

# REMEMBERING

# REMEMBERING

# REMEMBERING

# REMEMBERING

# REMEMBERING

## REMEMBERING

# REMEMBERING

# REMEMBERING

# REMEMBERING

# REMEMBERING

# REMEMBERING

# REMEMBERING

# REMEMBERING

# REMEMBERING

# REMEMBERING

# REMEMBERING

# REMEMBERING

# REMEMBERING

# REMEMBERING

# REMEMBERING

## REMEMBERING

# REMEMBERING

# REMEMBERING

## REMEMBERING

# REMEMBERING

# REMEMBERING

# REMEMBERING

# REMEMBERING

# REMEMBERING

# REMEMBERING

# REMEMBERING

# REMEMBERING

# REMEMBERING

# REMEMBERING

# REMEMBERING

# REMEMBERING

# REMEMBERING

# REMEMBERING

# REMEMBERING

# REMEMBERING

# REMEMBERING

# REMEMBERING

# REMEMBERING

## REMEMBERING

# REMEMBERING

www.ingramcontent.com/pod-product-compliance
Lightning Source LLC
Chambersburg PA
CBHW082057230426
43662CB00039B/2172